MECHANICS OF CONTINUOUS MEDIA AND ANALYSIS OF STRUCTURES

NORTH-HOLLAND SERIES IN

APPLIED MATHEMATICS AND MECHANICS

EDITORS:

E. BECKER
Institut für Mechanik
Technische Hochschule, Darmstadt

B. BUDIANSKY
Division of Applied Sciences
Harvard University

W. T. KOITER
Laboratory of Applied Mechanics
University of Technology, Delft

H. A. LAUWERIER
Institute of Applied Mathematics
University of Amsterdam

VOLUME 26

NORTH-HOLLAND PUBLISHING COMPANY
AMSTERDAM · NEW YORK · OXFORD

MECHANICS OF CONTINUOUS MEDIA AND ANALYSIS OF STRUCTURES

Roger VALID

Engineer graduate of Ecole Centrale de Paris
Docteur ès-Sciences Mathematiques
Laureate of the French Academy of Sciences

Scientific Assistant, Head of Structures Department
at the Office National d'Etudes et de Recherches Aerospatiales
(ONERA, France)

Professor at the Ecole Centrale des Arts et Manufactures of Paris

Foreword by

PROFESSOR PAUL GERMAIN

Member of the French Academy of Sciences

NORTH-HOLLAND PUBLISHING COMPANY
AMSTERDAM · NEW YORK · OXFORD

ISBN: 0 444 86150 5

Published by:

NORTH-HOLLAND PUBLISHING COMPANY
AMSTERDAM · OXFORD · NEW YORK

Sole distributors for the U.S.A. and Canada:
ELSEVIER NORTH-HOLLAND INC.
52 VANDERBILT AVENUE
NEW YORK, N.Y. 10017

Translation of: La Méchanique des Milieux Continus et le Calcul des Structures
Published by: Eyrolles, Paris (1977)
Copyright: Direction des Études et Recherches d'Électricité de France (1977)
Translated by: Tradunion, Versailles, France

Library of Congress Cataloging in Publication Data

Valid, Roger.
  Mechanics of continuous media and analysis of structures.

  (North-Holland series in applied mathematics and mechanics ; v. 26)
  Translation of Mécanique des milieux continus et le calcul des structures.
  Includes bibliographies and index.
  1. Structures, Theory of. 2. Continuum mechanics.
3. Finite element method. I. Title.
TA645.V3413   624.1'71   81-328
ISBN 0-444-86150-5   AACR1

PRINTED IN THE NETHERLANDS

INTRODUCTION

The basic characteristic of the treatise by Professor Valid is his systematic employment of modern coordinate-free analysis in the mechanics of continuous media. This approach is typical of a French school of applied mathematics and mechanics and it is not too familiar to the English speaking community in engineering science. Professor Germain's eloquent recommendation to engineers in his preface to the original French edition of this work therefore applies even more strongly outside France. The international world of engineering science is deeply indebted to Professor Valid for his willingness to prepare an English translation of his treatise in the North-Holland Series in Applied Mathematics and Mechanics.

W.T. KOITER

PREFACE

This is a course on Mechanics and not Mathematics. It is essentially designed for those who feel the need for certain theoretical developments in Mechanics, with a view to Applications in the field of the Analysis of Structures. Those who experience the quite legitimate need to base their calculations on precise theorems of existence and uniqueness, providing strict limits on conditions of application, hypotheses concerning data, solution spaces, and theorems of convergence, we would refer to specialized works on the subject.

However, the problems set by Mechanical Engineers have always been so numerous and complex, that the Mathematician still finds, and will continue to find for a considerable time, a wide scope for the exercise of his skills in this context. In fact the Engineer and Technician cannot wait, they must calculate and build. They are therefore engaged on two fronts - scientific and technical - but are also involved with Economics.

Having taken these precautions, we have divided the course into six main chapters, and after the first two, which are clearly differentiated, practical considerations and discretization alternate constantly with considerations of a more theoretical nature. A particular place has been given to the Finite Elements Method, which dominates the calculation of structures at the present time. However, this does not mean that other methods are of no interest, in particular the older Finite Differences Method, or the more recent Integral Equation method for the calculation of structures. Here again, the reader will refer to specialized works.

For the various subjects considered, a certain number of useful references are given for the reader seeking development in greater depth, but we frequently avoid the more conventional methods of presentation, in the hope of combining clarity with a more concise form. In addition, the annex contains all notations and definitions used in the text, and a number of relevant equations.

Furthermore, subjects which have not been dealt with or even referred to are considerably more numerous than those considered in this course. Should certain of the former be the subject of future addenda, this will be in order to make the course a little more complete, rather than to compete pointlessly with famous earlier or more recent works.

R. VALID

FOREWORD

In 1976, Monsieur Roger Valid had the honour of being asked to give a course at the Summer School that EDF[1], IRIA[2] and the CEA[3] organize each year. This school has a truly excellent reputation, founded on its judicious choice of subjects, highly qualified professors and most able audiences. Roger Valid subsequently decided to write up his notes in order to publish a book that would make his lectures available to a far wider audience than the privileged few who are lucky enough to have been to the summer school. This is the book I have the pleasure of introducing here.

Roger Valid's book gives the reader an insight into two of the most remarkable aspects of his scientific personality, easily recognized by those who are acquainted with his professional activities and the fine works he has produced. He is a research engineer and his main job is finding solutions to the problems of research departments and consulting engineers. This means that his solutions have to go as far as the actual calculations and clearly bring out the physical interpretations. It is therefore not surprising that his course includes a fund of information on the finite element method, which undoubtedly occupies first place amongst the methods nowadays used for structural analysis. However, he is also a scientist who cares a great deal about language and the most refined mathematical methods available: he attaches great importance to using the most efficient tools of contemporary mathematics in both his reasonings and his analytical developments.

The reader who takes up the book without being familiar with the concepts and notations that Roger Valid handles so masterfully will find this last aspect very striking - and perhaps a little disconcerting at first. But the very

---

(1) EDF - French Electricity Board
(2) IRIA - Institute of Research in 'Informatique et Automatisme'
(3) CEA - French Atomic Energy Commission

originality and merit of this course lie in the concise way the author presents results and formulae which are much more ponderously written up in the books usually intended for engineers. An appendix of formulae is attached to guide the reader who requires assistance in assimilating the concepts and symbols that he must know if he is to benefit to the full from his study of the six chapters of this course. He may have to spend a certain amount of time on this; but he is bound to find he has gone deeper and understood better afterwards. There will always be arguments as to how long engineers should wait before applying recent mathematical advances, given that it is considered wiser to wait a certain length of time for the final mathematical formulation to emerge: this will be the best adapted and the most often used of all the possible formulations suggested. However, if there were no pioneers, like Roger Valid, amongst the mechanical engineers to tell them they must keep on renewing their thinking habits, they would be very likely to go on lazily using very outdated mathematical frameworks.

The table of contents is detailed and explicit and it does not seem necessary to comment on it at length since the title of the book summarizes it perfectly: this is a course of mechanics of continuous media with a view to structural analysis. This explains why elastic behaviour occupies so much of the book. The important subjects discussed - finite element method, variational principles in elasticity, vibrations, buckling, shells - are all approached in the light of extensive knowledge of each, the methods are explained concisely but accurately, and some of the author's expositions and results are undoubtedly original.

To conclude, this is certainly not an easy book. Engineers and students of structural analysis, even well-informed, will find it interesting and useful because it presents subjects they know in a new light and is also a mine of precious information and references. Teachers and university and engineering school students with sufficient training in mechanics of continuous media, or at least in elasticity, and who want to start on structural analysis, will find this account of the subject fully up-to-date and written in modern terms: they will find it enriching and stimulating to study. We should be grateful to Roger Valid for writing a book which answers a definite need, now that advanced applied mathematical techniques are bringing profound renewal to structural analysis, one of the engineer's basic disciplines.

Paul GERMAIN
Professor at Ecole Polytechnique (France)

TABLE OF CONTENTS

CHAPTER VI : SHELL THEORY 207

1. - <u>GENERAL HYPOTHESES</u> - <u>HYPERELASTIC MEDIA</u> - EXTENSION TO ARBITRARY MEDIA

Taking a material medium occupying a domain $\Omega \subset E_3$ with a boundary Σ, a generic point M, at time t :

a) We assume that this medium is composed of identifiable molecules, namely with each molecule located at point M such that :

$$M = F(M_o, t),$$

M_o belongs to a reference domain, which can correspond to a state of the medium. M_o and t are LAGRANGE variables. The reference domain is called Ω_o, with a boundary Σ_o; $\Omega_o \subset E_3$. F is a mapping of an open set of $E_3 \times [o,T]$ in E_3, and $t \in [o,T] \subset \mathbb{R}$.

We assume that F is continuous, regular, differentiable, and with a regular derivative, which expresses the absence of cracking, cavitation or penetration.

In abridged form :
- cracking or cavitation $\Longrightarrow$ F discontinuous, or
$$\forall M_o, M'_o \in \Omega_o; \; M'_o \to M_o \not\Longrightarrow M' \to M \; ;$$

- penetration $\Longrightarrow$ F irregular, or
$$\exists M_o, \; M'_o \in \Omega_o; \; M_o, \; M'_o \Longrightarrow \text{same } M.$$

- F differentiable $\Longrightarrow \forall$ the field with value $dM_o \in \vec{E}_3$, at M_o

$\exists \; dM \in \vec{E}_3$, at M, such that $\quad dM = \dfrac{\partial M}{\partial M_o} \, dM_o = F'(M_o) dM_o \quad$ [see (A.28)].

- $\dfrac{\partial M}{\partial M_o} = F'(M_o)$ regular $\Longrightarrow \det\left(\dfrac{\partial M}{\partial M_o}\right) \neq 0$, in a state close to reference state $\left(\det\left(\dfrac{\partial M_o}{\partial M_o}\right) = 1\right)$; the body does not undergo any "overturning" effect.

b) Mass conservation : We assume that the mass of a set of adjacent molecules is conserved when the molecules are displaced. This hypothesis provides the equation of continuity.

Let ρ be the mass volume density at time t and at point M, and ρ_o be its value at the corresponding reference point M_o.

$\forall$ the fields of value d_1M_o, d_2M_o, d_3M_o at M, the mass δm of a volume element is constant, $\forall t$, so that [see (A.2)] :

$$\delta m = \rho \; \mathrm{vol}(d_1 M)(d_2 M)(d_3 M) = \rho_o \; \mathrm{vol}(d_1 M_o)(d_2 M_o)(d_3 M_o) = C^{te}$$

$$= \rho \; \det\left(\frac{\partial M}{\partial M_o}\right) \mathrm{vol}(d_1 M_o)(d_2 M_o)(d_3 M_o) = \rho_o \; \mathrm{vol}(d_1 M_o)(d_2 M_o)(d_3 M_o),$$

where :

$$(1.1) \qquad \rho \; \det\left(\frac{\partial M}{\partial M_o}\right) = \rho_o.$$

We can consider ρ and $\det\left(\frac{\partial M}{\partial M_o}\right)$ as functions of M and t (reciprocal images). M and t are called Euler variables. Under these conditions, if we derive (1.1) with respect to time, we obtain :

$$\frac{d}{dt}\left[\rho \; \det\left(\frac{\partial M}{\partial M_o}\right)\right] = 0,$$

(the derivative being material or with constant M_o), so that [see (A.36)]:

$$\frac{d\rho}{dt} \; \det\left(\frac{\partial M}{\partial M_\delta}\right) + \rho \; \frac{d}{dt} \; \det\left(\frac{\partial M}{\partial M_o}\right) = C$$

$$\frac{\partial \rho}{\partial t} + \frac{\partial \rho}{\partial M} \; \dot{M} + \rho T_r \left(\mathrm{Adj}\left(\frac{\partial M}{\partial M_o}\right) \frac{d}{dt} \; \frac{\partial M}{\partial M_o}\right) \frac{1}{\det\left(\frac{\partial M}{\partial M_o}\right)} = 0, \; \dot{M} = \frac{\partial M}{\partial t}.$$

$$\frac{\partial \rho}{\partial t} + \frac{\partial \rho}{\partial M} \; \dot{M} + \rho T_r \left(\frac{\partial M_o}{\partial M} \; \frac{\partial \dot{M}}{\partial M_o}\right) = 0$$

$$\frac{\partial \rho}{\partial t} + \frac{\partial \rho}{\partial M} \; \dot{M} + \rho T_r \left(\frac{\partial \dot{M}}{\partial M}\right) = 0$$

$$(1.2) \qquad \frac{\partial \rho}{\partial t} + \mathrm{div}\,[\rho \; \dot{M}] = 0.$$

(1.2) is another form of equation of continuity (1.1).

c) <u>Principle of energy localization</u>: We assume that there exists an internal volume density , depending on the state of the medium, at each point and at each instant of time. In the following paragraphs, thermodynamic phenomena are ignored, and transformations arc supposed to be isothermal. Furthermore, the state is assumed to depend at each point M, on the immediate vicinity of this point only,[1,10] and more precisely, we put for this type of medium, referred to as <u>hyperelastic</u> (in Lagrange variables):

$$\alpha = \underset{\sim}{\alpha} \, (M_o, \; \frac{\partial M}{\partial M_o}, \; \frac{\partial^2 M}{\partial M_o \, \partial M_o}, \; \ldots).$$

Finally, leaving aside the possible heterogeneity of the medium, we adopt a first gradient theory where α only depends on the first derivative of the transformation, namely :

$$\alpha = \underset{\sim}{\alpha} \, (\frac{\partial M}{\partial M_o}).$$

In fact we demonstrate that energy density α is only a function of endomorphism K of $\vec{E}_3$, such that :

$$K = \overline{\frac{\partial M}{\partial M_o}} \; \frac{\partial M}{\partial M_o}, \; \text{where the bar means transposition [see Annex § 6].}$$

(The tensor corresponding to K is called the Cauchy tensor (see Annex § 4.b)

Let us consider in fact two successive states $\mathcal{E}_1$ and $\mathcal{E}_2$, where $\mathcal{E}_2$ is obtained from $\mathcal{E}_1$, without deformation :

$$M_o \Longrightarrow M_1 \Longrightarrow M_2$$

$$\forall dM_o \text{ at } M_o \Longrightarrow dM_1 \text{ at } M_1 \Longrightarrow dM_2 \text{ at } M_2.$$

$$\overline{dM_2} \, dM_2 = \overline{dM_1} \, dM_1 \quad , \quad [\text{see (A.10)}],$$

the field of value dM_1, with $dM_2 = \dfrac{\partial M_2}{\partial M_1} \, dM_1 \quad , \quad [\text{see (A.28)}].$

Hence :

$$\overline{\frac{\partial M_2}{\partial M_1}} \; \frac{\partial M_2}{\partial M_1} = 1_{E_3} \Longrightarrow \alpha_2 = \alpha_1,$$

where α_2 and α_1 correspond to $\mathcal{E}_2$ and $\mathcal{E}_1$ respectively. But this equation also gives :

$$\overline{\frac{\partial M_2}{\partial M_o}} \; \overline{\frac{\partial M_o}{\partial M_1}} \; \frac{\partial M_2}{\partial M_o} \; \frac{\partial M_o}{\partial M_1} = 1_{E_3},$$

or :

$$\left[\overline{\frac{\partial M_1}{\partial M_o}}\right]^{-1} \; \overline{\frac{\partial M_2}{\partial M_o}} \; \frac{\partial M_2}{\partial M_o} \; \left[\frac{\partial M_1}{\partial M_o}\right]^{-1} = 1_{E_3},$$

and finally :

$$\frac{\overline{\partial M_2}}{\partial M_o} \frac{\partial M_2}{\partial M_o} = \frac{\overline{\partial M_1}}{\partial M_o} \frac{\partial M_1}{\partial M_o} = \ldots \implies \alpha_2 = \alpha_1 = \ldots$$

$$K_2 = K_1 = \ldots \implies \alpha_2 = \alpha_1 = \ldots$$

This demonstrates that :

$$\alpha = \underset{\sim}{\alpha}(K).$$

Thus the absence of deformation at point M in state $\mathcal{E}$, corresponds to :

$$\frac{\overline{\partial M}}{\partial M_o} \cdot \frac{\partial M}{\partial M_o} = 1_E \cdot \iff \overline{dM}\, dM = \overline{dM_o}\, dM_o, \quad \forall dM_o.$$

This leads to the definition of <u>deformation</u> D at M :

$$(1.3) \qquad D = \frac{1}{2}\left[\frac{\overline{\partial M}}{\partial M_o} \frac{\partial M}{\partial M_o} - 1_{E_3}\right] = \overline{D}.$$

D is an endomorphism of $\vec{E}_3$, corresponding to the so-called <u>Green tensor</u>. (See Annex § 4.b).We also define an endomorphism :

$$\frac{1}{2}\left[1_{E_3} - \frac{\overline{\partial M_o}}{\partial M} \frac{\partial M_o}{\partial M}\right]$$

which corresponds to the <u>Almansi tensor</u>.

2. - <u>STRESSES</u>

We can now put :

$$\alpha = \underset{\sim}{\alpha}\,(D)$$

the strain energy density at point M of the deformed body, so that the total deformation energy w is written:

$$(1.4) \qquad w = \int_{\Omega_o} \underset{\sim}{\alpha}\,(D)\,d\Omega_o,$$

with $\qquad d\Omega_o = \text{vol}(d_1 M_o)\,(d_2 M_o)\,(d_3 M_o).$

We now call Σ_F that part of boundary Σ of Ω on which we have a force surface density F, and Σ_U the complementary part of Σ on which we have a displacement field U_d $(\Sigma = \Sigma_F U \Sigma_U)$.

We shall use the expression <u>kinematically admissible displacement field</u> <u>U</u> in the deformed state, to describe a continuous displacement field in $\overline{\Omega}_o$, with derivative square summable in Ω_o (therefore such that its components $\in L^2(\Omega_o)$, and such that $U = U_d$ on Σ_{oU}.

We also use the expression <u>admissible virtual variation δU</u> to describe an admissible field of the previous type, but where :

$$\delta U = 0 \text{ on } \Sigma_{oU}$$

(δU can also be considered as a virtual velocity).

For a kinematically admissible (K.A.) virtual variation δU, the virtual variation of the deformation energy w, taking α to be differentiable (with α summable differential), is written:

$$\delta w = \int_{\Omega_o} \delta\alpha \, d\Omega_o$$

$$\delta w = \int_{\Omega_o} \frac{\partial\alpha}{\partial D} \, \delta D \, d\Omega_o .$$

Now $\frac{\partial\alpha}{\partial D}$ being linear with scalar value, and δD being Hermitian, $\exists$ a hermitian endomorphism C' such that [see (A.9)]

$$(1.5) \qquad \delta w = \int_{\Omega_o} T_r(C'\delta D) \, d\Omega_o, \quad C' = \overline{C'}.$$

C' and δD being two Hermitian endomorphisms of $\vec{E}_3$, the quantity $T_r(C'\delta D)$ is interpreted as a scalar product on the vectorial space of these endomorphisms [*]. C' is therefore an Euclidian vector space with dimension 6, which we call $\vec{E}_6$. If we call $\overset{\sim}{C'}$ the transpose of C' (Vector $\in \vec{E}_6$) in the sense of this scalar product, we have :

---

[*] Although the deformations, as functions of U, do not constitute a vector space, but only a subset of $\vec{E}_6$.

(1.6) $T_r(C'\delta D) = \tilde{C}'\delta D.$

We then obtain :

$$\frac{\partial \alpha}{\partial D} = \tilde{C}' \quad \epsilon \; \vec{E}_6 \quad *).$$

<u>C' is the Piola-Kirchhoff stress.</u>

Furthermore $\forall \delta M = \delta U$ C.A. (1.3) gives [see (A.35)]

$$\delta D = \frac{1}{2}\delta\left[\frac{\overline{\partial M}}{\partial M_o} \; \frac{\partial M}{\partial M_o}\right] = \frac{1}{2}\left[\frac{\overline{\partial M}}{\partial M_o}\delta \; \frac{\partial M}{\partial M_o} + \delta \; \frac{\overline{\partial M}}{\partial M_o} \; \frac{\partial M}{\partial M_o}\right].$$

Now $\delta \; \dfrac{\partial M}{\partial M_o} = \dfrac{\partial \delta M}{\partial M_o}$, as the "derivation" δ (at constant M_o) commutes with the space derivation $\dfrac{\partial}{\partial M_o}$. Thus :

$$\delta D = \frac{1}{2}\left[\frac{\overline{\partial M}}{\partial M_o} \; \frac{\partial \delta M}{\partial M_o} + \frac{\overline{\partial \delta M}}{\partial M_o} \; \frac{\partial M}{\partial M_o}\right]$$

$$= \frac{1}{2}\left[\frac{\overline{\partial M}}{\partial M_o} \; \frac{\partial \delta M}{\partial M} \; \frac{\partial M}{\partial M_o} + \frac{\overline{\partial M}}{\partial M_o} \; \frac{\overline{\partial \delta M}}{\partial M} \; \frac{\partial M}{\partial M_o}\right]$$

(1.7) $$\delta D = \frac{1}{2}\frac{\overline{\partial M}}{\partial M_o}\left[\frac{\partial \delta M}{\partial M} + \frac{\overline{\partial \delta M}}{\partial M}\right]\frac{\partial M}{\partial M_o}.$$

(1.5) is then written, with (1.7) :

$$\delta w = \int_{\Omega_o} T_r(\frac{1}{2} C' \; \frac{\overline{\partial M}}{\partial M_o}\left[\frac{\partial \delta M}{\partial M} + \frac{\overline{\partial \delta M}}{\partial M}\right]\frac{\partial M}{\partial M_o})d\Omega_o$$

$$= \int_{\Omega_o} T_r(\frac{1}{2} \; \frac{\partial M}{\partial M_o} \; C' \; \frac{\overline{\partial M}}{\partial M_o}\left[\frac{\partial \delta M}{\partial M} + \frac{\overline{\partial \delta M}}{\partial M}\right]) \; d\Omega_o, \; [\text{see (A.6)}].$$

$$\delta w = \int_{\Omega_o} T_r(\frac{\partial M}{\partial M_o} \; C' \; \frac{\overline{\partial M}}{\partial M_o} \; \frac{\partial \delta M}{\partial M}) \; d\Omega_o.$$

---

*) Although the deformations, as functions of U, do not constitute a vector space, but only a subset of $\vec{E}_6$.

Reverting to domain Ω by reciprocal images, we have :

$$\delta w = \int_\Omega T_r \left(\frac{\partial M}{\partial M_o} C' \frac{\overline{\partial M}}{\partial M_o} \frac{\partial \delta M'}{\partial M} \right) \frac{1}{\det\left(\frac{\partial M}{\partial M_o}\right)} \, d\Omega$$

If we now put :

$$(1.8) \qquad C = \frac{1}{\det\left(\frac{\partial M}{\partial M_o}\right)} \frac{\partial M}{\partial M_o} C' \frac{\overline{\partial M}}{\partial M_o} = \overline{C},$$

we have :

$$(1.9) \qquad \delta w = \int_\Omega T_r \left(C \frac{\partial \delta M}{\partial M} \right) d\Omega.$$

C is the <u>Cauchy stress</u>. It is the dual of the endomorphism of $\vec{E}_3$: $\frac{\partial \delta M}{\partial M}$ or $\frac{1}{2} \left[\frac{\partial \delta M}{\partial M} + \frac{\overline{\partial \delta M}}{\partial M} \right]$ [see $(A.11)_3$]. In (1.9) w is expressed in Euler variables.

<u>Extension to arbitrary media</u>: We shall generalize equation (1.9) for non-hyper-elastic media. This equation then enables us to define the stress, as stated above, as the dual of the quantity $\frac{1}{2} \left[\frac{\partial \delta M}{\partial M} + \frac{\overline{\partial \delta M}}{\partial M} \right]$, furthermore linear in δM.

As we shall see later, this generalization can be applied to the case of a non-Hermitian stress (see § 9).

Interpretation of the stress C is given in § 4.

3. – <u>DEFORMATION</u>

We have defined the deformation, at point M of the deformed state with respect to the reference state $\mathcal{E}_o$, by equation (1.3) :

$$(1.10) \qquad D = \frac{1}{2} \left[\frac{\overline{\partial M}}{\partial M_o} \frac{\partial M}{\partial M_o} - {}^1 E_3 \right].$$

As a function of displacement vector U, such that :

$$M = M_o + U,$$

this equation becomes :

$$(1.11) \qquad D = \frac{1}{2}\left[\frac{\partial U}{\partial M_o} + \overline{\frac{\partial U}{\partial M_o}}\right] + \frac{1}{2}\,\overline{\frac{\partial U}{\partial M_o}}\,\frac{\partial U}{\partial M_o}\,,$$

showing a linear part and a non-linear part, according to the vectorial variable U.

Its interpretation is simple and classical. If we consider a local unitary basis S_o at point M_o, assumed to originate from a map f of E_3, such that, ϑ being an open set of $\mathbb{R}^3$:

$$M_o = f(X) \qquad\qquad X \in \vartheta \subset \mathbb{R}^3$$

$$\frac{\partial M_o}{\partial X} = S_o \text{ and } \overline{S_o}\,S_o = 1_{\mathbb{R}^3},\ S_{oi} = \partial_i M_o \ (i = 1,\,2,\,3).$$

Referring back to (1.10), the components, here covariants of D in S_o, are written conventionally ε_{ij}, where ε is the matrix representing D in S_o. Thus :

$$\overline{S_o}\,D\,S_o = \varepsilon$$

$$\overline{S_{oi}}\,D\,S_{oj} = \varepsilon_{ij},\quad (i,\,j = 1,\,2,\,3).$$

$$\overline{S_{oi}}\,D\,S_{oj} = \frac{1}{2}\,\partial_i M_o\left[\overline{\frac{\partial M}{\partial M_o}}\,\frac{\partial M}{\partial M_o} - 1_{E_3}\right]\partial_j M_o$$

$$= \frac{1}{2}\left[\overline{\partial_i M}\,\partial_j M - \overline{\partial_i M_o}\ \partial_j M_o\right].$$

In particular :

$$\overline{S_{oi}}\,D\,S_{oi} = \varepsilon_{ii} = \frac{1}{2}\left[\overline{\partial_i M}\,\partial_i M - \overline{\partial_i M_o}\ \partial_i M_o\right]$$

$$S_{oi}\,D\,S_{oj} = \varepsilon_{ij} = \frac{1}{2}\,\overline{\partial_i M}\ \partial_j M\ ,\ i \neq j$$

$$\varepsilon_{ij} = \frac{1}{2}\,|\partial_i M|\,|\partial_j M|\ \cos\varphi,\ i \neq j,$$

where φ is the angle enclosed by the two vectors. If $\varphi = \frac{\Pi}{2}$, $\cos\varphi = 0$, the deformation occurs without angular distortion with respect to the reference value $\frac{\Pi}{2}$ for φ.

<u>Volume dilatation</u> : This is given by $\det(\frac{\partial M}{\partial M_o})$, such that :

$$d\Omega = \det(\frac{\partial M}{\partial M_o}) \, d\Omega_o \ .$$

Now :

$$\frac{\partial M}{\partial M_o} = \frac{\partial}{\partial M_o} \left[M_o + U \right] = 1_{E_3} + \frac{\partial U}{\partial M_o}.$$

If $\frac{\partial U}{\partial M_o}$ is small with respect to 1_{E_3}, we may linearize :

$$\frac{\partial M}{\partial M_o} = 1_{E_3} + \varepsilon \frac{\partial U}{\partial M_o}, \text{ with } \varepsilon^2 = 0^{22}.$$

Thus   [see (A.38) and (A.36)] :

$$\det(1_{E_3} + \varepsilon \frac{\partial U}{\partial M_o}) = \det(1_{E_3}) + \varepsilon T_r(\text{Adj}(1_{E_3}) \frac{\partial U}{\partial M_o})$$

$$= 1 + \varepsilon \underset{M_o}{\text{div}} U \ , \quad [\text{see (A.47)}].$$

The quantity $\underset{M_o}{\text{div}} U$ is therefore the linearized volume dilatation.

If we keep in D, only those terms which are linear in U, we have :

$$(1.12) \qquad D_L = \frac{1}{2} \left[\frac{\partial U}{\partial M_o} + \overline{\frac{\partial U}{\partial M_o}} \right].$$

$$(1.13) \qquad T_r(D_L) = T_r(\frac{\partial U}{\partial M_o}) = \underset{M_o}{\text{div}} U.$$

It is also possible to calculate the non-linearized volume dilatation as a function of U.

<u>Rotation</u> : We always have :

$$\frac{\partial M}{\partial M_o} = 1_{E_3} + \frac{\partial U}{\partial M_o}$$

$$(1.14) \qquad \frac{\partial U}{\partial M_o} = \frac{1}{2} \left[\frac{\partial U}{\partial M_o} + \overline{\frac{\partial U}{\partial M_o}} \right] + \frac{1}{2} \left[\frac{\partial U}{\partial M_o} - \overline{\frac{\partial U}{\partial M_o}} \right],$$

where $\frac{\partial U}{\partial M_o}$ is decomposed into its Hermitian and anti-Hermitian(skew symmetric) components.

(1.14) is then written :

(1.15) $\quad \dfrac{\partial U}{\partial M_o} = D_L + i \left(\dfrac{\text{rot } U}{2}\right)$, as $i(\text{rot } U) = \dfrac{\partial U}{\partial M_o} - \overline{\dfrac{\partial U}{\partial M_o}}$ $\quad$ (A.45).

Let us put :

(1.16) $\quad \Omega = \dfrac{1}{2} \text{ rot } U$

(no confusion with domain Ω is possible).

Ω is the vector of local rotation. Thus

(1.17) $\quad \dfrac{\partial U}{\partial M_o} = D_L + i(\Omega)$,

and in addition :

$$dM = \frac{\partial M}{\partial M_o}\, dM_o = \left[1_{E_3} + D_L + i(\Omega) \right] dM_o, \quad \forall dM_o.$$

Compatibility[2,9,12] : The deformation D has to meet, as an endomorphism of $\vec{E}_3$, compatibility conditions to ensure the existence of a field of vector U such that (1.11) be verified. In the non-linear case, we write that the Riemannian curvature of space E_3, expressed as a function of D, remains null at any point M.

In the linearized case, the calculation is much easier. In fact :

$$D_L = \frac{1}{2}\left[\frac{\partial U}{\partial M_o} + \overline{\frac{\partial U}{\partial M_o}} \right] = \overline{D_L}$$

$$\text{rot } D_L = \frac{1}{2} \text{ rot } \frac{\partial U}{\partial M_o}, \text{ as rot } \overline{\frac{\partial U}{\partial M_o}} = \text{rot grad } U = 0,$$

$$\text{[see (A.51) and (A.60)]}$$

$$= \frac{1}{2} \frac{\partial}{\partial M_o} \text{ rot } U = \frac{1}{2} \overline{\text{grad } [\text{ rot } U]}, \text{ [see (A.59)]}.$$

Hence :

(1.18) $\quad \text{rot } \overline{\text{rot } D_L} = 0$.

(1.18) is the compatibility condition precisely sought. A priori

$\mathrm{rot}\ \overline{\mathrm{rot}\ D_L}$ is an endomorphism of $\vec{E}_3$, represented by a square matrix of order 3, and which belongs consequently to a vector space of dimension 9. In fact $\mathrm{rot}\ \overline{\mathrm{rot}\ D_L} \in \vec{E}_6$ for :

$$(1.19) \qquad \mathrm{rot}\ \overline{\mathrm{rot}\ D_L} = \overline{\mathrm{rot}\ \overline{\mathrm{rot}\ D_L}}$$

In fact $\mathrm{rot}\ \overline{\mathrm{rot}\cdot D_L}$ can be calculated in any basis S_o at M_o, which is nevertheless constant, by :

$$\mathrm{rot}\ D_L = i(^i s^{-1})\partial_i D_L, \quad (i = 1,\ 2,\ 3),\quad [\text{see (A.63)}].$$

$$\overline{\mathrm{rot}\ D_L} = -\ \partial_i D_L\ i(\overline{^i s^{-1}}),\ \ \text{as}\ \ D_L = \overline{D_L}$$

$$\mathrm{rot}\ \overline{\mathrm{rot}\ D_L} = -\ i(\overline{^j s^{-1}})\partial_j \partial_i D_L\ i(\overline{^i s^{-1}}) = \overline{\mathrm{rot}\ \overline{\mathrm{rot}\ D_L}}\ .\qquad \text{Q.E.D.}$$

Reciprocally, if (1.18) is verified, $\exists$ in a simply connected open set, a vector field Ω defined to within any constant vector, such that :

$$(1.20) \qquad \mathrm{rot}\ D_L = \frac{\partial \Omega}{\partial M_o}\ \ (\text{Poincaré's theorem})\ \ [\text{see (A.4) and (A.60)}],$$

and by curvilinear integration on a curve C in this open set :

$$\Omega = \oint_{\substack{M_o}}^{M'_o} \mathrm{rot}\ D_L\ dM_o + \Omega_o\ \forall\ C\ \text{and}\ M'_o.$$

Furthermore, [see (1.17)], the form $D_L + i(\Omega)$ is a closed vector of degree 1, namely :

$$(1.21) \qquad \mathrm{rot}\ \overline{D_L + i(\Omega)} = 0.$$

In fact [see (A.19)] :

$$\mathrm{rot}\left[\overline{D_L} + \overline{i(\Omega)}\right] = \mathrm{rot}\left[D_L - i(\Omega)\right] = \mathrm{rot}\ D_L - \mathrm{rot}\ i(\Omega)$$

$$= \mathrm{rot}\ D_L - \left[\frac{\partial\Omega}{\partial M_o} - \mathop{\mathrm{div}}_{M_o}\Omega.\ 1_{E_3}\right]$$

$$= \mathop{\mathrm{div}}_{M_o}\Omega.\ 1_{E_3}\ ,\ \text{taking account of (1.20).}$$

Now :

$$\mathop{\mathrm{div}}_{M_o}\Omega = T_r\left(\frac{\partial\Omega}{\partial M_o}\right) = T_r(\mathrm{rot}\ D_L),\ \text{taking account of (1.20), and with}$$

the preceding representation [see (A.63)] :

$$T_r(\text{rot } D_L) = T_r(i(^iS^{-1})\partial_i D_L) = 0, \text{ as } D_L = \overline{D_L}. \qquad Q.E.D.$$

(1.21) therefore shows that there exists in the simply connected open set, a vector field U defined to within any constant vector, such that:

$$U = \oint_{M_o}^{M'_o} [D_L + i(\Omega)] dM_o + U_o.$$

4. - <u>EQUILIBRIUM EQUATIONS</u>

Let us consider a medium in a state of equilibrium under the action of a volume density f of given external forces in Ω, and a surface density F of given external forces on Σ_F, these two fields being square summable. We are also given :

$$U = U_d \text{ on } \Sigma_U;$$

Using (1.9), the equilibrium in the state considered is given by the principle of virtual work, thus :

$$(1.22) \qquad \int_\Omega T_r(C \frac{\partial \delta U}{\partial M}) d\Omega - \int_\Omega \overline{f}\delta U d\Omega - \int_{\Sigma_F} \overline{F}\delta U d\Sigma = 0, \ \forall \delta U \ \ K.A.$$

Furthermore, we assume $C = \overline{C}$, differentiable, with a derivative square summable in Ω and on Σ. (1.22) is written [see (A.50)] :

$$\int_\Omega \left[\text{div } [C\delta U] - \text{div } C.\delta U\right] d\Omega - \int_\Omega \overline{f}\delta U d\Omega - \int_{\Sigma_F} \overline{F}\delta U d\Sigma = 0, \forall \delta U \ K.A.,$$

or, by application of the Stokes formula (A.98) :

$$-\int_\Omega \left[\text{div } C + \overline{f}\right]\delta U d\Omega + \int_{\Sigma_F} \left[\ \overline{n}\ C\delta U - \overline{F}\delta U\right] d\Sigma = 0, \ \forall \delta U \ K.A.,$$

(where n is the unit normal to Σ, oriented externally). Thus we obtain the local equilibrium equations relative to each of domains Ω and Σ_F :

$$(1.23) \qquad \begin{cases} \text{div } C + \overline{f} = 0 \quad \text{in } \Omega \\[2ex] \overline{n}\ C = \overline{F} \qquad \text{on } \Sigma_F, \end{cases}$$

holding both when C is Hermitian or not. In case C is indeed Hermitian, these equations are equivalent to :

$$(1.24) \quad \begin{cases} \text{div } C + \overline{f} = 0 \quad \text{in } \Omega \\[2ex] C\, n = F \qquad\qquad \text{on } \Sigma_F, \text{ when } \quad C = \overline{C} . \end{cases}$$

It should be noted that if the virtual deformation energy is invariant under a rigid body rotation in any domain Ω', and $\delta\Omega$ be a virtual rotation independent of M, we have :

$$\int_{\Omega'} T_r (C\, \frac{\partial \delta U}{\partial M}) d\Omega = 0, \forall\, \delta\, U = i\,(\delta\Omega)\,(M), \forall\, \Omega'$$

$$\frac{\partial \delta U}{\partial M} = i\,(\delta\Omega), \forall\, \delta\, \Omega.$$

Hence   :

$$T_r (C\, i\,(\delta\Omega)) = 0\, \forall\, \delta\, \Omega \Longleftrightarrow C = \overline{C} .$$

Reciprocally (1.23), [or (1.24)], involves (1.22). Demonstration is immediate. Formulation (1.22) is called the weak formulation of the equilibrium equations, the terms of which are then assumed to belong to appropriate distributions spaces. In practice (1.22) gives an equilibrium in the mean).

Remarks : 1) In a dynamic problem, it is appropriate to add to the density f, a density of inertial forces, so that f is replaced by $f - \rho\ddot{M}$.

2) (1.24) supplies an immediate <u>interpretation of the stress C</u>, as :

$$Cn\, d\Sigma = F\, d\Sigma , \qquad \forall\, d\Sigma \subset \Sigma_F.$$

If we imagine a supplementary cut out $d\Sigma$ in Ω, we create a supplementary boundary element, and F is the local surface density required to balance the density Cn.

3) <u>Case of perfect fluids</u> : The stress C is decomposed into its scalar component and deviator, such that :

$$C_m = \frac{1}{3} T_r (C)$$

$$C_d = C - C_m \cdot 1_{E_3}$$

$$(1.25) \qquad C = C_m \cdot 1_{E_3} + C_d .$$

$C_m \cdot 1_{E_3}$ is the scalar (or hydrostatic) component, and C_d the deviator.
(1.25) gives :

$$(1.26) \qquad T_r(C_d) = 0 .$$

It should be noted that if we consider C, 1_{E_3} and $C_d \in \vec{E}_6$, we have :

$$C_m \cdot \overset{\sim}{1}_{E_3} \cdot C_d = T_r(C_m \cdot 1_{E_3} \cdot C_d) = C_m \, T_r(C_d) = 0 .$$

The two components of C thus defined, belong to supplementary orthogonal subspaces in $\vec{E}_6$. One has dimension 1, and is subtended by 1_{E_3}, and the other has dimension 5 and is the space of the deviators.

In the case of a perfect fluid, we have :

$$(1.27) \qquad C_d = 0$$

$$C = C_m \cdot 1_{E_3} .$$

C_m is normally called the pressure p.

Applying the same decomposition to the linearized virtual deformation :

$$\delta D_L = \frac{1}{2} \left[\frac{\partial \delta U}{\partial M} + \overline{\frac{\partial \delta U}{\partial M}} \right] = \delta D_{Lm} \cdot 1_{E_3} + \delta D_{Ld}$$

with :

$$\delta D_{Lm} = \frac{1}{3} T_r(\delta D_L) \; , \; T_r(\delta D_{Ld}) = 0 .$$

$$(1.13) \Longrightarrow \quad \delta D_{Lm} = \frac{1}{3} \, \underset{M}{\mathrm{div}} \; \delta U$$

Hence :

$$T_r(C \frac{\partial \delta U}{\partial M}) = T_r(C \, \delta D_L) = T_r \left(C_m \cdot 1_{E_3} \left[\frac{1}{3} \, \mathrm{div} \, \delta U . 1_{E_3} + \delta D_{Ld} \right] \right)$$

$$= \frac{1}{3} \, C_m T_r (\underset{M}{\mathrm{div}} \, \delta U . 1_{E_3})$$

$$= C_m \; \text{div} \; \delta U.$$
$$\quad\quad\; M$$

If we put :

$$\text{div} \; \delta U = \; \delta v,$$
$$\; M$$

the linearized virtual variation of volume, we obtain :

$$T_r(C \; \frac{\delta \partial U}{\partial M}) = p. \; \delta v.$$

5. – <u>HOMOGENEOUS AND ISOTROPIC LINEAR ELASTICITY</u>[13-15]

5.1 – <u>Isotropy</u>

If we consider a given mechanical medium, with symmetrical mechanical properties with respect to a plane P_o, and with an external force volume density f_o at point M_o, we obtain a resultant deformation D_o. Any direction V_o will be transformed to $D_o V_o$ by endomorphism D_o.

A symmetry with respect to plane P_o is characterized by a normal unitary endomorphism Q, such that :

$$\overline{Q} \; Q = 1_{E_3} = Q \; \overline{Q}$$

Any vector V_o is transformed to V, such that :

$$V = QV_o.$$

Taking a force density f, symmetrical with f_o with respect to P_o at M_o :

$$f = Qf_o.$$

This results in a deformation D at M_o, transforming vector V into a vector DV. Because of the property of symmetry, DV must be symmetrical with $D_o V_o$ with respect to P_o, therefore :

$$DV = QD_o V_o$$

or :

$$DQV_o = QD_o V_o, \; \forall V_o$$

As a result :

$$DQ = QD_o$$

$$D = QD_o Q^{-1}$$

namely:

(1.28) $D = QD_o \overline{Q}.$

In the case of an isotropic medium, this relation is true $\forall P_o$ and $\forall Q$, such that :

$$\overline{Q}\, Q = Q\, \overline{Q} = 1_{E_3}.$$

Now we saw in § 1 that in the case of a hyperelastic medium, the deformation energy volume density α was such that :

$$\alpha = \underset{\sim}{\alpha}\ (K) \quad \text{with } K = \overline{\frac{\partial M}{\partial M_o}}\ \frac{\partial M}{\partial M_o}.$$

If the medium is isotropic, the preceding argument demonstrates that this density is constant by transformation (1.28), namely :

(1.29) $\left[\underset{\sim}{\alpha}\ (K_o) = \underset{\sim}{\alpha}\ (Q\, K_o\, \overline{Q}), \quad \forall \text{ unitary } Q \right] \iff$ isotropy.

Under these conditions, we have

> Theorem : For an isotropic medium
>
> (1.30) $\alpha = f\Big(T_r(K_o),\ T_r(K_o^2),\ T_r(K_o^3)\Big),$
>
> or in other words the deformation energy only depends on the invariants of the Green tensor, namely the coefficients of its characteristic equation.

In fact :

$$K = Q\, K_o\, \overline{Q}\ ,\ (Q\, \overline{Q} = \overline{Q}\, Q = 1_{E_3}),$$

and K_o have the same characteristic equation, as [see (A.6)] :

(1.31) $T_r(K) = T_r(Q\, K_o\, \overline{Q}) = T_r(K_o),\ T_r(K^2) = T_r(K_o^2),\ T_r(K^3) = T_r(K_o^3).$

Reciprocally, if K and K_o have the same characteristic equation, $\forall K$, $\exists Q$ unitary such that :

$$K = Q \, K_o \, \overline{Q} \, , \quad \forall K \text{ satisfying} \quad (1.31)$$

In fact in the case where K and K_o have the same eigenvalues, and if we call Λ the diagonal matrix of these eigenvalues, S_o the basis of the eigenmodes of K_o, and S the basis of the eigenmodes of K, we have [see (1.16)]:

$$K_o = S_o \, \Lambda \, \overline{S_o} \quad \text{with} \quad \overline{S_o} \, S_o = 1_{R^3}$$

$$K = S \, \Lambda \, \overline{S} \quad \text{with} \quad \overline{S} \, S = 1_{R^3} \, .$$

Hence :

$$K = S \, \overline{S_o} \, K_o \, S_o \, \overline{S} = Q \, K_o \, \overline{Q} \, .$$

If we put :

$$S \, \overline{S_o} = Q .$$

Q correctly verifies that $\overline{Q} \, Q = Q \, \overline{Q} = 1_{E_3} \, .$

In conclusion, if α is invariant for endomorphisms K, having the same characteristic equation, α only depends on its invariants,   Q.E.D.

5.2 – Stress [5]

The deformation energy volume density, with respect to the reference volume, is given by (1.6), namely :

$$(1.32) \qquad T_r(C' \delta D) = T_r(C' \, \frac{\delta K}{2})$$

with :

$$K = \frac{\overline{\partial M}}{\partial M_o} \, \frac{\partial M}{\partial M_o}$$

Formula (1.32) is written , with (1.30) :

$$(1.33) \qquad T_r(C' \frac{\delta K}{2}) = \delta \alpha = \frac{\partial \alpha}{\partial T_r(K)} \, \delta T_r(K) + \frac{\partial \alpha}{\partial T_r(K^2)} \, \delta T_r(K^2) + \frac{\partial \alpha}{\partial T_r(K^3)} \, \delta T_r(K^3)$$

$$= f'_1 \, T_r(\delta K) + f'_2 \cdot 2 \, T_r(K \delta K) + f'_3 \cdot 3 \, T_r(K^2 \, \delta K),$$

putting :

$$f'_1 = \frac{\partial \alpha}{\partial T_r(K)} \;,\; f'_2 = \frac{\partial \alpha}{\partial T_r(K^2)}, \; f'_3 = \frac{\partial \alpha}{\partial T_r(K^3)}$$

evaluated at $M = M_o$. (1.33) is now written :

$$T_r(C'\cdot\frac{\delta K}{2}) = T_r\left(\left[f'_1\cdot 1_{E_3} + 2f'_2\cdot K + 3f'_3\cdot K^2\right]\delta K\right)$$

Hence :

(1.34) $$C' = 2\left[f'_1\cdot 1_{E_3} + 2f'_2\cdot K + 3f'_3\, K^2\right].$$

The Cauchy stress at M is obtained by (1.8), namely :

$$C = \frac{1}{\det(\frac{\partial M}{\partial M_o})} \; \frac{\partial M}{\partial M_o} \; C' \; \overline{\frac{\partial M}{\partial M_o}}.$$

(1.35) $$C = \frac{2}{\det(\frac{\partial M}{\partial M_o})} \left[f'_1\cdot \frac{\partial M}{\partial M_o}\overline{\frac{\partial M}{\partial M_o}} + 2f'_2\cdot\left[\frac{\partial M}{\partial M_o}\overline{\frac{\partial M}{\partial M_o}}\right]^2 + 3f'_3\left[\frac{\partial M}{\partial M_o}\overline{\frac{\partial M}{\partial M_o}}\right]^3\right].$$

This stress is therefore expressed as a function of $\frac{\partial M}{\partial M_o}\overline{\frac{\partial M}{\partial M_o}}$, referred to as anti-deformation.

5.3 - <u>Small deformation - linearization</u>

If the deformation D remains small, namely if $\frac{\partial M}{\partial M_o}\overline{\frac{\partial M}{\partial M_o}}$ is small with respect to 1_{E_3}, we can expand α by means of the Taylor-Maclaurin formula, starting from the value $D = 0$, namely [see (A.38)] :

(1.36) $$\alpha = \underset{\sim}{a}(D) = \underset{\sim}{a}(0) + \underset{\sim}{a}'(0)D + \frac{1}{2}\underset{\sim}{a}''(0)(D)(D) + 0(|D|^3)$$

with :

$$\frac{0(|D|^3)}{|D|^2} \to 0 \text{ when } D \to 0.$$

Generally, we adopt the following definition :

(1.37)   <u>Definition</u> : The expression <u>natural state</u> describes a state without external forces, from which we measure stresses and deformations, and from which we express or measure the constitutive law of the medium, namely the relation between stress and deformation.

If the natural state is taken as the reference state ($D = 0$), the above definition gives $C' = 0$ in this state, namely :

$$\underset{\sim}{\alpha}{}'(0) = 0$$

With the deformation energy defined to within any constant, (1.36) is written :

$$\alpha = \underset{\sim}{\alpha}(D) = \frac{1}{2}\underset{\sim}{\alpha}{}''(0)(D)(D).$$

disregarding terms above second order. Finally we put :

$$(1.38) \qquad \alpha = \frac{1}{2}\mathcal{A}(D)(D),$$

This formula is valid in any case of homogeneous, anisotropic linearization. $\mathcal{A}$ is a second order symmetrical tensor on $\vec{E}_6$, and α is a quadratic form of the deformation. We then have a clearly defined endomorphism A of $\vec{E}_6$, such that :

$$(1.39) \qquad \alpha = \frac{1}{2}\tilde{D} A D \quad \text{with} \quad A = \tilde{A}.$$

In the case of an <u>anisotropic medium</u>, A is expressed by a symmetrical matrix of R^6, still called $\mathcal{A}$, possessing $\dfrac{(6+1)6}{2} = 21$ coefficients[14,16].

In the case of an <u>isotropic medium</u>, as we have seen we can put :

$$\alpha = f(T_r(D),\ T_r(D^2),\ T_r(D^3)).$$

Let us replace D by εD provisionally, with $\varepsilon^3 = 0$ if D is small[22], and we develop this formula to the second order in ε, using the Taylor-Maclaurin formula. Thus :

$$\alpha = \varepsilon f_1'\, T_r(D) + \varepsilon^2 f_2'\, T_r(D^2) + \varepsilon^3 f_3'\, T_r(D^3)$$

$$+ \frac{1}{2}\varepsilon^2 f_1''\left[T_r(D)\right]^2 + \varepsilon^3 \left[\qquad\right] + \ldots$$

with : $\quad f_1' = \dfrac{\partial \alpha}{\partial T_r(D)}$, $\quad f_2' = \dfrac{\partial \alpha}{\partial T_r(D^2)}$, $\quad f_3' = \dfrac{\partial \alpha}{\partial T_r(D^3)}$,

$$f_1'' = \dfrac{\partial^2 \alpha}{\partial\left[T_r(D)\right]^2} , \text{ at point } D = 0.$$

The first order term for D is zero, as the reference state corresponds to $D = 0$, so that $f_1' = 0$.

Only retaining the second order terms, and changing the notations, we have :

$$(1.40) \qquad \alpha = \frac{1}{2} \left[\lambda \left[T_r(D) \right]^2 + 2 \mu \, T_r(D^2) \right]$$

λ and μ are called the Lamé coefficients.

For a kinematically admissible virtual variation of U, leading to a virtual variation δD of D, (1.40) and (1.6) give :

$$\delta \alpha = T_r \left(\left[\lambda \, T_r(D) \cdot 1_{E_3} + 2\mu \, D \right] \delta D \right) = T_r(C'\delta D) \cdot$$

Thus we have the law of linear isotropic behaviour, known as Hooke's law :

$$(1.41) \qquad C' = \lambda T_r(D) \cdot 1_{E_3} + 2 \mu D \cdot$$

Remarks : 1) In the linearized anisotropic case, we have :

$$\alpha = \frac{1}{2} \, \tilde{D} \, A \, D \quad , \qquad A = \tilde{A}$$

$$\delta \alpha = \tilde{D} \, A \, \delta D = \tilde{C}' \delta D = T_r(C'\delta D) \cdot$$

Hence :

$$(1.42) \qquad C' = AD \, , \, A = \tilde{A} \, .$$

2) In the linearized isotropic case, endomorphism A of $\vec{E}_6$ is represented by the following Cartesian coordinate matrix :

$$(1.43) \, \mathcal{A} = \begin{bmatrix} \lambda+2\mu & \lambda & \lambda & 0 & 0 & 0 \\ \lambda & \lambda+2\mu & \lambda & 0 & 0 & 0 \\ \lambda & \lambda & \lambda+2\mu & 0 & 0 & 0 \\ 0 & 0 & 0 & 2\mu & 0 & 0 \\ 0 & 0 & 0 & 0 & 2\mu & 0 \\ 0 & 0 & 0 & 0 & 0 & 2\mu \end{bmatrix}$$

5.4 – Equilibrium equations

The equilibrium equation $(1.24)_1$ can be written as a function of displacement U, taking account of the linear elastic constitutive law (1.41). If we linearize (1.3) and (1.8), retaining only the first order terms for U, we have (homogeneous medium) :

$$D_L = \frac{1}{2}\left[\frac{\partial U}{\partial M} + \overline{\frac{\partial U}{\partial M}}\right]$$

$$C' = C$$

$$\operatorname{div} C = \lambda \operatorname{div}\left[T_r\left(\frac{\partial U}{\partial M}\right)1_{E_3} + \mu\left[\frac{\partial U}{\partial M} + \overline{\frac{\partial U}{\partial M}}\right]\right]$$

$$= \lambda \frac{\partial}{\partial M}\left[T_r\left(\frac{\partial U}{\partial M}\right)\right] + \mu\left[\operatorname{div}\frac{\partial U}{\partial M} + \operatorname{div}\overline{\frac{\partial U}{\partial M}}\right]$$

$$\operatorname{div} C = \left[\lambda+\mu\right]\frac{\partial}{\partial M}\operatorname{div} U + \mu\operatorname{div}\left[\operatorname{grad} U\right].$$

$$(1.44)\qquad \overline{\operatorname{div} C} = \left[\lambda+\mu\right]\operatorname{grad}\operatorname{div} U + \mu\Delta U,$$

with : $\overline{\Delta U} = \operatorname{div}\left[\operatorname{grad} U\right].$

5.5 – Stability of material

In the case of an anisotropic linear medium, the specific energy is written by (1.39) :

$$\alpha = \frac{1}{2}\tilde{D} A D \text{ , with } D \in \vec{E}_6$$

and :

$$A = \tilde{A}.$$

This energy is positive $\forall D \neq 0$ for a stable material. As α is a positive definite form, the eigenvalues of A are real and positive. If λ_i is one of the 6 eigenvalues of A, and $D_i \in \vec{E}_6$ the corresponding eigenmode, we have :

$$AD_i = D_i\lambda_i \text{ with } \lambda_i > 0 \ (i = 1, 2, \ldots, 6),$$

but (1.42) gives the corresponding stress C'_1 , namely :

$$AD_i = D_i\lambda_i = C'_i.$$

In the isotropic case, (1.41) then gives :

$$C'_i = \lambda T_r(D_i).1_{E_3} + 2\mu D_i = D_i\lambda_i$$

namely :

(1.45) $\lambda T_r(D_i) \cdot 1_{E_3} + \left[2\mu - \lambda_i \right] D_i = 0,$

This formula demonstrates that $\left[2\mu - \lambda_i \right] D_i$ is a scalar endomorphism of $\vec{E}_3$. Two cases occur for λ_i :

a) $\lambda_i \neq 2\mu_i$; D_i is then scalar. If D_1 is the scalar mode concerned, and λ_1 the corresponding eigenvalue, (1.45) gives :

$$\left[3\lambda + 2\mu - \lambda_1 \right] D_1 = 0$$

and if $D_1 \neq 0$:

$$\lambda_1 = 3\lambda + 2\mu.$$

The corresponding deformation D_1 belongs to the one-dimensional subspace of $\vec{E}_6$, subtended by $1_{E_3} \in \vec{E}_6$. This is a deformation resulting from a volume variation only.

b) $\lambda_i = 2\mu . D_i$ is no longer scalar ; $i = 2, 3, \ldots 6$; 2μ is a quintuple eigenvalue. We know that D_i belongs to the subspace supplementary to the one-dimensional subspace already defined.

Therefore D_i is a deviator belonging to this deviator subspace of dimension 5, called $\vec{E}_5$. (1.45) is also correct if $\lambda \neq 0$:

$$T_r(D_i) = 0 \qquad i = 2, 3, \ldots, 5.$$

We also check that $D_i \perp D_1$ in $\vec{E}_6$:

$$\overset{\sim}{D_1} D_i = T_r(D_1 D_i) = D_1 T_r(D_i) = 0, \quad i = 2, 3, \ldots, 6.$$

Finally, the two distinct eigenvalues are :

$$\lambda_1 = 3\lambda + 2\mu > 0 \quad \text{and} \quad \lambda_i = 2\mu > 0, \quad (i = 2, 3, \ldots, 6).$$

<u>Remark</u> : (1.41) gives :

$$C' = C'_m + C'_d \quad \text{with} \quad C'_m = \frac{1}{3} T_r(C')$$

$$C'_m + C'_d = \lambda T_r(D) \cdot 1_{E_3} + 2\mu D$$

$$= \lambda T_r(D) \cdot 1_{E_3} + 2\mu \left[D_m 1_{E_3} + D_d \right]$$

$$= \left[3\lambda + 2\mu \right] D_m \cdot 1_{E_3} + 2\mu D_d$$

and taking the trace of the two members, we have :

$$\left[\begin{array}{l} C'_m = \left[3\lambda + 2\mu \right] D_m \\[2em] C'_d = 2\mu \, D_d \, . \end{array} \right.$$

These are equivalent formulae to (1.41), correctly giving the value of the coefficient of the volume dilatation according to the Lamé coefficients.

6. – VARIATIONAL PRINCIPLE FOR A LINEAR ELASTIC MEDIUM[17]

A medium of this type has a specific energy of deformation, given by (1.38), starting from the natural state :

$$(1.45) \qquad \alpha = \frac{1}{2} \mathcal{A}(D_L)(D_L) > 0, \ \forall \ D_L \neq 0,$$

where :

$$D_L = \underset{\sim}{D}_L(U) = \frac{1}{2} \left[\frac{\partial U}{\partial M_o} + \overline{\frac{\partial U}{\partial M_o}} \right] .$$

Under these conditions, we have :

> Principle : Solution U satisfies the following principle of minimum total potential energy :
>
> $$\text{Solution } U \Longleftrightarrow \mathcal{F}(\underset{\sim}{V}) \quad \text{min.} \quad \Big| \underset{\sim}{V} \text{ K.A.}$$
>
> with :
>
> (1.46)
>
> $$\mathcal{F}(\underset{\sim}{V}) = \int_{\Omega_o} \frac{1}{2} \mathcal{A}(D_L)(D_L) \, d\Omega_o - \int_{\underline{\Omega}_o} \overline{F} \, V \, d\overline{\Omega}_o \, .$$
>
> The last integral designates the work, in abridged from, of the given external forces.

This principle becomes a theorem, if we admit the principle of virtual work. In fact taking a kinematically admissible virtual displacement field δU, from solution $U : V = U + \delta U$.

$$\mathcal{F}(\underset{\sim}{U+\delta U}) - \mathcal{F}(\underset{\sim}{U}) = \int_{\Omega_o} \left[\frac{1}{2} \mathcal{A}(D_L + \delta D_L)(D_L + \delta D_L) - \frac{1}{2} \mathcal{A}(D_L)(D_L) \right] d\Omega_o$$

$$+ \int_{\underline{\Omega}_o} \left[\overline{F} \left[U + \delta U \right] - \overline{F} \, U \right] d\overline{\Omega}_o$$

$$= \int_{\Omega_o} \mathcal{A}\,(D_L)(\delta D_L) - \int_{\overline{\Omega_o}} \overline{F}\delta U d\overline{\Omega_o} + \int_{\Omega_o} \frac{1}{2}\mathcal{A}(\delta D_L)(\delta D_L)d\Omega_o$$

$$\underset{\sim}{\mathcal{J}}(U+\delta U) - \underset{\sim}{\mathcal{J}}(U) = \frac{1}{2}\int_{\Omega_o} \mathcal{A}\,(\delta D_L)(\delta D_L)d\Omega_o .$$

The first two integrals of the second member cancel each other, by application of the principle of virtual work, corresponding to equilibrium,
namely the stationarity of $\mathcal{J}(U)$. The first member is then positive by
virtue of (1.45). Q.E.D. $^{*)}$

7. - <u>THEOREM OF RECIPROCITY FOR A LINEARIZED HYPERELASTIC MEDIUM</u>

In a medium of this type, (1.39) and (1.42) apply :

$$\alpha = \frac{1}{2}\overset{\approx}{D}\,A\,D\,,\,A = \overset{\approx}{A}$$

$$C' = AD .$$

Let us consider two cases of loading, characterized by the subscripts 1
and 2 respectively, and the corresponding solutions, namely :

$$C_1' \,,\, D_1,\, U_1,\, f_1,\, F_1$$

$$C_2' \,,\, D_2,\, U_2,\, f_2,\, F_2,$$

with f_1 and f_2 in Ω_o, and F_1 and F_2 on Σ_{oF}, applying the principle of
virtual work referred to the reference state :

$$\int_{\Omega_o} T_r(C'\delta D)d\Omega_o - \int_{\Omega_o} \overline{f}\delta U d\Omega_o - \int_{\Sigma_{oF}} \overline{F}\delta U d\Sigma = 0 \quad \forall\,\delta\,U,\,K.A.$$

successively for each state of equilibrium, adopting a field $\delta U = U_2$ for
the first, and a field $\delta U = U_1$ for the second. We then have :

$$\int_{\Omega_o} T_r(C_1'D_2)d\Omega_o - \int_{\Omega_o} \overline{f}_1 U_2 d\Omega_o - \int_{\Sigma_{oF}} \overline{F}_1 U_2 d\Sigma_o = 0$$

$^{*)}$ In the case of non-linear deformations, and under certain hypotheses,
we shall see that the minimum total potential energy corresponds to
stability.

$$\int_{\Omega_o} T_r(C_2' D_1)\, d\Omega_o - \int_{\Omega_o} \overline{f}_2 U_1\, d\Omega_o - \int_{\Sigma_{oF}} \overline{F}_2 U_1\, d\Sigma_o = 0.$$

Then replacing C_1' by AD_1 and C_2' by AD_2, and subtracting, we have :

$$\int_{\Omega_o} \left[\overset{\circ}{D}_1\, AD_2 - \tilde{D}_2\, AD_1 \right] d\Omega_o - \int_{\Omega_o} \left[\overline{f}_1 U_2 - \overline{f}_2 U_1 \right] d\Omega_o$$

$$- \int_{\Sigma_{oF}} \left[\overline{F}_1 U_2 - \overline{F}_2 U_1 \right] d\Sigma_o = 0.$$

The first integral is zero, as $A = \tilde{A}$. We are left with :

$$(1.47) \qquad \int_{\Omega_o} \overline{f}_1 U_2\, d\Omega_o + \int_{\Sigma_{oF}} \overline{F}_1 U_2\, d\Sigma_o = \int_{\Omega_o} \overline{f}_2 U_1\, d\Omega_o + \int_{\Sigma_{oF}} \overline{F}_2 U_1\, d\Sigma_o,$$

this formula constituting the theorem of reciprocity already mentioned.

8. – STRESS FUNCTIONS[12, 18, 19]

Let us consider the problem of internally homogeneous equilibrium, namely that of a continuous medium occupying a domain Ω of E_3, and subjected only
to a surface distribution F of external forces over the part Σ_F of the
boundary Σ of Ω, the displacements being imposed on the complementary part
Σ_U. For a non-polarized medium, we have seen that the stress C satisfies the
internal equilibrium equations [see (1.23)] :

$$(1.48) \qquad \begin{bmatrix} \text{div } C = 0 \\[2mm] C = \overline{C}. \end{bmatrix}$$

If C is a field of endomorphism of class C^1, defined in a simply connected open set ω of Ω, we have the following result :

$$(1.49) \qquad \begin{bmatrix} \textbf{div } C = 0 \\[2mm] C = \overline{C} \end{bmatrix} \Longleftrightarrow \begin{bmatrix} \exists\, B = \overline{B} \ \text{ such that } C = \text{rot } \overline{\text{rot } B} \end{bmatrix},$$

where B is any Hermitian endomorphism of $\vec{E}_3$, the value of a class C^2 field
for example, defined in ω; B is the stress function, represented in any

basis of E_3, by 6 scalar functions. The second member of (1.49) is called the Beltrami representation.

In fact in any simply connected open set of Ω, $(1.48)_1$ and Poincaré's theorem give : $\exists$ endomorphism field A of $\vec{E}_3$, of class C^2, in this open set, such that [see (A.41), (A.48) and (A.50)] :

$$(1.50) \qquad C = \text{rot } A \ .$$

A is defined to within an irrotational endomorphism, namely to within any vector gradient U_1, therefore :

$$(1.51) \qquad A = A_1 + \text{grad } U_1 \ , \ (\text{grad } U_1 = \frac{\overline{\partial U_1}}{\partial M}) \ .$$

But $(1.48)_2$ then gives :

$$\text{rot } A = \overline{\text{rot } A}.$$

Therefore $\forall$ the constant vector fields V_1 and V_2 :

$$\overline{V}_1 \text{ rot } AV_2 = \overline{V}_2 \text{ rot } AV_1$$

or :

$$\overline{\text{rot } AV_1} \ V_2 = \overline{\text{rot } AV_2} \ V_1$$

so that [see (A.56)] :

$$\text{div } \left[i(AV_1)(V_2) - i(AV_2)(V_1) \right] = 0,$$

which is again a closure condition for the vector appearing between brackets.

Therefore in this same open set, $\exists$ a field of class C^3 of value $U \in \vec{E}_3$, such that :

$$i(AV_1)(V_2) - i(AV_2)(V_1) = \text{rot } U$$

and defined to within the gradient of any scalar u, such that :

$$U = U_2 + \text{grad } u, \ (\text{grad } u = \frac{\partial u}{\partial M}) \ .$$

Furthermore $\forall \ V_3 \in \vec{E}_3$:

$$\overline{V}_3 \left[i(AV_1)(V_2) + i(V_1)(AV_2) \right] = \overline{V}_3 \text{ rot } U$$

or, in succession :

$$\text{vol } (AV_1)(V_2)(V_3) + \text{vol } (V_1)(AV_2)(V_3) + \text{vol } (V_1)(V_2)(AV_3)$$

$$- \text{vol } (V_1)(V_2)(AV_3) = \overline{V_3} \text{ rot } U$$

$$T_r(A) \text{ vol}(V_1)(V_2)(V_3) - \text{vol } (V_1)(V_2)(AV_3) = \overline{V_3} \text{ rot } U$$

$$\overline{V_3} \, i(V_1)(V_2)T_r(A) - \overline{V_3} \, \overline{A} \, i(V_1)(V_2) = \overline{V_3} \text{ rot } U$$

$$\left[T_r(A) \cdot 1_{E_3} - \overline{A} \right] i(V_1)(V_2) = \text{ rot } U \cdot$$

But as rot U is linear for $i(V_1)(V_2)$, $\exists$ an endomorphism B of $\vec{E}_3$, the value
of a class C^3 field, defined in ω, such that :

$$\text{rot } U = \text{rot } B \, i(V_1)(V_2),$$

and defined to within any vector gradient, such that :

$$B = B_1 + \text{grad } U_3 .$$

Thus :

$$(1.52) \qquad T_r(A) \cdot 1_{E_3} - \overline{A} = \text{rot } B$$

now :

$$T_r(T_r(A) \cdot 1_{E_3} - \overline{A}) = T_r(\text{rot } B)$$

or :

$$T_r(A) = \frac{1}{2} T_r(\text{rot } B).$$

Therefore (1.52) becomes :

$$(1.53) \qquad \overline{A} = \frac{1}{2} T_r(\text{rot } B) \cdot 1_{E_3} - \text{rot } B.$$

Let us then decompose B into its Hermitian component B_S and anti-
Hermitian component B_A. We know that :

$$T_r(\text{rot } B_S) = 0 \text{ as } B_S = \overline{B_S}.$$

In addition, we have the property :

$$(1.54) \qquad \frac{1}{2} \, \text{rot} \left[T_r(\text{rot } B_A).1_{E_3} \right] = \text{rot } \overline{\text{rot } B_A}.$$

In fact since B_A is anti-Hermitian on $\vec{E}_3$, $\exists$ a single vector V, such that [see (A.20)] :

$$B_A = i(V).$$

Now :

$$\text{rot } B_A = \text{rot } i(V) = \frac{\partial V}{\partial M} - \text{div } V.1_{E_3} , \quad [\text{see } (A.58)]$$

and :

$$\frac{1}{2} \, T_r(\text{rot } B_A) = \frac{1}{2} \, T_r(\frac{\partial V}{\partial M}) - \frac{1}{2} \, T_r(\text{div } V.1_{E_3}) = - \, \text{div } V.$$

Therefore :

$$\frac{1}{2} \, \text{rot} \left[T_r(\text{rot } B_A).1_{E_3} \right] = -\text{rot} \left[\text{div } V.1_{E_3} \right].$$

But we also have :

$$\text{rot } \overline{\text{rot } B_A} = \text{rot } \frac{\partial V}{\partial M} - \text{rot} \left[\text{div } V.1_{E_3} \right] = - \, \text{rot} \left[\text{div } V.1_{E_3} \right],$$

which proves (1.54).

Then taking account of (1.54), (1.53) gives :

$$(1.55) \qquad \text{rot } A = - \, \text{rot } \overline{\text{rot } B_S} \ .$$

Finally, (1.55) and (1.50) correctly give the property previously indicated (1.49), or more precisely : $\exists$ some Hermitian endomorphism field B, defined on ω, of class C^3, such that :

$$C = \text{rot } \overline{\text{rot } B} \ , \ B = \overline{B} = B_1 + \text{grad } U_3 + \overline{\text{grad } U_3}, \ (B_1 = \overline{B_1}) \ .$$

It can be demonstrated that we can choose U_3, in certain cases, so as to have a stress function with three scalar components.

Remarks : 1) In the bidimensional case, we find that there exists in a simply connected open set ω, of E_2 (gauged and Euclidian), a scalar field u of class C^3, such that[12] :

$$\begin{bmatrix} \mathbf{div}\ C = 0 \\[1em] C = \bar{C} \end{bmatrix} \Leftrightarrow \left[C = i_2\ \mathbf{grad}\ \left[i_2\ \mathbf{grad}\ u \right] \right]$$

u is an Airy function.

2) This method makes it possible to find easily the conditions for global closure for a multiply connected domain Ω, using de Rham's theorem[12]. (See Appendix § 13).

9. – POLARIZED MEDIA[2,11,20]

In paragraph 4, we saw that while the virtual deformation energy density is constant under any rigid body rotation, the Cauchy stress C is Hermitian. This result fails in the case of polarized media.

A medium of this type is constituted by material points, each volume element of which is in a state of equilibrium under the action of forces and couples. For a virtual displacement field δU at M, the rotational vector of a volume element $d\Omega$ is equal to :

$$(1.57) \qquad \delta\theta = \frac{1}{2}\ \mathrm{rot}\ \delta U.$$

In fact each differential vector dM undergoes a virtual displacement :

$$\delta dU = d\delta U = i(\delta\theta)(dM).$$

Hence :

$$\frac{\partial \delta U}{\partial M} - \overline{\frac{\partial \delta U}{\partial M}} = i(\mathrm{rot}\ \delta U) = 2\ i(\delta\theta) \qquad (Q.E.D.)$$

Let us consider a domain Ω with a boundary Σ, occupied by a medium of this type, and an element Ω' with a boundary Σ', subjected to a force volume density f, a force surface density F', a couple volume density m, and a surface couple density m'. The complement of Ω' in Ω in fact exerts actions F' and m' on Σ'.

If we call $\delta \mathcal{C}'_i$ the virtual work of the internal forces on $\overline{\Omega'}$ in a displacement δU, the principle of virtual work gives the equilibrium of this element, under the action of the virtual work of the external forces δ_e,

namely :

$$\delta\mathcal{C}_i + \delta\mathcal{C}_e = 0, \forall \delta U \; K.A.$$

or again :

$$(1.58) \quad \delta\mathcal{C}_i + \int_{\Omega'} \overline{f}\delta U d\Omega' + \int_{\Sigma'} \overline{F}\delta U d\Sigma' + \int_{\Omega'} \overline{m}\delta\theta d\Omega' + \int_{\Sigma'} \overline{m}'\delta\theta \; d\Sigma' = 0, \forall \delta U \; K.A.$$

with (1.57).

Now the equilibrium of element $d\Sigma'$ of Σ', of unit normal n, implies the existence of an endomorphism $\overline{C}$ *), such that :

$$\overline{C} \, n = F'$$

and an endomorphism $\overline{C}_c$, such that :

$$\overline{C}_c \, n = m'$$

C and C_c are the stress and couple stress respectively, defined at any point of Ω, as Ω' is indeterminate.   (1.58) is then written :

$$\delta\mathcal{C}_i + \int_{\Omega'} \overline{f}\delta U d\Omega' + \int_{\Omega'} \overline{m} \, \delta\theta d\Omega' + \int_{\Sigma'} \left[\overline{\overline{C}}.n\delta U + \overline{\overline{C}}_c \, n\delta\theta \right] d\Sigma' = 0$$

$\forall \delta U$ K.A., with (1.57).

By application of the Stokes formula :

$$\delta\mathcal{C}_i + \int_{\Omega'} \left[\mathrm{div}\left[C\delta U \right] + \mathrm{div}\left[C_c\delta\theta \right] + \overline{f}\delta U + \overline{m} \, \delta\theta \right] d\Omega' = 0$$

$$(1.59) \quad \left[\delta\mathcal{C}_i + \int_{\Omega'} \left[\left[\mathrm{div} \, C + \overline{f} \right]\delta U + \left[\mathrm{div} \, C_c + \overline{m} \right]\delta\theta + T_r(C\frac{\partial\delta U}{\partial M}) + T_r(C_c\frac{\partial\delta\theta}{\partial M}) \right] d\Omega' \right.$$

$$\left. \forall\delta U, \; K.A., \; \text{with (1.57)}. \right.$$

Let us consider a constant vector field δU in Ω'. Under these conditions, $\delta\mathcal{C}_i = 0$ and $\delta\theta = 0$. As Ω' is indeterminate in Ω, we find the following conditions required for equilibrium :

$$(1.60) \quad \mathrm{div} \, C + \overline{f} = 0 \; \text{in} \; \Omega.$$

---

*) The hypothesis whereby this mapping is an endomorphism, does not lead
   to any subsequent contradiction.

(1.59) therefore becomes :

$$\left[(1.61)\quad \left[\delta \mathcal{C}_i + \int_{\Omega'}\left[\left[\mathrm{div}\ C_c + \overline{m}\right]\delta\theta + T_r(C\ \frac{\partial\delta U}{\partial M}) + T_r(C_c\ \frac{\partial\delta\theta}{\partial M})\right]d\Omega' = 0, \forall \delta U\ \mathrm{K.A.},\right.\right.$$

with (1.57).

Let us now take a field δU, such that $\delta\theta$ is constant in Ω', then $\delta\mathcal{C}_i = 0$ and :

$$\delta U = i(\delta\theta)(M),\quad d\delta U = i(\delta\theta)dM,\quad \frac{\partial\delta U}{\partial M} = i(\delta\theta).$$

In this case, (1.61) becomes :

$$(1.62)\qquad \int_{\Omega'}\left[\left[\mathrm{div}\ C_c + \overline{m}\right]\delta\theta + T_r[C\ i(\delta\theta)]\right]d\Omega' = 0,\quad \forall \delta\theta = c^t$$

now :

$$T_r(C\ i(\delta\theta)) = T_r(C_A\ i(\delta\theta))\ \mathrm{with}\ C_A = \frac{C - \overline{C}}{2}.$$

Let us put, provisionally, [see (A.20)]

$$C_A = i(W)\quad \mathrm{where}\ W = i^{-1}(C_A)$$

$$T_r(C\ i(\delta\theta)) = T_r(i(W)\ i(\delta\theta)) = -2\ \delta\theta\ W = -2\ \overline{\delta\theta}\ i^{-1}(C_A).$$

And (1.62) then gives another equilibrium equation :

$$\overline{\mathrm{div}\ C_c} - 2\ i^{-1}(C_A) + m = 0\ \mathrm{in}\ \Omega,$$

which, taken with (1.60), therefore gives the system :

$$(1.63)\qquad \left[\begin{array}{l}\overline{\mathrm{div}\ C} + f = 0 \\[2mm] \overline{\mathrm{div}\ C_c} - 2\ i^{-1}(C_A) + m = 0\end{array}\right\}\ \mathrm{in}\ \Omega.$$

(1.63) can be written with C_A eliminated, as $(1.63)_2$ gives :

$$\frac{1}{2}\ i\ \overline{(\mathrm{div}\ C_c + m)} = C_A,$$

$$\mathrm{div}\ C_A = \frac{1}{2}\ \mathrm{div}\left[i\ \overline{(\mathrm{div}\ C_c + m)}\right] = \frac{1}{2}\ \mathrm{rot}\left[\overline{\mathrm{div}\ C_c + m}\right],$$

which in $(1.63)_1$, gives, with $(1.63)_2$:

$$(1.64) \quad \begin{cases} \overline{\text{div } C_S} + \frac{1}{2} \text{rot}\left[\overline{\text{div } C_c} + m\right] + f = 0, \quad C_S = \frac{C + \overline{C}}{2}, \\[2ex] \overline{\text{div } C_c} - 2\, i^{-1}(C_A) + m = 0 \end{cases}$$

Equation (1.61), taken with $(1.63)_2$, enables us to calculate $\delta\mathcal{C}_1$. We find :

$$(1.65) \quad \delta\mathcal{C}_i = - \int_{\Omega'} T_r(C_S \frac{\partial \delta U}{\partial M}) + T_r(C_c \frac{\partial \text{ rot } \delta U}{\partial M}) \quad \forall\, \delta U \text{ K.A.}$$

This formula, associated with (1.58), written in Ω and on Σ, enables us to write a weak formulation, also given by (1.64) and the boundary conditions, in which the anti-symmetrical component C_A no longer intervenes in Ω.

In the case where $C_c = m = f = 0$, we find $C_A = 0$, but we can assume this last condition to be satisfied in Ω in the mean, and not locally. In this case, there exists in a simply connected open set an endomorphism field A of $\vec{E}_3$, of class C^2, such that [see (1.50)] :

$$C = \text{rot } A .$$

A is a first order stress function, which has been used in this perspective[21].

CHAPTER II

THE FINITE ELEMENT METHOD

1. THE DISPLACEMENT METHOD IN STATIC PROBLEMS
2. OTHER TYPES OF ELEMENTS AND STATIC PROBLEMS
3. OTHER TYPES OF PROBLEMS

1. – THE DISPLACEMENT METHOD IN STATIC PROBLEMS

1.1. – General[1-12]

The Finite Element Method is a discretization method, whose purpose is to find an approximate solution to a physical problem, set in a variational form, namely a weak approximate solution to the problem.

The method consists in partitioning the medium into finite elements, limited by boundaries, writing the global functional to be minimized as a sum of the relative functionals for each element.

In each element, the unknown fields are represented by a Ritz method, using a functional basis particular to each element, the basis functions of which are called interpolation functions (or shape functions).

Points, called nodes, are placed on the boundaries of the elements, and in their domain where appropriate, and the coefficients of the Ritzian representation are expressed according to the nodal values of the unknown fields, these values being called degrees of freedom, constituting the unknown quantities of the discretized problem.

The assembly of adjacent elements is an operation required to express the global functional as the sum of elementary functionals. This assembly is achieved by writing the equality of the unknown nodal values at the common nodes of the elements. Once assembly has been carried out, there only remain the independent nodal unknowns, which constitute the definitive degrees of freedom of the problem.

The interpolation functions are chosen, in principle, so as to achieve continuity of the unknown fields, when crossing the common boundary between elements (conformity).

The boundary of the global domain is approached by joining the elementary boundaries of the elements concerned.

In conclusion, the method is presented as a piecewise Ritz method, designed to convert the problem into algebraic form using nodal unknowns.

For example, let us take a two-dimenstional elastic medium under static
load. The medium occupies a domain Ω with a boundary Σ. It is given a
force volume density field f in Ω, and a force surface density field F on
part Σ_F of Σ. On the complementary part Σ_U of Σ, a displacement field U_d
is given. Here is an elementary description of the method.

We consider the linearized problem, and we call X a point of $\Omega \subset \mathbb{R}^2$, and
in Cartesian coordinates, we call σ the matrix representing the stress C,
and ε the matrix representing the deformation D, such that :

$$X = \begin{bmatrix} x \\ y \end{bmatrix} ; \quad U = \underset{\sim}{U}(X) = \begin{bmatrix} u \\ v \end{bmatrix} ; \quad \sigma = \begin{bmatrix} \sigma_x & \sigma_{xy} \\ \sigma_{xy} & \sigma_y \end{bmatrix} ; \quad \varepsilon = \begin{bmatrix} \varepsilon_x & \varepsilon_{xy} \\ \varepsilon_{xy} & \varepsilon_y \end{bmatrix}$$

$$(2.1) \qquad \varepsilon = \frac{1}{2} \left[\frac{\partial U}{\partial X} + \overline{\frac{\partial U}{\partial X}} \right] .$$

U must satisfy the equation [see (1.22)] :

$$(2.2) \qquad \int_\Omega T_r(\sigma \delta \varepsilon) d\Omega - \int_\Omega \overline{f} \delta U d\Omega - \int_{\Sigma_F} \overline{F} \delta U d\Sigma = 0, \ \forall \delta U \ \text{K.A.}$$

where :

$$\delta \varepsilon = \frac{1}{2} \left[\frac{\partial \delta U}{\partial X} + \overline{\frac{\partial \delta U}{\partial X}} \right] ,$$

and where σ is related to ε by a linear constitutive equation. σ
and ε are second order matrices of $\mathbb{R}^2$, which can be considered as vectors
(columns) of $\mathbb{R}^3$, putting for example :

$$(2.3) \qquad \sigma = \begin{bmatrix} \sigma_x \\ \sigma_y \\ \sigma_{xy} \end{bmatrix} \quad \varepsilon = \begin{bmatrix} \varepsilon_x \\ \varepsilon_y \\ 2\varepsilon_{xy} \end{bmatrix} = \begin{bmatrix} \dfrac{\partial u}{\partial x} \\ \dfrac{\partial v}{\partial y} \\ \dfrac{\partial u}{\partial y} + \dfrac{\partial v}{\partial x} \end{bmatrix} .$$

Thus :

$$T_r(\sigma\delta\varepsilon) = \overline{\sigma\delta\varepsilon}$$

The linear constitutive equation is then written :

(2.4) $\sigma = \mathcal{A}\varepsilon$, where $\mathcal{A}$ is a matrix of R^3,

with :

$$\mathcal{A} = \overline{\mathcal{A}}$$

We look for an approximate solution. Decomposing Ω into elements which can be triangular for example, numbered e, and in each triangle we represent the unknown solution U by a linear combination of basis functions $T_i(X)$, with finite number m, so that :

(2.5)
$$U = \underset{\sim}{U}(X) = \begin{bmatrix} \underset{\sim}{u}(x,y) \\ \\ \underset{\sim}{v}(x,y) \end{bmatrix} = \begin{bmatrix} \underset{\sim}{u}_1(x,y).{}^1\alpha + \underset{\sim}{u}_2(x,y).{}^2\alpha + \ldots + \underset{\sim m}{u}(x,y).{}^m\alpha \\ \\ \underset{\sim}{v}_1(x,y).{}^1\beta + \underset{\sim}{v}_2(x,y).{}^2\beta + \ldots + \underset{\sim m}{v}(x,y).{}^m\beta \end{bmatrix}.$$

The nodes being naturally placed at the apexes of each triangle, and we shall take as unknown nodal values, the displacements of each node. We number the three nodes of a triangle provisionally 1, 2 and 3. Their coordinates and displacements will then be, respectively :

$$\begin{bmatrix} x_1 \\ y_1 \end{bmatrix}, \begin{bmatrix} q_{u1} \\ q_{v1} \end{bmatrix}, \begin{bmatrix} x_2 \\ y_2 \end{bmatrix}, \begin{bmatrix} q_{u2} \\ q_{v2} \end{bmatrix}, \begin{bmatrix} x_3 \\ y_3 \end{bmatrix}, \begin{bmatrix} q_{u3} \\ q_{v3} \end{bmatrix}.$$

Then limiting ourselves to 2x3 basis functions, we have the following equations :

(2.6)
$$\begin{bmatrix} q_{u1} = \underset{\sim}{u}_1(x_1,y_1).{}^1\alpha + \underset{\sim}{u}_2(x_1,y_1).{}^2\alpha + \underset{\sim 3}{u}(x_1,y_1).{}^3\alpha \\ \\ q_{v1} = \underset{\sim}{v}_1(x_1,y_1).{}^1\beta + \underset{\sim}{v}_2(x_1,y_1).{}^2\beta + \underset{\sim 3}{v}(x_1,y_1).{}^3\beta \\ \\ q_{u2} = \underset{\sim}{u}_1(x_2,y_2).{}^1\alpha + \underset{\sim}{u}_2(x_2,y_2).{}^2\alpha + \underset{\sim 3}{u}(x_2,y_2).{}^3\alpha \\ \\ q_{v2} = \underset{\sim}{v}_1(x_2,y_2).{}^1\beta + \underset{\sim}{v}_2(x_2,y_2).{}^2\beta + \underset{\sim 3}{v}(x_2,y_2).{}^3\beta \\ \\ q_{u3} = \underset{\sim}{u}_1(x_3,y_3).{}^1\alpha + \underset{\sim}{u}_2(x_3,y_3).{}^2\alpha + \underset{\sim 3}{u}(x_3,y_3).{}^3\alpha \\ \\ q_{v3} = \underset{\sim}{v}_1(x_3,y_3).{}^1\beta + \underset{\sim}{v}_2(x_3,y_3).{}^2\beta + \underset{\sim 3}{v}(x_3,y_3).{}^3\beta \end{bmatrix}$$

System (2.6) can be written, for element e :

$$(2.7) \qquad \begin{bmatrix} {}^e q_u \\ \\ {}^e q_v \end{bmatrix} = \mathcal{C} \begin{bmatrix} \alpha \\ \\ \beta \end{bmatrix}$$

with obvious notations, where $\mathcal{C}$ is a matrix of R^6, assumed to be regular, whence :

$$(2.8) \qquad \begin{bmatrix} \alpha \\ \\ \beta \end{bmatrix} = \mathcal{C}^{-1} \begin{bmatrix} {}^e q_u \\ \\ {}^e q_v \end{bmatrix}$$

(2.5) can then be written, taking account of (2.8) :

$$U = \underset{\sim}{U}(X) = \begin{bmatrix} \underset{\sim}{T}_{u1}(X) & 0 & \underset{\sim}{T}_{u2}(X) & 0 & \underset{\sim}{T}_{u3}(X) & 0 \\ \\ 0 & \underset{\sim}{T}_{v1}(X) & 0 & \underset{\sim}{T}_{v2}(X) & 0 & \underset{\sim}{T}_{v3}(X) \end{bmatrix} \begin{bmatrix} {}^e q_{u1} \\ {}^e q_{v1} \\ {}^e q_{u2} \\ {}^e q_{v2} \\ {}^e q_{u3} \\ {}^e q_{v3} \end{bmatrix}$$

or again :

$$(2.9) \qquad U = \underset{\sim}{T}(X).{}^e q$$

Taking account of (2.3) and (2.9), the deformation ε for a point X of element e is written :

$$(2.10) \qquad \varepsilon = \underset{\sim}{\mathcal{B}}(X).{}^e q \quad , \quad \delta\varepsilon = \underset{\sim}{\mathcal{B}}(X).\delta^e q.$$

Then, using (2.10) and (2.4) :

$$\sigma = \mathcal{A}\,\underset{\sim}{\mathcal{B}}(X).{}^e q$$

whence :

$$(2.11) \qquad \overline{\sigma\delta\varepsilon} = \overline{{}^e q\,\underset{\sim}{\mathcal{B}}(X)}\,\mathcal{A}\underset{\sim}{\mathcal{B}}(X)\,\delta^e q.$$

For each element e, the left hand side member of (2.2) is then written :

$$(2.12) \qquad \left[\overline{{}^{e}q} \int_{\Omega_e} \overline{\underset{\sim}{\mathcal{B}}(X)} \mathcal{A}\underset{\sim}{\mathcal{B}}(X)\,d\Omega - \int_{\Omega_e} \overline{\underset{\sim}{f}(X)} \underset{\sim}{\tau}(X)\,d\Omega - \int_{\Sigma_{Fe}} \overline{\underset{\sim}{F}(X)} \underset{\sim}{\tau}(X)\,d\Sigma \right] \delta {}^{e}q .$$

In (2.12), Σ_{Fe} can be empty if $\Sigma_{Fe} \cap \Sigma_{F} = \emptyset$. The calculation of (2.12) is generally performed by using an appropriate and convenient local mapping for the element e considered.

Formula (2.12) can be written :

$$(2.13) \qquad \left[\overline{{}^{e}q}\, K_e - \overline{{}^{e}Q} \right] \delta {}^{e}q \qquad \text{avec } K_e = \overline{K_e} .$$

The matrix K_e is called the <u>elementary stiffness matrix</u>, and ${}^{e}Q$ is the column of generalized external forces, relative to the unknown nodal values ${}^{e}q$.

For <u>assembly</u>, the most natural method consists in representing the unknown nodal displacements ${}^{e}q$, in a reference system common to all elements, then writing the equality of the common node displacements.

Let us then call q the column formed by the set of preceding sub-columns ${}^{e}q$, the nodes and elements being numbered in an appropriate order, which we will indicate in due course. Let us call S' the general change of reference system matrix, and q' the column of non-assembled nodal displacements representing q in S', giving :

$$(2.14) \qquad q = S'q'.$$

or, for one element e :

$${}^{e}q = {}^{e}S'q'$$

S', or sub-matrices ${}^{e}S$, are obviously regular. It is then easy to write the equation for the components of q', relative to the common nodes, giving the linear relation :

$$(2.15) \qquad q' = S''q'',$$

where q" this time represents the column of nodal unknowns after assembly, this column having a dimension less than that of q'. S" is therefore a rectangular matrix.

$$q'' \in \mathbb{R}^n$$

if n is the number of the final independent unknowns.

Calculation of (2.2), discretized, is obtained by summing all elementary energies, such as (2.12), so that :

$$\sum_e \left[\overline{^eq}\ K_e - \overline{^eQ} \right] \delta\, ^eq$$

or otherwise :

$$\left[\overline{q}\ K - \overline{Q} \right] \delta q$$

with :

$$K = \begin{bmatrix} K_1 & & & & \\ & \ddots & & & \\ & & K_e & & \\ & & & \ddots & \\ & & & & K_E \end{bmatrix}$$

Then (2.2) becomes, by (2.14) and (2.15) :

$$\left[\overline{q'}\ \overline{S'}\ K\ S' - \overline{Q}\ S' \right] \delta q' = 0, \quad \forall \delta q' \quad \text{satisfying (2.15).}$$

So :

$$\left[\overline{q''}\ \overline{S''}\ \overline{S'}\ K\ S'S'' - \overline{Q}\ S'\ S'' \right] \delta q'' = 0, \quad \forall \delta q''.$$

Namely :

$$\left[\overline{S''}\ \overline{S'}\ K\ S'\ S'' \right] q'' = \overline{S''}\ \overline{S'}\ Q,$$

and finally :

$$(2.16) \quad \begin{bmatrix} K''q'' = Q'' \quad, \quad K'' = \overline{K''} \\ \text{with :} \\ K'' = \overline{S''}\ \overline{S'}\ K\ S'\ S'', \ Q'' = \overline{S''}\ \overline{S'}\ Q. \end{bmatrix}$$

K" is the global stiffness matrix for the assembled system, q" is the
column of the degrees of freedom for the assembled system, and Q" is the
column of generalized external loads relative to the unknown quantity q".

The static problem thus discretized and converted to algebraic form,
comes down, as we see, to the resolution of a matrix equation for which
appropriate algorithms exist, as we know, the matrices in question being
symmetrical.

1.2. - <u>Remarks and complements</u>

1) <u>Plane problems</u>

The problem taken as an example was bidimensional. It provides the resolution of plane problems with plane stresses or plane deformations, the
calculation of the stress or deformation relative to the third dimension
then being made very simply, as appropriate, by writing that one or other
of these quantities is zero, using the three dimensional constitutive law.

2) <u>Interpolation functions</u>

The choice of interpolation functions is generally based on the following
observation.

A Ritz method applied globally to the complete structure, requires the
adoption of a carefully chosen functional basis, to allow for the representation of the solution in sufficiently accurate fashion, with the minimum of
terms in the space of the kinematically admissible displacement fields.

It is clear that, for given kinematic and static conditions, and which
are more often highly varied, the choice of such a basis implies a prior
knowledge of at least the general form of the solution, or its main characteristics, and this can only be envisaged in very special cases.

On the contrary, the subdivision of the domain into sufficiently small
elements has the effect of decomposing the solution into parts, the simplicity of which is inversely proportional to the size of the elements. It is
then easy to imagine, that in each element, a polynominal representation
(linear, quadratic, cubic, etc.) is sufficient to provide a correct repre-

sentation of varied solutions, each part being connected to the other at
common boundaries, so that the representation be continuous.

It is also clearly understandable that the degree of the polynomial interpolation should be higher, as the dimensions of the elements are larger.

Going back to the previous example, with the three-node triangular elements, we can adopt for example the following linear representation with
six α and β parameters, for (2.5) :

$$U = \begin{bmatrix} \underset{\sim}{u}(x,y) \\ \\ \underset{\sim}{v}(x,y) \end{bmatrix} = \begin{bmatrix} \alpha_1 + \alpha_2 x + \alpha_3 y \\ \\ \beta_1 + \beta_2 x + \beta_3 y \end{bmatrix},$$

the number of which is compatible with the number of nodal degrees of freedom.　This type of representation leads to particularly simple calculations,
but at the cost of a partition into a correspondingly high number of elements.

We also observe that this linear representation gives constant deformations in each element, and consequently constant stresses in linear elasticity. The field of deformation is then discontinuous for the complete structure.

3) Conformity

As indicated previously, we are restricting ourselves to finding a continuous solution U in $\overline{\Omega}$, under conditions which have been specified, and
where theorems of existence and uniqueness exist.

It is important to know whether the preceding method is capable of
providing a continuous approximate solution. This is precisely the case if
the approximate unknown field U is continuous, after assembly, where the
common interelement boundaries are crossed. These elements are then described as conforming ones.

In the previous example of linear interpolation, the displacement along
one side of a triangle is also linear, and its representation will therefore be unique according to displacement of the two edge and nodes. Assembly
of the elements has the effect of equalizing the displacement of nodes
located on the sides common to two adjacent triangles, and the displacement

will indeed be continuous all along this side. The preceding triangular
element is indeed conforming .

The utilization of non-conforming elements is nevertheless possible, and
even inevitable as we shall see, in certain types of element. This non-
conformity introduces an error which can be evaluated with respect to an
energy norm. It is sufficient to calculate the total work of the boundary
stresses found for each element.

Non-conformity has an influence on the type of convergence, according to
the number of elements introduced [*].

4) <u>Assembly of elements and numbering</u>

Let us consider a structure comprising two finite elements, having a part
of their boundary common, and let us call $\begin{bmatrix} {}^1q \\ {}^2q \end{bmatrix}$ the column of the degrees of
freedom of element e_1, and $\begin{bmatrix} {}^3q \\ {}^4q \end{bmatrix}$ the column for element e_2 before assembly.

We assume that ${}^2q$ is the sub-column of the degrees of freedom for the
nodes of e_1 on the common side, with ${}^3q$ the corresponding sub-column for e_2.

The elastic force of the two non-assembled elements is written :

$$Kq = \begin{bmatrix} {}^1K_1 & {}^1K_2 & & \\ \overline{{}^1K_2} & {}^2K_2 & & 0 \\ & & {}^3K_3 & {}^3K_4 \\ 0 & & \overline{{}^3K_4} & {}^4K_4 \end{bmatrix} \begin{bmatrix} {}^1q \\ {}^2q \\ {}^3q \\ {}^4q \end{bmatrix} .$$

Assembly consists in writing :

$${}^2q = {}^3q,$$

---

which gives the change of variables :

$$
q = \begin{bmatrix} {}^1q \\ {}^2q \\ {}^3q \\ {}^4q \end{bmatrix} = \begin{bmatrix} \underline{1} & 0 & 0 \\ 0 & \underline{1} & 0 \\ 0 & \underline{1} & 0 \\ 0 & 0 & \underline{1} \end{bmatrix} \begin{bmatrix} {}^1q \\ {}^2q \\ {}^4q \end{bmatrix} = S'' \, q''
$$

with the preceding notations.

The elastic force for the assembled system is then written with :

$$
K'' = \overline{S''} \, K \, S''
$$

$$
K''q'' = \begin{bmatrix} {}^1K_1 & {}^1K_2 & 0 \\ \overline{{}^1K_2} & {}^2K_2 + {}^3K_3 & {}^3K_4 \\ 0 & \overline{{}^3K_4} & {}^4K_4 \end{bmatrix} \begin{bmatrix} {}^1q \\ {}^2q \\ {}^4q \end{bmatrix} .
$$

By a judicious numbering of the degrees of freedom, the structure of the global stiffness matrix has a diagonal band, allowing the use of appropriate inversion algorithms in an economic manner. In this case, we generally seek to reduce the mean or maximum width of the band. However, modern methods of computer storage data have now largely modified this requirement.

To reduce the bandwidth, the numbering should be such that all coupled elements have degrees of freedom the numbering of which is as close as possible. Certain relatively general rules exist, but the case of multiply connected domains remains a matter of delicate optimization.

5) Convergence[13-18]

Under certain hypotheses, and where a unique exact solution exists, there are cases where the convergence of the method can be proved with respect to the fineness of the modelling, namely when the number of elements of a

given type increases indefinitely, each element having dimensions tending
towards zero.

Convergence in the energy sense consists in taking the deformation
energy (assumed to be quadratic) as the square of the norm of the solution.

In this case, it is clear that the convergence will be monotonic if, for
finer and finer idealizations, each partition contains the preceding one.
In fact, the solution satisfies the principle of minimum total potential
energy. This minimum is the lower, the greater the number of degrees of
freedom, in other words, the constraints will be looser and looser for the
same class of approximate solutions, namely of interpolation. This is true
for a conforming element idealization. Non-conformity compromises this
result.

Nevertheless, it should be noted that the asymptotic tendency of the
solution towards a certain value, provides no guarantee of proximity or
convergence towards the exact solution. A necessary criterion normally
adopted, consists in utilizing elements possessing the property of admitting
constant deformations, and in particular null for rigid body displacements,
in each element. This is justified if we consider elements the dimension of
which tends towards zero.

Apart from problems of the convergence of the approximate solution towards the exact solution, it is also appropriate to study the associated
problems of the increase in the error committed for a given idealization on
the one hand, and the rapidity of convergence on the other, according to the
dimension of the elements.

A solution to the first of these problems is found in the separate utilization of two, primal and dual methods[5,19,20,31,] as we shall see. For
the second, contemporary research[13,17,18] has made it possible, in certain
cases, to establish useful theorems and formulae giving the approximate
value of the error, according to certain characteristic dimensions and shape
parameters of the elements.

6) <u>Overestimation of the stiffness in the displacement method</u>

If we consider as before a conforming element idealization of a linear
elastic system, the approximate global stiffness is overestimated with

respect to the exact stiffness.

To simplify matters, we will consider a discrete system for which the total potential energy is written, for a given generalized force $Q \in R^n$:

$$W = \frac{1}{2} \overline{q} \, K \, q - \overline{Q} \, q \, . \, , \quad K = \overline{K} \, .$$

The exact solution q_{ex} makes W minimum, and satisfies the equation :

$$Kq_{ex} = Q.$$

Hence :

$$W_{min} = \frac{1}{2} \overline{Q} \, q_{ex} - \overline{Q} \, q_{ex} = - \frac{1}{2} \overline{Q} \, q_{ex} < 0.$$

For a "less fine" discretization, giving the approximate solution q_a,[*] the corresponding minimum total potential energy will be greater by reason of the constraints introduced. Therefore :

$$- \frac{1}{2} \overline{Q} \, q_a \geqslant - \frac{1}{2} \overline{Q} \, q_{ex}$$

For a special load $Q = 1_i \, {}^iQ$, (i fixed), relative to the same degree of freedom in both cases, we have :

$$- \frac{1}{2} \, {}^iQ \, {}^iq_a \geqslant \frac{1}{2} \, {}^iQ \, {}^iq_{ex.} \, ,$$

Whence :

$$ {}^iq_a \leqslant {}^iq_{ex}.$$

The conclusions are inverted with the dual method (see next chapter).

7) <u>Rigid body displacements</u>

When a structure is deformed, one or more elements are liable to displace as rigid bodies, implying zero deformation in certain areas.

In principle this eventuality requires, for each element, the possibility of admitting rigid body displacements. It is clear that the elementary stiffness matrices K_e will then be singular. This local singularity disappears after assembly, but the global stiffness matrix will be singular nevertheless, as the generalized unknown q" could take rigid body mode

---

[*] We assume the configuration space extended, so as to have the same number of dimensions as before ; q_a being correctly calculated on this extension.

values. Obviously, static problems must be at least globally isostatic. In
the free structure dynamic problems which we shall consider later, we must
take account of these rigid body modes in structural analysis programmes.

8) <u>Static reduction methods</u>

It may be useful, or even essential, according to the programme or computer available, to decompose the structure into substructures, or establish
super-elements corresponding to the assembly of a number of finite elements.
In all cases, a "reduction method" consists in eliminating carefully selected degrees of freedom. For example, we eliminate the degrees of freedom
internal to the substructures, the others being used for assembly of the
substructures, or in a set of several elements, only retaining the super-
element boundary degrees of freedom.

For example, let us consider a discretized linear structure, with unknown
q and loaded by generalized force Q. If we split the degrees of freedom of
the structure into a sub-column ^1q of degrees of freedom (boundary degrees
of freedom, for example), and a sub-column ^2q of internal degrees of freedom,
we obtain the equation :

$$\begin{bmatrix} ^1K_1 & ^1K_2 \\ \hline ^1K_2 & ^2K_2 \end{bmatrix} \begin{bmatrix} ^1q \\ ^2q \end{bmatrix} = \begin{bmatrix} ^1Q \\ ^2Q \end{bmatrix}$$

We can then eliminate ^2q by elementary calculation :

$$^1K_1 \, ^1q + {}^1K_2 \, ^2q = {}^1Q$$

$$^2K_1 \, ^1q + {}^2K_2 \, ^2q = {}^2Q .$$

Hence :

$$^2q = {}^2K_2^{-1} \left[^2Q - \overline{^1K_2} \, ^1q \right] ,$$

giving the final equation :

$$\left[^1K_1 - {}^1K_2 \, ^2K_2^{-1} \, \overline{^1K_2} \right] . \, ^1q = {}^1Q - {}^1K_2 \, ^2K_2^{-1} . {}^2Q .$$

This method is therefore particularly interesting when $^2Q = 0$.

This method of reduction is systematized as a "frontal" method, where each
element is assembled successively with the preceding elements, and where the
degrees of freedom corresponding to certain nodes are eliminated by a Gauss

substitution at each stage. These nodes are those for which assembly has
been completely achieved at each stage of progression of the front of
assembly. Storage of these data makes it possible to go back to the eliminated unknowns[32,33].

Dynamic reduction methods also exist, and we shall consider these at a
later stage.

9) <u>Nature of the approximate equilibrium in a weak formulation</u>

As we can easily observe, a variational formulation, whether exact or
approximate, expresses an equilibrium obtained <u>in the mean</u> on the domain
occupied by the structure, and in relation to a given class of unknown
fields. The approximate weak solution does not correspond to a local equilibrium, in contrast to a strong solution which would satisfy the local
equilibrium equations. The solution can then belong to a function space
larger than any strong solution, namely a less regular function space,
consequently easier to calculate taking account of the properties of regularity of the geometrical, kinematic and dynamic data.

It is for this reason that a solution $\underset{\sim}{U} \in C°(\overline{\Omega})$ is generally sought, the
derivative of which is only scalar square summable. The consequence is that
the stress obtained only achieves mean equilibrium, and being a direct
function of deformation, it generally presents discontinuities where the
element boundaries are crossed. The "displacement" method is therefore
characterized by a poor description of the stresses. Hence it is appropriate
to envisage methods for smoothing the stresses.

Furthermore, for the displacement to be kinematically admissible, it must
satisfy the following equation :

$$U = U_d \text{ on } \Sigma_U .$$

In the preceding method, we assumed implicitly that the representation
basis $\underset{\sim}{T}(X)$ [see (2.9)] met this condition, in other words it subtended a
linear subspace of kinematically admissible vector   functions. Obviously
we cannot select a kinematically admissible basis in each problem, and
generally the polynomial interpolations adopted would not be so.

We therefore consider only vector fields U_d which are compatible with
the selected interpolations. After discretization and assembly, this corresponds to breaking down the unknown q into given degrees of freedom, ^2q
corresponding to U_d on Σ_U, and with ^1q unknown. The unknown generalized
reaction forces ^2Q then correspond to ^2q. This gives the following system :

$$
\begin{bmatrix} {}^1K_1 & {}^1K_2 \\ \overline{{}^1K_2} & {}^2K_2 \end{bmatrix}
\begin{bmatrix} {}^1q \\ {}^2q \end{bmatrix}
=
\begin{bmatrix} {}^1Q \\ {}^2Q \end{bmatrix}
$$

where ^1q and ^2Q are unknown, and ^2q and ^1Q are known.

The first equation gives ^1q, and the second can be used to calculate the reaction ^2Q. It should be noted that if we take $U_d = 0$ on Σ_U, this corresponds to cancellation of a certain column ^2q. The equation to be solved is then :

$$
^1K_1 \; ^1q = {}^1Q
$$

A more precise justification consists in satisfying the equation $U = U_d$ on Σ_U, in a weak form, using a Lagrangian multiplier, interpreted naturally as a constraint force. After discretization, this new nodal unknown is precisely the preceding unknown ^2Q.

10) <u>More general Lagrange functions and independent variables</u>

In the preceding variational principle, the Lagrange function took the form :

$$
L = \underset{\sim}{L}\left(U, \frac{\partial U}{\partial X}\right) \in \mathbf{R}, \quad X \in \Omega
$$

and the stationarity equation gave :

$$
\int_\Omega \left[\frac{\partial L}{\partial U}\, \delta U + \frac{\partial L}{\underset{\sigma}{\partial U}}\, \frac{\partial \delta U}{\partial X} \right] d\Omega , \forall \delta U \; K.A.
$$

$$
= \int_\Omega \left[\frac{\partial L}{\partial U}\, \delta U - \mathrm{div}\; C.\delta U \right] d\Omega + \int_\Sigma \overline{\delta U\; C}\; n \; d\Sigma .
$$

This expression demonstrates that on Σ, or parts of Σ, it is appropriate to give either the vector field U, or the force field $\overline{C}\, n$ (where n is still the external unit normal to Σ).

In certain problems, we can encounter Lagrange functions which depend on the unknown by higher derivation orders. For example :

$$\int_\Omega \mathcal{L}(U, \frac{\partial U}{\partial X}, \frac{\partial^2 U}{\partial X \partial X}) d\Omega$$

whence :

$$(2.17) \quad \delta \int_\Omega \mathcal{L}(U, \frac{\partial U}{\partial X}, \frac{\partial^2 U}{\partial X \partial X}) d\Omega$$

$$= \int_\Omega \left[\frac{\partial L}{\partial U} \delta U + \frac{\partial L}{\partial \frac{\partial U}{\partial X}} \frac{\partial \delta U}{\partial X} + \frac{\partial L}{\partial \frac{\partial^2 U}{\partial X \partial X}} \frac{\partial^2 \delta U}{\partial X \partial X} \right] d\Omega \, , \forall \, \delta U \; \text{K.A.}$$

To facilitate interpretation, and to avoid the introduction of more complex operators, we write the preceding formula using any fixed basis, giving, in succession :

$$\int_\Omega \left[L_i \, \delta^i U + L_i^j \, \delta^i U_{,j} + L_i^{jk} \, \delta^i U_{,jk} \right] d\Omega$$

$$= \int_\Omega \left[L_i \, \delta^i U - L_{i,j}^j \, \delta^i U - L_{i,j}^{jk} \, \delta^i U_{,k} \right] d\Omega$$

$$+ \int_\Sigma \left[n_j \, L_i^j \, \delta^i U + n_j \, L_i^{jk} \, \delta^i U_{,k} \right] d\Sigma$$

$$= \int_\Omega \left[L_i \, \delta^i U - L_{i,j}^j \, \delta^i U + L_{i,jk}^{jk} \, \delta^i U \right] d\Omega$$

$$+ \int_\Sigma \left[n_j \, L_i^j \, \delta^i U + n_j \, L_i^{jk} \, \delta^i U_{,k} - n_k \, L_{i,j}^{jk} \, \delta^i U \right] d\Sigma .$$

In this formula, i,j,k = 1,2,3. The tensors with components L_i^j and L_i^{jk} represent the dual stresses of quantities $\frac{\partial \delta U}{\partial X}$ and $\frac{\partial^2 \delta U}{\partial X \partial X}$. In addition, n_j is a covariant component of the normal to Σ.

Classic examples in the mechanical field are those of capillarity[21], or shells[22]. In the first case, the stress with components L_i^{jk} is associated with a surface tension, and in the second this is associated with the bending and torsional moments corresponding to variations of the curvature of the middle surface *. If the edge Σ of Ω contains lines of singularity, or of line loading, these lines must be considered as non-empty edges of Σ. If we take $\mathcal{C}$ as the union of these lines, a new partial integration then gives the final formula as follows :

$$\int_\Omega \left[L_i \, \delta^i U - L_{i,j}^j \, \delta^i U + L_{i,jk}^{jk} \, \delta^i U \right] d\Omega$$

$$+ \int_\Sigma \left[n_j \, L_i^j \, \delta^i U - \left[n_j \, L_i^{jk} \right]_{;k} \, \delta^i U - n_k \, L_{i,j}^{jk} \, \delta^i U \right] d\Sigma$$

$$+ \int_{\mathcal{C}} \nu_k \, n_j \, L_i^{jk} \, \delta^i U \, ds + \int_\Sigma \left[-\left[\widehat{\operatorname{div}} \, \pi A \right]_i \, \delta^i U + n_k n_j L_i^{jk} \frac{\partial \delta^i U}{\partial n} \right] d\Sigma$$

where $n_j L_i^{jk} = {}^k A_i \Longrightarrow A = f(X) \in \mathcal{L} \, (\vec{E}_3, \vec{E}_3)$, $X \in \Sigma$; $i,j,k = 1,2,3$. [for $\widehat{\operatorname{div}} \, \pi A$: see (A.96)].

In this formula, the volume, surface or curvilinear tensors are the covector "coefficients" of the δU. In the line integral, the quantities ν_k are covariant components of the normal to   in the plane tangent to Σ, and oriented canonically. Furthermore, the divergence operations appear clearly. Finally, the independent unknown still remains ${}^i U$, but also $\frac{\partial^i U}{\partial n}$ on Σ.

It should be noted that in this second gradient theory, it is possible to achieve equilibrium of external force line densities **). This theory can be generalized[23].

The virtual variation (2.17) shows that it is advisable to find a continuous solution U, with continuous derivative and such that components L_i, L_i^j, L_i^{jk} and ${}^i U_{,jk}$ be square summable.

---

*) In the case of shells, integrations and derivations are made on a
    Riemannian differentiable manifold of dimension 2 (See Chapter VI).
**) This is not possible in a first gradient theory.

We can however envisage another method, at least in its principle, considering the two unknowns U and V associated by the constraint :

$$V = \frac{\partial U}{\partial X} \ .$$

The canonical formulation then gives :

$$\delta \int_{\Omega} \underset{\sim}{L}(W, \frac{\partial W}{\partial X}) d\Omega$$

where :

$$W = \begin{bmatrix} U \\ \\ V \end{bmatrix} \text{ with } V = \frac{\partial U}{\partial X}$$

The constraint is introduced by means of a Lagrangian multiplier Λ, considered as an independent unknown. We always look for a continuous solution W, the partial derivatives of which have a summable square in Ω, together with the stress and multiplier components. Although the number of unknowns increased, the interest of the method essentially lies in the fact that the constraint is only satisfied in the mean, authorizing interpolations of a lower degree than in the preceding method, with consequent compensation in the order of the discretized system.

11) <u>Patch test</u>

The patch test proposed by Irons in 1972, is a convergence criterion in the use of non conforming elements. According to Strang and Fix[13], it appears that this is a necessary and sufficient condition of convergence, which can thus be formulated for an element with linear deformations.

If σ_o is a constant stress field in element e, and ε is the deformation field corresponding to the displacement field U, such that for this field the element is not conforming, the criterion of convergence is written :

$$\int_{\Omega_e} T_r(\sigma_o \varepsilon) \, d\Omega_e = 0 \ , \ \forall \sigma_o C^t.$$

The criterion is equivalent to :

$$\int_{\Sigma_{Fe}} \overline{\sigma_o n} \, Ud\Sigma = 0 \, , \, \forall \, \sigma_o \, C^t$$

This criterion will be justified later (see chapter III).

2. - OTHER TYPES OF ELEMENTS AND STATIC PROBLEMS

2.1. - Axisymmetrical three-dimensional bodies with axisymmetrical loading

This case is highly analogous with the bidimensional case. The finite elements considered are toroidal, elements with symmetry of revolution, the meridian section of which is triangular, for example. A point on this section will be :

$$X = \begin{bmatrix} r \\ z \end{bmatrix},$$

where r is the distance from the axis, and z is the ordinate on the axis. The displacement U is :

$$U = \underset{\sim}{U}(r,z) = \begin{bmatrix} \underset{\sim}{u}(r,z) \\ \underset{\sim}{v}(r,z) \end{bmatrix},$$

where u is the component in the meridian plane, orthogonal to the axis, and v is the component in the direction of the axis.

The deformation ε has the same components as in the bidimensional case, plus a circumferential component designated ε_φ, due to the variation of r with axisymmetrical deformation. This gives :

$$\varepsilon = \begin{bmatrix} \varepsilon_z \\ \varepsilon_r \\ \varepsilon_\varphi \\ 2\varepsilon_{rz} \end{bmatrix} \begin{bmatrix} \dfrac{\partial v}{\partial z} \\ \dfrac{\partial u}{\partial r} \\ \dfrac{u}{r} \\ \dfrac{\partial u}{\partial z} + \dfrac{\partial v}{\partial r} \end{bmatrix}.$$

We can adopt a linear interpolation for U, and verify the basic criteria
of convergence.

We then have [see (2.10)] :

$$\varepsilon = \underset{\sim}{\mathcal{B}} (X).^e q .$$

To integrate $\underset{\sim}{\mathcal{B}} (X)$ in the element, we can take $\mathcal{B}_m$ as the mean value at
the centre of gravity of the triangle, and put :

$$\mathcal{B} = \mathcal{B}_m + \underset{\sim}{\mathcal{B}} (r,z) .$$

The latter term gives integrals of the type :

$$\int \frac{dr}{r} \, dz \, , \quad \int \frac{z}{r} \, drdz, \quad \int \frac{z^2}{r} \, drdz.$$

Generalized nodal forces are applied along a parallel equivalent to a
node. As their distribution is constant along the paralle due to the axi-
symmetry, the two components at point $X = \begin{bmatrix} r \\ z \end{bmatrix}$ will be $2 \pi r \, F_r$ and $2 \pi r \, F_z$.

We can also introduce strains of a thermal origin. For a temperature
variation θ, the corresponding deformation will be :

$$\varepsilon_\theta = \begin{bmatrix} \alpha\theta \\ \alpha\theta \\ \alpha\theta \\ 0 \end{bmatrix} \, , \quad ou \ \varepsilon_\theta = \begin{bmatrix} \alpha_z \theta \\ \alpha_r \theta \\ \alpha_r \theta \\ 0 \end{bmatrix}$$

with coefficients of thermal dilatation corresponding to an isotropic or
orthotropic thermal behaviour.

2.2. – <u>Improved elements in the bidimensional case</u>

<u>Quadrilateral elements</u>

Quadrilateral elements can be constructed from triangular elements,
giving a greater number of degrees of freedom, and possibly improved
accuracy :

- either by combining two triangles, which gives half the number of elements.
  If we take the average for the stresses of the quadrilateral element, we
  obtain better result than with the previous triangular element, namely less
  oscillation, even with a rough subdivision.

- or taking the average of the stiffness values obtained by the two possible
  triangular partitioned configurations. Again, the results are improved.

Rectangular elements

There are two degrees of freedom, u and v, for each of the four nodes,
allowing a second degree interpolation for u and v, but containing a maximum
of four terms. We select the supplementary quadratic term to ensure continuity of displacement on the interfaces, namely :

$$u = \alpha_1 + \alpha_2 x + \alpha_3 y + \alpha_4 xy$$

$$v = \beta_1 + \beta_2 x + \beta_3 y + \beta_4 xy.$$

Linearity of displacement along each side provides continuity.

Isoparametric quadrilateral elements

To obtain the same result as in the case of a rectangle, using a quadrilateral element, we parameterize the plane, using a map F, so that :

$$X = \begin{bmatrix} x \\ y \end{bmatrix} = F\begin{pmatrix} \xi \\ \eta \end{pmatrix}$$

so that each side of the quadrilateral is given by a constant value of the
new co-ordinates, thus $\xi = \pm 1$ and $\eta = \pm 1$. The angles are indexed by i,j,k,
and the corresponding displacements are respectively u_i, v_i ; u_j, v_j etc.
The displacement of point X, for a quadratic interpolation of the preceding
type, is written :

$$\left[\begin{aligned} u &= \frac{1}{4}\left[(1-\xi)(1-\eta)u_i + (1+\xi)(1-\eta)u_j + (1+\xi)(1+\eta)u_k + (1-\xi)(1+\eta)u_\ell \right] \\ v &= \frac{1}{4}\left[(1-\xi)(1-\eta)v_i + (1+\xi)(1-\eta)v_j + (1+\xi)(1+\eta)v_k + (1-\xi)(1+\eta)v_\ell \right]. \end{aligned} \right.$$

Along a straight line, where ξ (or η) is constant, x and y vary linearly in η (or ξ). Furthermore, for $\xi = \pm1$, $\eta = \pm1$ we obtain the co-ordinates x_i, y_i ; x_j, y_j, etc. for the nodes, giving the following change of variables :

$$\left[\begin{array}{l} x = \frac{1}{4} \left[(1-\xi)(1-\eta)x_i + (1+\xi)(1-\eta)x_j + (1+\xi)(1+\eta)x_k + (1-\xi)(1+\eta)x_\ell \right] \\[2mm] y = \frac{1}{4} \left[(1-\xi)(1-\eta)y_i + (1+\xi)(1-\eta)y_i + (1+\xi)(1+\eta)y_k + (1-\xi)(1+\eta)y_\ell \right]. \end{array} \right.$$

The formulae for the coordinate transformation are identical to the interpolation formulae for the displacement, hence the designation "isoparametric element", conforming like the rectangular element.

The presence of the Jacobian determinant in the transformation, in the calculation of the integrals, makes the techniques of numerical integration necessary in practice.

We shall generalize the technique of isoparametric elements, presented here in an elementary case, the use of which, although not economic, reveals to be useful in cases of delicate and precise idealizations.

<u>6-node triangle</u>

If we place an additional node, for example in the middle of each side of a triangular element, we obtain a 6-node triangle, the 12 degrees of freedom of which authorize 6 polynomial terms in the interpolation of point X displacement, namely :

$$\left[\begin{array}{l} u = \alpha_1 + \alpha_2 x + \alpha_3 y + \alpha_4 x^2 + \alpha_5 xy + \alpha_6 y^2 \\[2mm] v = \beta_1 + \beta_2 x + \beta_3 y + \beta_4 x^2 + \beta_5 xy + \beta_6 y^2. \end{array} \right.$$

On one side, parameterized by curvilinear abscissa s, the displacement component u (or v) will be parabolic, giving :

$$\underset{\sim}{u}(s) = a + bs + cs^2,$$

the determination of which will be unique, according to the displacements of the three nodes of this side. The assembly of these three nodes with those of the side common to an adjacent triangle will therefore make this displacement identical for both triangles, along their common side. The representation is therefore conforming. This quadratic interpolation gives linear deformations and consequently stresses, which are therefore more accurate than in the case of the 3-node triangle initially considered. These results are also more accurate than if we consider the four triangles, each with 3 nodes, obtained with the preceding 6 nodes, as each of these four triangles, with a linear interpolation, would give constant deformations, and therefore discontinuous at the interfaces.

2.3. - Three-dimensional elements and problems

All the above can be generalized without theoretical difficulty to three-dimensional problems, by the introduction of a supplementary co-ordinate z, and a supplementary displacement component w.

For example, 8-node elements can result from the assembly of five tetrahedrons, in two possible ways. The case of the triangle with 6 nodes here becomes that of the 10-node tetrahedron, giving complete conformity and enabling us to obtain linear stresses.

2.4. - Plate bending

The linearized theory of elastic plates is a special case of the linearized shell theory, in which two basic simplifications are introduced :

1) The bending problems are not coupled with the membrane problems.

2) The derivatives are evaluated in a plane, and therefore in a linear space, and are therefore ordinary and non-covariant.

Anticipating a little with respect to the shell theory, to be presented later, let us consider a thin shell with constant thickness h, generated by a segment of which the mid-point m describes a surface Σ_m, while remaining normal to this surface. We call C_m the edge of Σ_m. Point M of the shell, in the deformed state or not, and situated on the normal at m, is given by :

$$M = m + \overrightarrow{mM},$$

or

$$M = m + Nz, \quad \overline{N} N = 1$$

where N is the unit normal at m, and z is the normal co-ordinate of M with respect to m.

A virtual displacement (or linearized real displacement) of M is therefore written :

$$\delta M = \delta m + \delta N z + N \delta z.$$

A simple surface theory consists in assuming that, for a thin shell, the thickness does not change as an initial approximation, so that :

$$(2.18) \qquad \delta z = 0.$$

Whence :

$$(2.19) \qquad \delta M = \delta m + \delta N z$$

In this formula, δm and δN are virtual surface fields by hypothesis, namely only depending on m $\in \Sigma_m$.

Another kinematic hypothesis consists in assuming that the normal to Σ_m, as a material normal, remains normal to Σ_m after deformation. Thus for any vector field dm tangent to Σ_m at m, the condition of orthogonality :

$$\overline{N} dm = 0$$

continues to hold during deformation (virtual or real), namely :

$$\delta \left[\overline{N} dm \right] = 0 \Rightarrow \overline{\delta N}\, dm + \overline{N} \delta dm = 0, \forall\, dm \text{ tangent.}$$

Hence :

$$(2.20) \qquad \overline{\delta N} = - \overline{N}\, \frac{\partial \delta m}{\partial m} \quad \text{or} \quad \delta N = - \overline{\frac{\partial \delta m}{\partial m}}\, N$$

The two conditions (2.18) and (2.20) constitute the two kinematic hypotheses of Kirchhoff-Love.

Furthermore, as the normal N is unitary, in the virtual (or linearized real) displacement, the condition :

$$\overline{N}\, N = 1$$

gives :

$$\overline{\delta N}\, N = 0.$$

δN is therefore a vector tangent to Σ_m.

In the context of more general hypotheses, it should be noted that the two independent vector fields correspond to virtual variations δm and δN.

In the case of the preceding hypothesis, δN is not an independent variable, as demonstrated by (2.20). Thus, expressed in Cartesian coordinates in the case of a plate :

$$m = \begin{bmatrix} x \\ y \end{bmatrix}, \quad \delta m = \begin{bmatrix} \delta u \\ \delta v \\ \delta w \end{bmatrix} \in \mathbb{R}^3, \quad N = \begin{bmatrix} 0 \\ 0 \\ 1 \end{bmatrix}$$

$$\frac{\partial \delta m}{\partial m} = \begin{bmatrix} \dfrac{\partial \delta u}{\partial x} & \dfrac{\partial \delta u}{\partial y} \\[2ex] \dfrac{\partial \delta v}{\partial x} & \dfrac{\partial \delta v}{\partial y} \\[2ex] \dfrac{\partial \delta w}{\partial x} & \dfrac{\partial \delta w}{\partial y} \end{bmatrix} \qquad \delta N = - \begin{bmatrix} \dfrac{\partial \delta w}{\partial x} \\[2ex] \dfrac{\partial \delta w}{\partial y} \end{bmatrix}.$$

Incidentally, we immediately find that the virtual rotation vector of the normal is written :

$$\delta \theta = \begin{bmatrix} \dfrac{\partial \delta w}{\partial y} \\[2ex] - \dfrac{\partial \delta w}{\partial x} \end{bmatrix}.$$

Before giving some indications concerning the finite elements for bending of plates, it is appropriate to identify the independent boundary unknowns.

To write the equilibrium equations for a shell, we see that these can be deduced from the three-dimensional principle of virtual work, by developing the virtual energy of deformation [see (1.9)] :

$$\delta W = \int_\Omega T_r (C \, \frac{\partial \delta M}{\partial M}) d\Omega,$$

and taking account of (2.19), we have :

$$\delta W = \int_\Omega T_r (C \frac{\partial}{\partial M} [\delta m + \delta N \ z]) d\Omega.$$

Decomposing the integration in Ω in classical way, into an integration with respect to z and an integration on Σ_m, we put δW in the following form:

$$(2.21) \qquad \delta W = \int_{\Sigma_m} T_r \ (f_1 \frac{\partial \delta m}{\partial m} + f_2 \frac{\partial \delta m}{\partial m}) \ d\Sigma.$$

Partial integrations then give a virtual energy in the form :

$$(2.22) \qquad \delta W = \int_{\Sigma_m} \left[\overline{g_1} \ \delta m + \overline{g_2} \ \delta N \right] d\Sigma + \int_{C_m} \left[\overline{h_1} \ \delta m + \overline{h_2} \ \delta N \right] ds$$

where h_1, h_2, g_1, g_2 are the dual forces of δm and δN respectively and s is the curvilinear abscissa of C_m.

But as N is not an independent variable, account must be taken for the constraint (2.20), and a new partial integration executed. In the first integral, the only independent virtual unknown remaining is obviously δm. But the situation is not the same on the boundary. In fact :

$$\overline{\delta N} = - \overline{N} \frac{\partial \delta m}{\partial m} = - \frac{\partial \overline{N} \delta m}{\partial m} + \overline{\delta m} \ \frac{\partial N}{\partial m}.$$

Thus δN is expressed as a function of δm, and the derivative of its normal component, which we have called δw. Therefore :

$$\overline{\delta N} = - \frac{\partial \delta w}{\partial m} + \overline{\delta m} \ \frac{\partial N}{\partial m} .$$

The last term of (2.22) is then written :

$$(2.23) \qquad \int_{C_m} \overline{\delta N} \ h_2 \ ds = \int_{C_m} \left[- \frac{\partial \delta w}{\partial m} + \overline{\delta m} \ \frac{\partial N}{\partial m} \right] h_2 \ ds .$$

The quantity $\frac{\partial \delta w}{\partial m}$, the surface derivative of δw, must be calculated on C_m. For example, we can write its components in the form $\frac{\partial \delta w}{\partial s}$ and $\frac{\partial \delta w}{\partial v}$, where v is the unit normal to C_m, in the plane tangent to Σ_m (and oriented externally to Σ_m).

It is clear that in (2.23), the term for $\frac{\partial \delta w}{\partial s}$ gives rise to a partial integration on C_m, leaving only the independent variable w. On the other hand, the term $\frac{\partial \delta w}{\partial \nu}$ remains as such, constituting a boundary independent variable with δm.

The kinematic explanation is as follows : $\frac{\partial \delta w}{\partial \nu}$ represents variation δN in rotation of the normal round the tangent to C_m at m (namely a variation in direction ν). The latter is free at the edge, while the Kirchhoff–Love hypothesis corresponding to constraint (2.20), implies that normal N, which should remain normal to Σ_m, remains normal to C_m in particular.

In conclusion, as the problem of bending is uncoupled from the membrane problem, in linearized thin plate problems, the only independent unknowns of the plate remaining at the edge, are the following (in virtual expression form) :

$$\delta w \quad \text{and} \quad \frac{\partial \delta w}{\partial \nu}$$

or :

$$w \quad \text{and} \quad \frac{\partial w}{\partial \nu}$$

as real variables. In particular, at any singular point of C_m where two distinct normals exist, two values of $\frac{\partial w}{\partial \nu}$ must be distinguished. This is precisely the case for the intersection node of two sides of a finite plate element. For example we will take the following case :

Bending of rectangular plates

Let q be the column of the degrees of freedom for the four nodes, indexed 1, 2, 3 and 4. In Cartesian coordinates, the degrees of freedom and the dual force at node i will be :

$$
{}^i q = \begin{bmatrix} w \\ \dfrac{\partial w}{\partial y} \\ \dfrac{\partial w}{\partial x} \end{bmatrix}_i
\quad , \quad
{}^i Q = \begin{bmatrix} Q_w \\ Q_{\theta_x} \\ Q_{\theta_y} \end{bmatrix}_i
$$

and :

$$q = \begin{bmatrix} {}^1q \\ {}^2q \\ {}^3q \\ {}^4q \end{bmatrix} \quad , \quad Q = \begin{bmatrix} {}^1Q \\ {}^2Q \\ {}^3Q \\ {}^4Q \end{bmatrix}$$

for the complete plate, corresponding to 12 degrees of freedom.

At any point $X = \begin{bmatrix} x \\ y \end{bmatrix}$ of Σ_m, only the normal deflection q is independent. We therefore select an interpolation of w with 12 parameters, namely :

$$\underset{\sim}{w}(x,y) = \alpha_1 + \alpha_2 x + \alpha_3 y + \alpha_4 x^2 + \alpha_5 xy + \alpha_6 y^2 +$$
$$+ \alpha_7 x^3 + \alpha_8 x^2 y + \alpha_9 xy^2 + \alpha_{10} y^3$$
$$+ \alpha_{11} x^3 y + \alpha_{12} xy^3,$$

eliminating certain 4th degree terms. The preceding model offers the advantage of providing cubic interpolation for w, when m runs along one side of the rectangle, namely with x or y constant, therefore with a curve defined by four parameters. Now on a side, for example defined by nodes 1 and 2, we have 4 nodal degrees of freedom, namely :

$$w_1, \quad \left[\frac{\partial w}{\partial \nu}\right]_1 \quad , \quad w_2, \quad \left[\frac{\partial w}{\partial \nu}\right]_2,$$

which define this cubic interpolation in unique manner. <u>w is therefore continuous at the interfaces.</u>

But in addition, $\frac{\partial w}{\partial \nu}$ also varies parabolically on a side where, for example taking the side joining nodes 1 and 2, we only have two degrees of freedom, $\left[\frac{\partial w}{\partial \nu}\right]_1$ and $\left[\frac{\partial w}{\partial \nu}\right]_2$. A parabolic interpolation of $\frac{\partial w}{\partial \nu}$ on each side, requiring three parameters, is therefore not defined in a unique manner for two adjacent elements. We therefore have a discontinuity of the rotation $\frac{\partial w}{\partial \nu}$ at the boundary. <u>The element is not conforming.</u>

Calling again C the matrix for passage from α to q [see (2.7)] :

$$^i q = {}^i C \alpha \ , \ q = C \alpha \ , \ \alpha = {}^{-1} q$$

$$\underset{\sim}{w}(x,y) = \begin{bmatrix} 1 & x & \ldots\ldots & xy^3 \end{bmatrix} \mathcal{C}^{-1} q = \underset{\sim}{T}(X) \cdot q \, .$$

In shell theory, we shall see that the bending deformation energy for a shell can be written [see $(6.58)_1$] :

$$\delta W_{f1.} = \int_{\Sigma_m} T_r(m \, \delta \, \kappa) d\Sigma$$

where κ represents flexural deformation, and **m** the corresponding dual moments. These are Hermitian endomorphisms of the plane tangent to Σ_m, which can be represented in cartesian coordinates by :

$$m = \begin{bmatrix} M_x & M_{xy} \\ \\ M_{xy} & M_y \end{bmatrix} \qquad \kappa = \begin{bmatrix} \kappa_x & \kappa_{xy} \\ \\ \kappa_{xy} & \kappa_y \end{bmatrix} .$$

For a plate, we find :

$$m = \begin{bmatrix} M_x \\ \\ M_y \\ \\ M_{xy} \end{bmatrix} \qquad \kappa = \begin{bmatrix} - \dfrac{\partial^2 w}{\partial x^2} \\ \\ - \dfrac{\partial^2 w}{\partial y^2} \\ \\ \dfrac{2\partial^2 w}{\partial x \partial y} \end{bmatrix} .$$

$\mathbf{m} = \mathcal{A}\kappa$ with :

$$\mathcal{A} = \frac{Eh^3}{12(1-\nu^2)} \begin{bmatrix} 1 & \nu & 0 \\ \nu & 1 & 0 \\ 0 & 0 & \frac{1-\nu}{2} \end{bmatrix} \text{ or } \mathcal{A} = \begin{bmatrix} \mathcal{A}_x & \mathcal{A}_1 & 0 \\ \mathcal{A}_1 & \mathcal{A}_y & 0 \\ 0 & 0 & \mathcal{A}_{xy} \end{bmatrix} ,$$

depending on whether the plate is isotropic or orthotropic, where h is the thickness, and E and ν are Young's modulus and Poisson's ratio of elasticity respectively. The preceding interpolation gives :

$$\kappa = \underset{\sim}{\mathcal{B}}(X) \cdot q$$

and calculations are made using the general method, integration preferably being executed by numerical procedures.

The three-dimensional deformation ε at M, again taking the linearized case, is written :

$$\varepsilon = z \begin{bmatrix} \kappa_x & \kappa_{xy} & 0 \\ \kappa_{xy} & \kappa_y & 0 \\ 0 & 0 & 0 \end{bmatrix}$$

The three-dimensional stress σ at M is therefore also proportional to coordinate z. We find :

$$\sigma = \frac{6}{h^2} z \begin{bmatrix} M_x & M_{xy} & 0 \\ M_{xy} & M_y & 0 \\ 0 & 0 & \frac{h^2}{6z}\sigma_z \end{bmatrix}$$

<u>Important remark</u>: The adoption of hypothesis (2.18) is equivalent to assuming the variation in thickness to be negligible but it certainly implies the existence of an indeterminate reaction stress σ_z, relative to this constraint. In fact it is more realistic to assume the deformation energy relative to a variation $\sigma_z \neq 0$, with respect to the others, as negligible, which in no way modifies the preceding calculations. On the other hand, practical experience demonstrates that it is rather σ_z which can frequently be considered negligible leading to a state of plane, or tangential, stresses. The deformation ε_z is then calculated using the three-dimensional constitutive law.

Bending of triangular plates

A polynomial development for w must include 9 terms, if a 3-node triangle is adopted. Now a development to the 3rd degree includes 10 terms. To eliminate one, while maintaining symmetry, we can put $\alpha_8 = \alpha_9$. But in addition, the matrix for passage from the α unknowns to nodal unknowns becomes singular, for certain configurations of the triangle. It is preferable to abandon this method.

A method using areal coordinates was proposed by Zienkiewicz and
Cheung[6], where a point X of the triangle is defined by the areas of the
three triangles generated by the segments joining this point to the angles,
areas the sum of which is equal to that of the triangular element in
question, giving three independent co-ordinates. We then select cubic
interpolations for w, which as before ensures continuity of displacements,
but not of rotations. In addition, the "constant deformation" criterion is
not satisfied. We then add corrector functions, also cubic, to the interpolation functions, according to the areal co-ordinates, these being zero
and with zero slope at the nodes, allowing constant deformations.

Another type of triangular bending element consists in adopting a 4th
degree development for w, as for the preceding rectangular element. We
then add 3 supplementary nodes to the mid-points of the sides, and a carefully selected supplementary degree of freedom on each of same. This gives
the total of 12 degrees of freedom required by the development in w.

This degree of freedom is the rotation $\left[\dfrac{\partial w}{\partial \nu}\right]$. Thus the rotation $\dfrac{\partial w}{\partial \nu}$ being
parabolic according to s, along a side, is correctly defined in unique man-
ner as a function of the three $\dfrac{\partial w}{\partial \nu}$ rotation degrees of freedom at the 3 nodes
located on the side.

In this model devised by Sander and Fraeijs de Veubeke, distinction must
be made in the programme between nodes with 3 degrees of freedom, and nodes
with a single degree of freedom. This is a drawback, but the element is
fully conforming.

Numerous types of bending elements have been suggested, and the reader
should refer to the very prolific specialized literature on the subject.

2.5.- Straight beams in bending

The linear theory of straight beams employs similar methods to those for
plates. If we ignore the coplanar deformation energy of the cross-sections,
the virtual deformation energy is written in the following form :

$$\delta W = \int_{C_m} \left[\overline{T} \frac{d\delta m}{dx} + \overline{M} \frac{d\delta\theta}{dx} + \overline{T}\, i(t)(\delta\theta) \right]\ dx$$

with :

$$\mathbf{T} = \int_{\Sigma_m} Ct\,d\Sigma, \quad \mathbf{M} = \int_{\Sigma_m} i\,(\overrightarrow{mM})\,(Ct)\,d\Sigma \quad , \quad \forall M \in \Sigma_m.$$

In these formulae, t is the unit tangent at m to the center line C_m, para-meterized by x, and orthogonal to the cross section Σ_m at m. δm is the virtual displacement of m, and $\delta\theta$ the virtual rotation of Σ_m.

Furthermore, if we assume that the cross sections remain orthogonal to C_m after deformation, the virtual deformation energy can be written :

$$\delta W = \int_{C_m} \left[T_x \frac{d\delta u}{dx} + M_x \frac{d\delta\theta_x}{dx} - M_y \frac{d^2\delta w}{dx^2} + M_z \frac{d^2\delta v}{dx^2} \right] dx.$$

In this formula, where the transverse shear T_y, T_z has disappeared, T_x is the component on t of $\mathbf{T}$, θ_x is the component of rotation round t, and M_x, M_y, M_z and δu, δv, δw, the components of M and δm respectively, in a unit-ary basis $\begin{bmatrix} t & S \end{bmatrix}$, where $S = \begin{bmatrix} S_y & S_z \end{bmatrix}$ sub-tends the plane of Σ_m orthogonal to t.

In both cases, and for similar reasons to those indicated in the case of plates, the independent degrees of freedom at a boundary point of C_m, for example for plane bending in the plane (x,y), are the displacement w and the rota-tion $\theta_y (= -\frac{dw}{dx})$. This leads to a cubic interpolation in x for the variable w, for an element with two nodes. The standard hypothesis of uniaxial stress Ct, represented by σ_{xx}, easily leads to the curvilinear constitutive law in plane bending :

$$M_y = - E\,I_y \frac{d^2 w}{dx^2}$$

where I_y is the moment inertia of Σ_m with respect to S_y, the bending axis, in a case where the transverse shear is neglected, as also the extension $\frac{du}{dx}$ of any point M of Σ_m, in comparison with the extension $-z \frac{d^2 w}{dx^2}$ due to rotation θ_y.

2.6.- <u>Shells as assembly of flat elements</u>

We can represent a shell intuitively, by flat finite elements. We obtain generalized plane and bending forces. For a flat element, the corresponding linearized deformations are independent, as already stated. Triangular elements are generally best suited for shells of arbitrary shape, although rectangles and quadrilaterals are suitable for the idealization of certain geometric configurations. Non-coplanar quadrangular elements have also been created.

It is naturally appropriate to represent the distribution of external forces by generalized forces at the nodes, and also solve problems of representation in a common reference system, in order to achieve assembly.

Nevertheless, plane elements are no longer coplanar, destroying the continuity of the unknowns at the interfaces, and compromising convergence.

Furthermore, the assembly of non-coplanar flat elements gives rise to problems of conditioning for the global stiffness matrices. If we consider a shell comprised of flat elements, not working in the membrane mode, each element presents a zero stiffness in the case of displacements normal to its plane. Now the elements must be finally represented using a common reference system, to achieve assembly of all the nodes, and each node must have all its degrees of freedom, namely a displacement vector with three components, in order to achieve the three-dimensional assembly.

Taking then a flat membrane finite element among those composing the shell, any nodal displacement will have an in-plane component, and a normal component in a local reference system. The stiffness of the element in the normal direction is zero, but the adjacent element will provide a non-negligible contribution in this direction, if its plane forms a sufficient angle with the first.

On the other hand, if the dihedral angle is small, the contribution made by the adjacent element will be very small in the direction normal to the first. This results in a quasi-singularity of the global stiffness matrix (also independent of the reference system).

The same argument applies for flat bending elements. The only bending stiffnesses relate to normal displacements and in-plane rotations with respect to the elements. Here again, assembly makes it necessary to consider, at each node, in addition to a complete displacement vector, a complete "normal" rotation vector which will have, locally, in-plane and normal components for each element. No membrane stiffness of the pure bending element corresponds to the in-plane displacement component, or to the normal component of the (normal) rotation vector. If the dihedral angle of the adjacent element is large, the stiffness contribution provided by this element, relative to these components, will be non-negligible (although two adjacent elements have separate normals, the rotation vector is common to both normals). If on the other hand, the dihedral angle is very small, the corresponding stiffnesses, which will be zero for one of the elements, will be negligible in the adjacent element. This again gives a quasi-singularity of the global stiffness matrix.

Practically in the case of plate bending elements modelling a shell, work in the membrane mode is always associated with the bending one, and the stiffnesses corresponding to in-plane displacements are non-zero.

To obviate the shortcomings of poor conditioning, in the case of small dihedral angles between elements, methods based on a same principle have been proposed : local stiffnesses are modified slightly, to a sufficient degree to obtain a conditioning compatible with the degree of accuracy available on the computer used, but not enough to modify the solution.

For this purpose, local reference systems can be used for the nodes concerned, one axis of the system being close to the normals to the neighbouring planes. We then eliminate the equations relating to degrees of freedom with quasi-zero stiffnesses[24]. Alternatively, one can add small stiffnesses on the degrees of freedom concerned, but which are adequate as mentioned above[11]. With both methods, indeed, is modified the global stiffness, but the second procedure is easier to use.

However, it should be noted that the possibility of representing stiffness matrices in a programme, by means of local reference systems selected at will, makes certain operations very easy, such as the analysis of structures presenting repetition symmetries for example.

To avoid these difficulties, and also to improve the idealization, curved finite elements were created[25], but in this case the difficulties increase, as also the number of degrees of freedom. Although the number of elements for a given level of accuracy can fall, the corresponding gain is not clear.

Finally, it should be noted that a flat element idealization of a shell introduces artificial imperfections, which can compromise the buckling analysis of certain shells, as we shall see[26];

2.7.- Axisymmetrical shells

In this case, the finite elements are conical frustra. As before, we consider the tangential forces and bending forces for axisymmetrical loading. The element is represented in a meridian plane, by a straight segment limited by two nodes, representing two parallels. A point of the element is defined by its curvilinear abscissa s, and its displacement in a meridian plane, is given by its meridian component u, and its component w normal to the meridian segment :

$$U = \begin{bmatrix} \underset{\sim}{u}(s) \\ \\ \underset{\sim}{w}(s) \end{bmatrix}.$$

In contrast to the case of plates, tangential deformation ε_s in the meridian direction, and ε_θ in the circumferential direction, and the curvature deformation χ_s and χ_θ, are coupled. Calling φ the half-angle at the apex of the element, we find :

$$\varepsilon = \begin{bmatrix} \varepsilon_s \\ \varepsilon_\theta \\ \chi_s \\ \chi_\theta \end{bmatrix} \begin{bmatrix} \dfrac{du}{ds} \\ \left[w \cos\varphi + u \sin\varphi \right]\dfrac{1}{r} \\ \dfrac{d^2w}{ds^2} \\ \dfrac{-\sin\varphi}{r}\ \dfrac{dw}{ds} \end{bmatrix},$$

where r is the radius of the parallel at the point concerned. There are therefore 4 corresponding surface stresses :

$$\sigma = \begin{bmatrix} N_s \\ N_\theta \\ M_s \\ M_\theta \end{bmatrix}$$

linked to the deformations by the elastic constitutive law of the shell.
In the isotropic case, we find :

$$\sigma = \mathcal{A}\varepsilon \quad \text{with} \mathcal{A} = \frac{Eh}{1-\nu^2} \begin{bmatrix} 1 & \nu & 0 & 0 \\ \nu & 1 & 0 & 0 \\ 0 & 0 & \dfrac{h^2}{12} & \dfrac{\nu h^2}{12} \\ 0 & 0 & \dfrac{\nu h^2}{12} & \dfrac{h^2}{12} \end{bmatrix},$$

where h is the thickness of the element.

The independent degrees of freedom on each node, will be the displacement, and as we have seen, the normal rotation (on the boundary) of the
normal, or θ_s, (corresponding to $\dfrac{\partial w}{\partial s}$ at an internal point). On a node with
index i, the degrees of freedom will therefore be :

$$q^i = \begin{bmatrix} u \\ w \\ \theta_s \end{bmatrix}_i$$

We can adopt the following interpolations :

$$U = \underset{\sim}{U}(s) = \begin{bmatrix} \underset{\sim}{u}(s) \\ w(s) \end{bmatrix} \begin{bmatrix} \alpha_1 + \alpha_2 s \\ \beta_1 + \beta_2 s + \beta_3 s^2 + \beta_4 s^3 \end{bmatrix},$$

which indeed give 6 degrees of freedom per element. The element is characterized geometrically by its angle φ, and its length L. For the remainder,
the general method is applied.

In the case of non-axisymmetrical loading, we must introduce three
displacement components, u, v, and w, where v is the circumferential
component (0 in the preceding case).

In this case there are 6 surface deformation components, and 6 corresponding stresses, or :

$$\varepsilon = \begin{bmatrix} \varepsilon_s \\[1ex] \varepsilon_\theta \\[1ex] 2\varepsilon_{s\theta} \\[1ex] \chi_s \\[1ex] \chi_\theta \\[1ex] \chi_{s\theta} \end{bmatrix} = \begin{bmatrix} \dfrac{\partial u}{\partial s} \\[2ex] \dfrac{1}{2r}\dfrac{\partial v}{\partial \theta} + \left[w\cos\varphi + u\sin\varphi \right]\dfrac{1}{r} \\[2ex] \dfrac{1}{r}\dfrac{\partial u}{\partial \theta} + \dfrac{\partial v}{\partial s} - v\dfrac{\sin\varphi}{r} \\[2ex] -\dfrac{\partial^2 w}{\partial s^2} \\[2ex] -\dfrac{1}{r^2}\dfrac{\partial^2 w}{\partial \theta^2} + \dfrac{\cos\varphi}{r^2}\dfrac{\partial v}{\partial \theta} - \dfrac{\sin\varphi}{r}\dfrac{\partial w}{\partial s} \\[2ex] 2\left[-\dfrac{1}{r}\dfrac{\partial^2 w}{\partial s\partial \theta} + \dfrac{\sin\varphi}{r^2}\dfrac{\partial w}{\partial \theta} + \dfrac{\cos\varphi}{r}\dfrac{\partial v}{\partial s} - \dfrac{\sin\varphi\cos\varphi}{r^2}v \right] \end{bmatrix} , \sigma = \begin{bmatrix} N_s \\[1ex] N_\theta \\[1ex] N_{s\theta} \\[1ex] M_s \\[1ex] M_\theta \\[1ex] M_{s\theta} \end{bmatrix}$$

To reduce the general problem with two dimensions, s and θ, to a problem with a single dimension s, we can develop the loading in Fourier series, according to the azimut θ. For example, the surface density can be :

$$F = \begin{bmatrix} F_u(s,\theta) \\[1ex] F_w(s,\theta) \end{bmatrix} = \sum_n \begin{bmatrix} F_{u,n}(s)\,\cos n\theta \\[1ex] F_{w,n}(s)\,\cos n\theta \end{bmatrix} .$$

Also when expanded in Fourier series, the solution satisfies uncoupled problems for each value of index n, if :

$$U = \begin{bmatrix} u \\[1ex] v \\[1ex] w \end{bmatrix} = \sum_n \begin{bmatrix} u_n(s)\,\cos n\theta \\[1ex] v_n(s)\,\sin n\theta \\[1ex] w_n(s)\,\cos n\theta \end{bmatrix} .$$

For the component $v_n(s)$, we must adopt an interpolation which ensures continuity of the slope (rotation of the normal), for example linear.

2.8.– <u>Structures under elastoplastic behaviour</u>

For the purpose of demonstrating application of the finite element method, to mechanical problems of a more general nature than linear elastic problems, we will take the example of the analysis of structures under partial or total elastoplastic behaviour, and for which we will give a few brief indications.

According to the most frequently accepted and best verified theory[27,28,22], we know that a volume element enters the plastic domain, namely a plastic deformation exists, if the stress σ satisfies the following plastic flow criterion, for very small increments $d\sigma$:

$$\underset{\sim}{F}(\sigma) = 0 \text{ et } \frac{\partial F}{\partial \sigma} d\sigma > 0 \cdot$$

F depends on the hardening parameters, and is referred to as the "plastic potential". On the other hand, if condition :

$$\underset{\sim}{F}(\sigma) < 0 \text{ ou } \frac{\partial F}{\partial \sigma} d\sigma < 0$$

is met, there is no plastic deformation, and the medium remains elastic at the point considered.

A flow law, associated with the plastic potential, enables us to write a constitutive law of the hypoelastic type :

$$(2.24) \qquad d\sigma = \left[\mathcal{A} - \alpha \underset{\sim}{\mathcal{A}}^{*}(\sigma) \right] d\varepsilon$$

between the increments in stress and deformation, which are considered here, as before, as columns of $\mathbb{R}^{6}$. $\mathcal{A}$ is the matrix of classical elastic behaviour, and α is a scalar equal to 1 or 0, depending on whether the plastic flow criterion is satisfied or not. The preceding incremental law is written for small displacements. Furthermore, it is also of the same type in the case of a perfectly plastic elastic medium (no hardening). The determination of this law is based on experimental tests.

The finite element method can easily be applied to this case, to determine the increments of the generalized displacement dq, corresponding to the generalized load increment dQ. By application of (2.2), where U is replaced by dU, and δU by δdU, we obtain a discretized equation of the type :

$$\left[K - \alpha K^*(q)\right]dq = dQ .$$

The solution dq is then calculated by iteration for each increment dQ. The most efficient method is the modified Newton-Raphson method : at step n+1 of the iteration, we easily find :

$$dq_{n+1} = \alpha K^*(q+dq_n) + dQ .$$

This method is ordinarily referred to as the "initial stress method", by reason of the presence of the term $\alpha K^*(q+dq)$, resulting from the stresses which must be subtracted from the elastic stresses, to reestablish elasto-plastic equilibrium.

It must be admitted that this simple and efficient method is based on iteration of the explicit type, and does not give the exact solution. More accurate algorithms of the implicit type, based on hypotheses of convexity which are always met physically, have recently been created[29].

2.9.- <u>Improvement of elements</u>[6]

To improve the elements from the point of view of continuity, certain writers propose the addition of correction functions to the interpolation functions. For example, for a plate bending element, cubic development for the variable w, along a side defined by two nodes, led to a parabolic variable $\frac{\partial w}{\partial \nu}$ non-continuous along this side. The continuity of rotation, and consequently slope, along the side was not realized. In this example, the method consists in adding a function ϕ for each side, such that :

$\phi = 0$ on all sides.

$\frac{\partial \phi}{\partial \nu} = 0$ on all sides except one, where $\frac{\partial \phi}{\partial \nu}$ varies parabolically.

ϕ and $\frac{\partial \phi}{\partial \nu}$ are continuous throughout the element.

Thus we can ensure a total linear variation of variable $\frac{\partial w}{\partial \nu}$ on each side, or again a parabolic variation, by imposing this quantity on an intermediate node.

Another improvement proposed consists in imposing <u>supplementary unknowns</u>, in addition to the unknowns necessary for definition of the model, in order to ensure the continuity of certain quantities, for example certain derivatives, thus increasing the number of degrees of freedom.

For example, for a triangle loaded in its plane, we can take the following as nodal unknowns on each node i :

$$
{}^i q = \begin{bmatrix} u \\ v \\ \dfrac{\partial u}{\partial x} \\ \dfrac{\partial v}{\partial x} \end{bmatrix}_i \qquad \text{or better} \quad {}^i q = \begin{bmatrix} u \\ v \\ \dfrac{\partial u}{\partial y} - \dfrac{\partial v}{\partial x} \end{bmatrix}_i
$$

the latter ensure the continuity of the rotation.

To improve the idealization, particularly on the boundaries, <u>curved elements</u> are proposed, these being obtained by local transformation of elements with a simple boundary. If X represents a point on a linear boundary element, x, a point on the curved element, is deduced from the first by a regular, non-linear transformation :

$$x = F(X)$$

Interpolation of the first is :

$$U = \underset{\sim}{T}(X)\, q,$$

and that of the second becomes :

$$u = \underset{\sim}{T}(F^{-1}(x))\, q = \underset{\sim}{t}(x) q.$$

A polynomial interpolation no longer remains so.

A highly advantageous method consists in adopting a transformation which ensures the continuity of the displacement for the transformed element, when that of the first is also ensured. We are led to <u>isoparametric elements</u>[11,17,30], based on the following representation :

$$
\left[\begin{array}{l} x = \underset{\sim}{T}(X) \cdot q, \quad U = \underset{\sim}{T}(X) \cdot q \\[2em] u = \underset{\sim}{T}(X) \cdot q\,. \end{array} \right.
$$

For example : $X \in \mathbb{R}^2$, $U \in \mathbb{R}^3$, $x \in \mathbb{R}^3$, $\left({}^k x \in \left[{}^k x_0, {}^k x_1 \right] , k = 1, 2 \right)$

or : $X \in \mathbb{R}^3$, $U \in \mathbb{R}^3$, $x \in \mathbb{R}^3$, $\left({}^k x \in \left[{}^k x_0, {}^k x_1 \right] , k = 1, 2, 3 \right)$.

The transformed element therefore has a manifold structure. As for the nodes with index i, the nodal unknown ${}^i q$ is equal to the value of the unknown field U for this point, we have :

$$\underset{\sim}{U}(X_i) = \underset{\sim i}{T}(X_i) \, {}^i q = {}^i q$$

$$\underset{\sim i}{T}(X_i) = \underline{1} \, , \, \underset{\sim j}{T}(X_i) = 0 \ \ \forall i \neq j.$$

From this, we deduce both continuity at the interfaces of the representation, and continuity of the unknown field u, provided the continuity of U is ensured.

Integrations are generally carried out numerically, taking account of the Jacobian determinant of the transformation. It is appropriate to check that the latter is indeed regular, as regularity can be compromised in the case of highly distorted transformed elements.

It is also necessary to ensure that the convergence criteria are satisfied.

If the transformation uses different interpolations for the geometry and for the unknown field, super-parametric or sub-parametric elements can be defined[11].

3. – OTHER TYPES OF PROBLEMS[6]

As mentioned at the start of this chapter, the finite element method can be applied to all problems based on a variational principle. This is the case in particular with problems of diffusion (conduction, mixing, flow in porous media, etc.), flow of irrotational perfect fluids, electrostatics, magnetism, etc.

The media can be isotropic or not, and homogeneous or not. As a simple example, let us consider a problem of conduction. Depending on whether we are concerned with electrical or thermal conduction, the unknown field is a voltage or temperature scalar field u.

Taking the case of thermal conduction, for example and to establish our language, if q_c is the vector of thermal current intensity, per unit of time, and Q_c the quantity of heat leaving domain Ω of boundary Σ, we have :

$$\int_\Sigma \overline{q_c}\, nd\Sigma = \int_\Omega \operatorname{div} q_c d\Omega = \int_\Omega - c\rho\, \frac{\partial u}{\partial t}\, d\Omega = \frac{-\partial Q_c}{\partial t}$$

where c is the specific heat and ρ is the mass density in Ω, and n the unit normal external to Σ, and Q_c the heat quantity flowing outwards. Hence :

$$\operatorname{div} q_c = -\rho\ c\ \frac{\partial u}{\partial t} .$$

The constitutive law linking q_c and u is the linear Fourier law :

$$q_c = A\ \operatorname{grad}\ u.$$

A is an endomorphism, which becomes χ (scalar) in the isotropic case, referred to as "thermal conductivity".

The internal equilibrium equation is written classically :

$$\operatorname{div} A\ \operatorname{grad}\ u + c\rho\ \frac{\partial u}{\partial t} = h, \quad \text{in } \Omega,$$

where h is a volume density of heat per unit of time, due to internal sources.

The boundary conditions express, on the part Σ_H of Σ, the equilibrium between outgoing heat $\overline{n}\, q_c\, d\Sigma$, the losses by convection or radiation, assumed to be linearized, namely $\alpha u\, d\Sigma$, and the given quantity of heat $Hd\Sigma$, giving the equation :

$$\overline{n}\ A\ \operatorname{grad}\ u + \alpha u - H = 0 \quad \text{on } \Sigma_H.$$

The temperature u is assumed to be given on the complementary part Σ_u of Σ.

This results in the following variational principle :

$$\delta\left[\int_\Omega \left[\frac{1}{2}\ \overline{\operatorname{grad}\ u}\ A\ \operatorname{grad}\ u + \left[h - c\rho\, \frac{\partial u}{\partial t}\right] u\right]\, d\Omega + \left[\int_{\Sigma_H}\left[\frac{\alpha}{2}\, u^2 - Hu\right] d\Sigma\right]\right] = 0$$

$\forall \delta$ u admissible.

Application of the finite element method consists in partioning the medium into elements, for example triangular elements in a bidimensional problem, the unknown temperature field u being discretized by its nodal values. If we adopt a linear interpolation, we have :

$$u = \alpha_1 + \alpha_2 \, x + \alpha_3 \, y$$

and finally, if q is the column of nodal temperatures, we obtain the following matrix equation :

$$Kq + B\dot{q} = Q, \text{ with } \dot{q} = \frac{dq}{dt} ,$$

We can solve this equation as a standard differential equation, or using a finite difference method. If we decompose the time into short intervals Δt, a simple method consists in assuming $\dot{q}$ to be linear with t in the interval Δt, which gives :

$$q_t = q_{t-\Delta t} + \frac{1}{2}\left[\dot{q}_{t-\Delta t} + \dot{q}_t\right]\Delta t$$

whence :

$$\dot{q}_t = -\dot{q}_{t-\Delta t} + \left[q_t - q_{t-\Delta t}\right]\frac{2}{\Delta t} ,$$

This finally gives an equation for q_t, with a second member obtained at step $t-\Delta t$. This algorithm has the advantage of being both stable and accurate[34].

CHAPTER III

VARIATIONAL PRINCIPLES IN LINEAR ELASTICITY

1. PRINCIPLE OF POTENTIAL ENERGY
2. HU-WASHIZU PRINCIPLE, OR THREE FIELD PRINCIPLE
3. HELLINGER-REISSNER PRINCIPLE, OR TWO FIELD PRINCIPLE
4. A FRAEIJS DE VEUBEKE TWO FIELD PRINCIPLE
5. PRINCIPLE OF COMPLEMENTARY ENERGY
6. TWO FIELD HYBRID PRINCIPLE OF PIAN
7. PRINCIPLE OF VIRTUAL DISPLACEMENTS AND PRINCIPLE OF VIRTUAL STRESSES
8. APPLICATION OF VARIATIONAL PRINCIPLES

1. – PRINCIPLE OF POTENTIAL ENERGY

The principle of the minimum of total potential energy of a conservative linear system, is also referred to as the "primal principle". This principle was explained in paragraph 6 of chapter I, and is reminded below. In the case of a linearized hyperelastic medium, under linear deformation, the solution U satisfies the following principle :

(3.1)

$$U \text{ solution} \Longleftrightarrow \min_{V \ \{V \ K.A.\}} \mathcal{J}(\underset{\sim}{V}) \ , \ V = \underset{\sim}{V}(M_o) \ , \ M_o \in \overline{\Omega}_o$$

where :

$$\mathcal{J}(\underset{\sim}{V}) = \int_{\Omega_o} \frac{1}{2} \underset{\sim}{\alpha} \ (D_L)(D_L) d\Omega_o - \int_{\Omega_o} \overline{f}.V d\Omega_o - \int_{\Sigma_{oF}} \overline{F} \ V d\Sigma_o$$

$$D_L = \frac{1}{2} \left[\frac{\partial V}{\partial M_o} + \overline{\frac{\partial V}{\partial M_o}} \right] = \underset{\sim}{D}_L(V) = \overline{D_L} \ .$$

$\mathcal{J}(\underset{\sim}{V})$ is the total potential energy, Ω_o with boundary Σ_o is the domain occupied by the medium in the natural reference state, f is the volume density for given external loads, and F is surface density for given external loads on a part Σ_{oF} of boundary Σ_o ; $\alpha = \frac{1}{2} \underset{\sim}{\alpha} \ (D_L)(D_L)$ is the volume density of the elastic deformation energy, quadratic in D_L.

The necessary condition of stationarity is deduced from (3.1) :

(3.2)

$$\delta \mathcal{J}(\underset{\sim}{V}) = \int_{\Omega_o} \underset{\sim}{\alpha} \ (D_L)(\delta D_L) d\Omega_o - \int_{\Omega_o} \overline{f} \delta U d\Omega_o - \int_{\Sigma_{oF}} \overline{F} \delta U d\Sigma_o = 0$$

with :

$$D_L = \frac{1}{2} \left[\frac{\partial U}{\partial M_o} + \overline{\frac{\partial U}{\partial M_o}} \right], \ \forall \delta U \ K.A.$$

Condition (3.2) is sufficient, as we have seen, in the case of linear deformation.

The principle of potential energy originates from the more general principle of virtual work, concerning which reference can be made to item 1 in the bibliography.

Numerous other principles can be deduced therefrom. We mention the more important of them, but many complements will be found in reference 2.

2.- HU-WASHIZU PRINCIPLE, OR THREE FIELD PRINCIPLE[3]

In principle (3.1), vector field V is the only unknown field. It must be kinematically admissible, satisfying the following condition in particular :

$$(3.3) \qquad V = U_d \quad \text{on} \quad \Sigma_{oU}$$

or again with :

$$V = U + \delta U$$

$$\delta U = 0 \quad \text{on} \quad \Sigma_{oU},$$

U is the exact solution, and Σ_{oU} is the part of Σ_o where U_d is given.

We can then also take D_L as an independent variable, and introduce constraint (3.3) in principle (3.1), namely the two constraints :

$$(3.4) \qquad \left[\begin{array}{l} D_L - \dfrac{1}{2}\left[\dfrac{\partial V}{\partial M_o} + \overline{\dfrac{\partial V}{\partial M_o}} \right] = 0 \quad \text{in} \quad \Omega_o \\[2em] V - U_d = 0 \qquad\qquad\qquad \text{on} \quad \Sigma_{oU} \end{array} \right.$$

in the principle, using two independent Lagrangian multipliers [see (A.102)], which we shall call respectively t for $(3.4)_1$ and T for $(3.4)_2$. Multiplier t is a Hermitian endomorphism, by reason of the hermiticity of D_L, while multiplier T is vectorial. Using the scalar products already introduced in Chapter I on space $\vec{E}_6$, (3.1) gives the equation :

$$(3.5) \qquad \left[\delta\left[\int_{\Omega_o} \alpha\, d\Omega_o - \int_{\Omega_o} \overline{f}\, U d\Omega_o - \int_{\Sigma_{oF}} \overline{F}\, U d\Sigma_o \right.\right.$$

$$\left.\left. + \int_{\Omega_o} \tilde{t}\left[\frac{1}{2}\left[\frac{\partial U}{\partial M_o} + \overline{\frac{\partial U}{\partial M_o}} \right] - D_L \right] d\Omega_o + \int_{\Sigma_{oU}} \overline{T}\left[U - U_d \right] d\Sigma_o \right] = 0 \right.$$

(3.5)
continued $\quad\left|\begin{array}{l} \forall\,\delta D_L,\ \delta U,\ \delta t,\ \delta T \text{ to within conditions of derivability and} \\ \text{summability, and with } t = \bar{t}. \end{array}\right.$

It should be noted that the four variables D_L, U, t, and T are now independent. Furthermore, we have replaced V by U without any problem, to make the equation more convenient.

Now using formula (A.50) :

$$(3.6) \qquad \tilde{t}\left[\frac{1}{2}\left[\frac{\partial U}{\partial M_o} + \frac{\overline{\partial U}}{\partial M_o}\right]\right] = T_r\left(t\,\frac{\partial U}{\partial M_o}\right) = \mathrm{div}\,\left[t\ U\right] - \mathrm{div}\ t\ U,$$

with partial integration, (3.5) gives :

$$
\left[\begin{array}{l}
\displaystyle\int_{\Omega_o}\left[\left[\frac{\partial\alpha}{\partial D_L} - \tilde{t}\right]\delta D_L - \left[\mathrm{div}\ t + \bar{f}\right]\delta U + \delta\tilde{t}\left[\frac{1}{2}\left[\frac{\partial U}{\partial M_o} + \frac{\overline{\partial U}}{\partial M_o}\right] - D_L\right]\right]\,d\Omega_o \\[20pt]
\quad + \displaystyle\int_{\Sigma_{oF}\cup\Sigma_{oU}}\overline{tn}\,\delta U\,d\Sigma_o - \int_{\Sigma_{oF}}\bar{F}\,\delta U\,d\Sigma_o + \int_{\Sigma_{oU}}\left[\overline{\delta T}\left[U_d - U\right] - \bar{T}\,\delta U\right]d\Sigma_o = 0 \\[20pt]
\quad \forall\,\delta D_L,\ \delta U,\ \delta t,\ \delta T\,,
\end{array}\right.
$$

or again :

$$
(3.7)\quad
\left[\begin{array}{l}
\displaystyle\int_{\Omega_o}\left[\left[\frac{\partial\alpha}{\partial D_L} - \tilde{t}\right]\delta D_L - \left[\mathrm{div}\ t + \bar{f}\right]\delta U + \delta\tilde{t}\left[\frac{1}{2}\left[\frac{\partial U}{\partial M_o} + \frac{\overline{\partial U}}{\partial M_o}\right] - D_L\right]\right]\,d\Omega_o \\[20pt]
\quad + \displaystyle\int_{\Sigma_{oF}}\left[\overline{tn} - \bar{F}\right]\delta U\,d\Sigma_o + \int_{\Sigma_{oU}}\left[\overline{tn} - \bar{T}\right]\delta U\,d\Sigma_o + \int_{\Sigma_{oU}}\overline{\delta T}\left[U_d - U\right]d\Sigma_o = 0 \\[20pt]
\quad \forall\,\delta D_L,\ \delta U,\ \delta t,\ \delta T.
\end{array}\right.
$$

Equation (3.7) is equivalent to (3.2) and (3.4), but as is natural, it also provides supplementary equations, making it possible to calculate the multipliers, considered here as independent variables. (3.7) then gives the following local equations :

$$(3.8) \quad \left[\begin{array}{ll}
\tilde{t} = \dfrac{\partial \alpha}{\partial D_L} & \\[4pt]
\text{div } t + \bar{f} = 0 & \\[4pt]
D_L - \dfrac{1}{2}\left[\dfrac{\partial U}{\partial M_o} + \dfrac{\overline{\partial U}}{\partial M_o} \right] = 0 & \left. \right\} \quad \text{in } \Omega_o \\[10pt]
U - U_d = 0 & \text{on } \Sigma_{oU} \\[4pt]
tn = F & \text{on } \Sigma_{oF} \\[4pt]
tn = T & \text{on } \Sigma_{oU}
\end{array} \right.$$

From $(3.8)_2$ and $(3.8)_5$ we deduce that <u>t is a statically admissible stress,</u> namely balancing the vector field f and F, but then $(3.8)_1$ indicates that t is the true stress C, deduced from the deformation through the constitutive law. We therefore have rigourously :

$$t = C .$$

Finally T is the surface density of reaction forces due to constraint $(3.8)_4$ on Σ_U. This reaction is due to the stress field t by $(3.8)_6$, and can be eliminated.

Retaining only the three independent fields of values D_L, U and t, equation (3.7) then gives the following variational principle :

$$(3.9) \quad \left[\begin{array}{l}
\delta \left[\displaystyle\int_{\Omega_o} \left[\dfrac{1}{2}\, \underset{\sim}{\alpha}\, (D_L)(D_L) + \tilde{t}\left[\dfrac{1}{2}\left[\dfrac{\partial U}{\partial M_o} + \dfrac{\overline{\partial U}}{\partial M_o} \right] - D_L \right] - \bar{f}\, U \right] d\Omega_o \right. \\[14pt]
\qquad\qquad \left. - \displaystyle\int_{\Sigma_{oF}} \bar{F}\, U d\Sigma_o + \int_{\Sigma_{oU}} \overline{tn}\left[U_d - U \right] d\Sigma_o \right] = 0 \\[14pt]
\forall\, D_L,\ \delta U,\ \delta t .
\end{array} \right.$$

It should be noted that (3.9) is a weak formulation of the problem, in which the "stress" t is statically admissible in the mean, the constitutive law $(3.8)_1$ is satisfied in the mean, and the kinematic conditions (3.4) are also satisfied in the mean.

<u>Important remark</u> : The elimination of the multiplier T by means of $(3.8)_6$, implies that the independent variable U, associated with $(3.8)_6$ on Σ_{oU}, must disappear from functional (3.5) in the domain constituted by the boundary Σ_{oU}.

In fact, given a functional :

$$y = \chi \ (x_1, \ x_2, \ \ldots, \ x_p)$$

with p independent variables, a principle of stationarity for y gives the following Euler equations :

$$\frac{\partial y}{\partial x_j} = 0 \ , \qquad j = 1, \ 2, \ \ldots, \ p$$

If we require variables x_j to satisfy one of the equations strictly, for example :

$$\frac{\partial y}{\partial x_i} = 0$$

the principle becomes :

$$\delta y = \frac{\partial y}{\partial x_1} \ \delta x_1 + \frac{\partial y}{\partial x_2} \ \delta x_2 + \ \ldots \ + \frac{\partial y}{\partial x_{i-1}} \ \delta x_{i-1} + \frac{\partial y}{\partial x_{i+1}} \ \delta x_{i+1} + \ \ldots \ +$$

$$+ \ \frac{\partial y}{\partial x_p} \ \delta x_p = 0$$

$\blacktriangledown \ x_1, \ x_2, \ \ldots, \ x_p$ satisfying the equation $\frac{\partial y}{\partial x_i} = 0$.

This signifies that y no longer depends on x_i (in the domain where this equation is assumed to hold).

This condition must be satisfied, if necessary, by modifying the principle so that variable x_i is genuinely independent in the different disconnected domains where it is liable to occur. This check being made, it is permissible to remodify the principle again.

In the preceding case, (3.7) gives, with $(3.8)_6$, the modified form :

$$\left[\delta \left[\int_{\Omega_o} \left[\alpha - t\tilde{D}_L - [\operatorname{div} t + \overline{f}] \, U \right] d\Omega_o + \int_{\Sigma_{oF}} [\overline{tn} - \overline{F}] \, U d\Omega_o + \int_{\Sigma_{oU}} \overline{tn} \, U_d d\Omega_o \right] = 0 \atop \forall D_L, \ U, \ t, \right.$$

where we check that variable U no longer appears on Σ_{oU}, as an independent variable (in particular independent of its value in other domains Ω_o and Σ_{oF}).

We note that the elimination of certain independent variables (in this case T), this elimination being aimed at simplification, is not always possible in a functional, the dual variable (in this case U), remaining in the principle (in the domain considered). This is so in certain non-linear mixed principles[25].

As a further interesting example, let us consider the primal principle where the constraint :

$$U = U_d \ , \quad \text{on } \Sigma_{oU},$$

is introduced by means of the vectorial multiplier T.

The principle is then written

$$\left[\int_{\Omega_o} T_r \left(C \frac{\partial \delta U}{\partial M_o} \right) d\Omega_o - \int_{\Omega_o} \overline{f} \delta U d\Omega_o - \int_{\Sigma_{oF}} \overline{F} \delta U d\Sigma_o - \delta \int_{\Sigma_{oU}} \overline{T} \left[U - U_d \right] d\Sigma_o = 0, \forall \delta U, \delta T \right.$$

where $C = \underset{\sim}{C}(U)$ is given by the constitutive law.

We write, in classic form :

$$\left[\int_{\Omega_o} - \left[\mathrm{div} \left[\underset{\sim}{C}(U) \right] + \overline{f} \right] \delta U d\Omega_o + \int_{\Sigma_{oF}} \left[\overline{\underset{\sim}{C}(U).n} - \overline{F} \right] \delta U d\Sigma_o + \int_{\Sigma_{oU}} \left[\left[\overline{\underset{\sim}{C}(U).n} - \overline{T} \right] \delta U \right. \right.$$

$$\left. \left. - \overline{\delta T} \left[U - U_d \right] \right] \right] d\Sigma_o = 0$$

$$\forall \delta U, \ \delta T .$$

If we wish to eliminate T by the Euler equation :

$$T = \underset{\sim}{C}(U).n, \quad \text{on } \Sigma_{oU},$$

we find :

$$\left[\int_{\Omega_o} - \Big[\text{div}[\underset{\sim}{C}(U)] + \overline{f} \Big] \delta U d\Omega_o + \int_{\Sigma_{oF}} \Big[\overline{\underset{\sim}{C}(U).n} - \overline{\overline{F}} \Big] \delta U d\Omega_o - \int_{\Sigma_{oU}} \delta \Big[\overline{\underset{\sim}{C}(U).n} \Big]\Big[U-U_d \Big] d\Sigma_o = 0 \right.$$

$$\left. \forall \delta U \right.$$

where we see that the independent variable U remains in the integral on Σ_{oU}.

Nevertheless a functional can be found formally in that case. However, the term $\underset{\sim}{C}(U).n$ cannot be kept, for it will be written as a function of the derivative $\frac{\partial U}{\partial n}$ on Σ_{oU} ; this derivative cannot be integrated by parts and has to be taken as an independent variable. We must then either retain the multiplier T, or respect the constraint $U = U_d$ on Σ_{oU}

To conclude, it should be noted that simultaneous elimination of two independent variables, by means of a single equation, is not automatic.

3. – <u>HELLINGER–REISSNER PRINCIPLE, OR TWO FIELD PRINCIPLE</u>[4]

We can set ourselves the task of finding a solution which satisfies strictly and locally, certain of the equations $(3.8)_1$ – $(3.8)_5$. Taking for example equation $(3.8)_1$, we have :

$$\underset{\sim}{t} = \frac{\partial \alpha}{\partial D_L} = \underset{\sim}{C}.$$

In this case, the term in δD_L disappears in (3.7). The corresponding equation $(3.8)_1$, used to calculate the independent variable D_L, is absent, and we must naturally eliminate this variable, precisely by means of the local equation $(3.8)_1$. With the notations of chapter I, we had :

$$\alpha = \frac{1}{2} \underset{\sim}{\alpha} (D_L)(D_L) = \frac{1}{2} D_L \underset{\sim}{A} D_L \ , \ A = \underset{\sim}{A}$$

$$(3.10) \qquad t = AD_L = C.$$

Whence :

$$D_L = A^{-1} t = Bt \text{ with } A^{-1} = B = \underset{\sim}{B}.$$

Thus :

$$\beta = \frac{1}{2} \underset{\sim}{\beta}(t)(t) = \frac{1}{2} t \underset{\sim}{B} t,$$

β is the elastic complementary energy.

Thus we have :

$$(3.11) \qquad \tilde{D}_L = \frac{\partial \beta}{\partial t} = \underset{\sim}{\beta}(t) = \tilde{t}B \; ; \; D_L = Bt$$

α and β are positive definite quadratic forms.

Now putting (3.11) in (3.7) (where T is replaced by tn), we obtain :

$$(3.12) \qquad \begin{aligned} &\int_{\Omega_o} \left[-\left[\operatorname{div} t + \overline{f}\right]\delta U + \delta\tilde{t}\left[\frac{1}{2}\left[\frac{\partial U}{\partial M_o} + \frac{\overline{\partial U}}{\partial M_o}\right] - Bt\right]\right] d\Omega_o \\ &\qquad + \int_{\Sigma_{oF}} \left[\overline{tn} - \overline{F}\right]\delta U d\Sigma_o + \int_{\Sigma_{oU}} \overline{\delta tn}\left[U_d - U\right] d\Sigma_o = 0 \\ &\forall \delta U, \delta t \end{aligned}$$

or

$$\begin{aligned} &\int_{\Omega_o} - \operatorname{div} t . \delta U d\Omega_o + \int_{\Sigma_{oF}} \overline{tn\delta U} d\Sigma_o + \int_{\Sigma_{oU}} \overline{tn\delta U} d\Sigma_o - \int_{\Sigma_{oU}} \overline{tn\delta U} d\Sigma_o \\ &\quad + \int_{\Omega_o} \delta\tilde{t}\left[\frac{1}{2}\left[\frac{\partial U}{\partial M_o} + \frac{\overline{\partial U}}{\partial M_o}\right] - Bt\right] d\Omega_o - \int_{\Omega_o} \overline{f}\delta U d\Omega_o - \int_{\Sigma_{oF}} \overline{F}\delta U d\Sigma_o \\ &\quad + \int_{\Sigma_{oU}} \overline{\delta tn}\left[U_d - U\right] d\Sigma_o = 0, \forall \delta t, \delta U. \end{aligned}$$

Finally we obtain the following <u>principle with two independent fields</u> :

$$(3.13) \qquad \begin{aligned} \delta &\left[\int_{\Omega_o} \left[\tilde{t}\frac{1}{2}\left[\frac{\partial U}{\partial M_o} + \frac{\overline{\partial U}}{\partial M_o}\right] - \frac{1}{2}\underset{\sim}{\beta}(t)(t) - \overline{f}\, U\right] d\Omega_o \right. \\ &\qquad \left. - \int_{\Sigma_{oF}} \overline{F}U d\Sigma_o + \int_{\Sigma_{oU}} \overline{tn}\left[U_d - U\right] d\Sigma_o\right] = 0 \end{aligned}$$

$\forall t, U.$ (t is stress C which verifies (3.10)).

Certain research work has been devoted to convergence studies concerning this principle[23], or one of its variants[24].

4. - <u>A FRAEIJS DE VEUBEKE TWO FIELD PRINCIPLE</u>[5]

Reverting to (2.7) and (3.8), we will take the field of multiplier t in the class of statically admissible fields, namely satisfying the equilibrium equations $(3.8)_2$ and $(3.8)_5$.   (3.7) then becomes (with T again replaced by tn) :

$$(3.14) \quad \begin{cases} \displaystyle\int_{\Omega_o} \left[\left[\frac{\partial \alpha}{\partial D_L} - \widetilde{t} \right] \delta D_L + \delta\widetilde{t} \left[\frac{1}{2}\left[\frac{\partial U}{\partial M_o} + \frac{\overline{\partial U}}{\partial M_o} \right] - D_L \right] \right] d\Omega_o \\[2ex] \qquad\qquad + \displaystyle\int_{\Sigma_{oU}} \overline{\delta tn} \left[U_d - U \right] d\Sigma_o = 0 \\[3ex] \forall\, \delta D_L \quad \text{and} \quad \forall \delta t \quad \text{S.A.} \end{cases}$$

The coefficients of δU are zero, and the vector field U should no longer appear as an unknown.   In fact $(3.8)_2$ and $(3.8)_2$ give :

$$(3.15) \quad \begin{cases} \text{div } \delta t = 0 \quad \text{in} \quad \Omega_o \\[2ex] \delta tn = 0 \qquad \text{on } \Sigma_{oF} \end{cases}$$

This means that the virtual field δt must be "statically admissible homogeneously" (S.A.H.).

Taking account of (3.15) and (3.6), and after simplification, (3.14) is therefore written :

$$(3.16) \quad \begin{cases} \delta \left[\displaystyle\int_{\Omega_o} \left[\frac{1}{2} \underset{\sim}{\alpha} \, (D_L)(D_L) - \widetilde{t}D_L \right] d\Omega_o + \displaystyle\int_{\Sigma_{oU}} \overline{tn} \; U_d \, d\Sigma_o \right] = 0 \\[3ex] \forall\, D_L \text{ and } \forall t \text{ S.A.} \end{cases}$$

As was to be expected, the variable U is eliminated. (3.16) represents
the Fraeijs de Veubeke two field principle.

5. - <u>PRINCIPLE OF COMPLEMENTARY ENERGY</u>[6]

This principle has one unknown field, the stress field, and is also designated as the "dual principle". It can be deduced from the Hellinger-
Reissner principle discussed in Chapter 3.

In fact considering equation (3.12), and taking the multiplier t, also
equal to the stress C as we have seen, in the class of statically admissible
stresses, namely satisfying $(3.8)_2$ and $(3.8)_5$, we obtain :

$$\int_{\Omega_o} \delta \tilde{t} \left[\frac{1}{2} \left[\frac{\partial U}{\partial M} + \overline{\frac{\partial U}{\partial M}} \right] - Bt \right] d\Omega_o + \int_{\Sigma_{oU}} \overline{\delta tn} \left[U_d - U \right] d\Sigma_o = 0$$

$$\forall t \text{ S.A.} \quad \text{and} \quad \forall \delta t \text{ S.A.H.}$$

We know that the variable U must be eliminated. In fact, taking account
of (3.6), and after simplification, the equation becomes :

$$(3.17) \quad \int_{\Omega_o} - \delta \tilde{t} \, Bt \, d\Omega_o + \int_{\Sigma_{oU}} \overline{\delta tn} \, U_d \, d\Sigma_o = 0 \ \forall t \text{ S.A.}$$

Hence the dual principe, with one unknown stress field :

$$(3.18) \quad \delta \left[\int_{\Omega_o} \frac{1}{2} \underset{\sim}{\beta}(t)(t) d\Omega_o - \int_{\Sigma_{oU}} \overline{tn} \, U_d \, d\Sigma_o \right] = 0, \, \forall t \text{ S.A.}$$

We then obtain the following result, referred to as the "principle of
complementary energy".

If we call "complementary energy" the quantity :

$$(3.19) \quad \mathfrak{D}(\underset{\sim}{t}) = - \int_{\Omega_o} \frac{1}{2} \underset{\sim}{\beta}(t)(t) d\Omega_o + \int_{\Sigma_{oU}} \overline{tn} \, U_d \, d\Sigma_o \quad ,$$

the exact solution C is that which maximizes the complementary energy calculated for the class of statically admissible stress fields t. Or otherwise :

$$(3.20) \qquad \left[\begin{array}{l} \text{Solution } C \Longleftrightarrow \text{Max } \mathcal{D}(\underset{\sim}{t}) \\ \qquad\qquad t \in \{t \text{ S.A.}\} \\[2mm] \text{with :} \\[2mm] \mathcal{D}(\underset{\sim}{t}) = - \int_{\Omega_o} \frac{1}{2} \underset{\sim}{\beta}(t)(t) d\Omega_o + \int_{\Sigma_{oU}} \overline{tn}\, U_d\, d\Sigma_o . \end{array} \right.$$

In fact let us consider the solution C, and any increment δC, so that :

$$t = C + \delta C$$

$$\mathcal{D}(\underset{\sim}{C+\delta C}) - \mathcal{D}(\underset{\sim}{C}) = - \int_{\Omega_o} \frac{1}{2} \underset{\sim}{\beta}(C+\delta C)(C+\delta C) d\Omega_o + \int_{\Sigma_{oU}} \overline{[C+\delta C]n}\, U_d\, d\Sigma_o$$

$$+ \int_{\Omega_o} \frac{1}{2} \underset{\sim}{\beta}(C)(C) d\Omega_o - \int_{\Sigma_{oU}} \overline{Cn}\, U_d\, d\Sigma_o$$

$$= - \int_{\Omega_o} \underset{\sim}{\beta}(C)(\delta C) d\Omega_o + \int_{\Sigma_{oU}} \overline{\delta Cn}\, U_d\, d\Sigma_o$$

$$- \int_{\Omega_o} \frac{1}{2} \underset{\sim}{\beta}(\delta C)(\delta C) d\Omega_o .$$

Now as C meets the condition of stationarity of $\mathcal{D}(\underset{\sim}{t})$ by hypothesis, we have:

$$\mathcal{D}(\underset{\sim}{C+\delta C}) - \mathcal{D}(\underset{\sim}{C}) = - \int_{\Omega_o} \frac{1}{2} \underset{\sim}{\beta}(\delta C)(\delta C) d\Omega_o \leqslant 0 ,$$

as β is a positive definitive quadratic form.

6. – <u>TWO FIELD HYBRID PRINCIPLE OF PIAN</u>[7]

This principle can be deduced easily from the Hellinger-Reissner principle with two fields, t and U, described in paragraph 3.

Considering equation (3.12), and taking the stress t in the class of fields satisfying internal equilibrium only, namely $(3.8)_2$, equation (3.12) immediately demonstrates that we must expect the variable U to be eliminated in Ω_o. Taking account of (3.6), equation (3.12) is written, in succession :

$$\int_{\Omega_o} \delta \tilde{t} \left[\frac{1}{2} \left[\frac{\partial U}{\partial M_o} + \overline{\frac{\partial U}{\partial M_o}} \right] - Bt \right] d\Omega_o + \int_{\Sigma_{oF}} \left[\overline{tn} - \overline{F} \right] \delta U d\Sigma_o$$

$$+ \int_{\Sigma_{oU}} \overline{\delta tn} \left[U_d - U \right] d\Sigma_o = 0$$

$\forall t$ S.A. in Ω_o and $\forall U$.

$$\int_{\Omega_o} \left[-\left[\text{div} \delta t \cdot U + \delta \tilde{t} Bt \right] d\Omega_o + \int_{\Sigma_{oF}} \overline{\delta tn} \, U d\Sigma_o + \int_{\Sigma_{oU}} \overline{\delta tn} \, U \, d\Sigma_o \right.$$

$$+ \int_{\Sigma_{oF}} \left[\overline{tn} - \overline{F} \right] \delta U \, d\Sigma_o + \int_{\Sigma_{oU}} \overline{\delta tn} \left[U_d - U \right] d\Sigma_o = 0$$

$\forall t$ S.A. in Ω_o and $\forall U$.

But δt also meets $(3.15)_1$, therefore becoming, after simplification :

$$(3.21) \quad \left[\begin{array}{l} \displaystyle\int_{\Omega_o} - \delta \tilde{t} \, Bt \, d\Omega_o + \int_{\Sigma_{oF}} \left[\overline{tn} - \overline{F} \right] \delta U d\Sigma_o + \int_{\Sigma_{oU}} \overline{\delta tn} \, U_d \, d\Sigma_o + \\[2ex] \displaystyle\hspace{6cm} + \int_{\Sigma_{oF}} \overline{\delta t \, n} \, U d \, \Sigma_o = 0 \\[2ex] \forall \, t \text{ S.A.} \hspace{3cm} (\text{ in } \Omega_o), \, \forall U \text{ on } \Sigma_{oF} . \end{array} \right.$$

From this, we deduce the following two field principle :

$$(3.22) \quad \left[\begin{array}{l} \delta \left[\displaystyle\int_{\Omega_o} - \frac{1}{2} \, \beta(t) \, (t) d\Omega_o + \int_{\Sigma_{oF}} \left[\overline{tn} - \overline{F} \right] U \, d\Sigma_o \right. \\[2ex] \displaystyle\hspace{5cm} \left. + \int_{\Sigma_{oU}} \overline{tn} \, U_d \, d\Sigma_o \right] = 0 \\[2ex] \forall t \text{ S.A. in } \Omega_o , \, \forall U \text{ on } \Sigma_{oF} . \end{array} \right.$$

It is clear that if the unknown stress field t is totally statically admissible, formula (3.21) again gives (3.18) and the principle of complementary energy.

7. - PRINCIPLE OF VIRTUAL DISPLACEMENTS AND PRINCIPLE OF VIRTUAL STRESSES

We have seen that in any deformed state, the Cauchy stress C, satisfying
equilibrium, must verify the principle of virtual displacements (1.22),
namely :

$$(3.23) \quad \left[\int_\Omega T_r \left(C\, \frac{\partial \delta U}{\partial M} \right) d\Omega - \int_\Omega \bar{f}\delta U \, d\Omega - \int_{\Sigma_F} \bar{F}\delta U d\Sigma = 0 \right.$$

$$\left. \forall \, \delta U \text{ K.A.} \right.$$

It is in fact from this linearized principle, in linear elasticity and
linear deformation, starting from the natural reference state, that it has
been posssible to deduce the other principles.

However, we can refer all the fields appearing in (3.23) to an earlier
state with subscript zero, taken as the reference state, by reciprocal
images. We then have :

$$(3.24) \quad \left[\begin{array}{l} \displaystyle\int_{\Omega_o} T_r(C'\delta D)\, d\Omega_o - \int_{\Omega_o} \bar{f}\delta U \det\left(\frac{\partial M}{\partial M_o}\right) d\Omega_o \\[3ex] \qquad\qquad - \displaystyle\int_{\Sigma_{oF}} \bar{F} \det\left(\frac{\partial M}{\partial M_o}\right)_2 \delta U d\Sigma_o = 0 \\[3ex] \forall \, U \text{ K.A., with } D = \underset{\sim}{D}(U); \end{array} \right.$$

C' is the Piola-Kirchhoff stress in Ω_o, where U is the value of the actual
displacement field between the two states. If these two states are assumed
to be close, we can linearize (3.24) from the reference state, then obtaining :

$$(3.25) \quad \left[\begin{array}{l} \displaystyle\int_{\Omega_o} T_r(C'\delta D_L) d\Omega_o - \int_{\Omega_o} \bar{f}\delta U d\Omega_o - \int_{\Sigma_{oF}} \bar{F}\delta U \, d\Sigma_o = 0 \\[3ex] \forall \, U \text{ K.A. with } D_L = \dfrac{1}{2}\left[\dfrac{\partial U}{\partial M_o} + \overline{\dfrac{\partial U}{\partial M_o}} \right]. \end{array} \right.$$

As in paragraph 2, we can then introduce the two constraints :

$$(3.26) \quad \begin{bmatrix} \dfrac{1}{2}\left[\dfrac{\partial U}{\partial M_o} + \overline{\dfrac{\partial U}{\partial M_o}}\right] - D_L = 0 & \text{in} & \Omega_o \\[2em] U - U_d = 0 & & \text{on} \quad \Sigma_{oU} \end{bmatrix}$$

using two Lagrangian multipliers, t and T, and taking D_L as an independent unknown, while including constraint $(3.26)_2$ in a weak form. By a calculation similar to that indicated in paragraph 2, we find :

$$(3.27) \quad \begin{bmatrix} \displaystyle\int_{\Omega_o} \left[\left[\tilde{C}'-\tilde{t}\right]\delta D_L - \left[\text{div } t + \overline{f}\right]\delta U + \delta\tilde{t}\left[\dfrac{1}{2}\left[\dfrac{\partial U}{\partial M_o} + \overline{\dfrac{\partial U}{\partial M_o}}\right] - D_L\right]\right]d\Omega_o \\[2em] + \displaystyle\int_{\Sigma_{oF}} \left[\overline{tn} - \overline{F}\right]\delta U \; d\Sigma_o \\[2em] + \displaystyle\int_{\Sigma_{oU}} \overline{\delta tn}\left[U_d - U\right] d\Sigma_o = 0 \\[2em] \forall \, \delta D_L, \; \delta t, \; \delta U. \end{bmatrix}$$

This gives the following local equations :

$$(3.28) \quad \begin{bmatrix} C' - t = 0 & \\[1em] \text{div } t + \overline{f} = 0 & \left.\begin{array}{c} \\ \\ \\ \end{array}\right\} \quad \text{in} \quad \Omega_o \\[1em] \dfrac{1}{2}\left[\dfrac{\partial U}{\partial M_o} + \overline{\dfrac{\partial U}{\partial M_o}}\right] - D_L = 0 & \\[2em] tn - F = 0 & \text{on} \quad \Sigma_{oF} \\[1em] U_d - U = 0 & \text{on} \quad \Sigma_{oU}. \end{bmatrix}$$

If we select the t fields in the class of fields verifying $(3.28)_1$, $(3.28)_2$ and $(3.28)_4$, (3.27) becomes :

$$(3.29) \quad \left[\begin{array}{l} -\displaystyle\int_{\Omega_o} \delta\tilde{t}\, D_L\, d\Omega_o + \int_{\Sigma_{oU}} \overline{\delta t n}\, U_d\, d\Sigma_o = 0 \\[2ex] \forall\, \delta t \text{ S.A. is homogeneous, satisfying :} \\[2ex] \text{div } \delta t = 0 \text{ in } \Omega_o, \quad (\text{with } \delta t = \overline{\delta t}) \\[2ex] \delta t n = 0 \text{ on } \Sigma_{oF}. \end{array} \right.$$

We note that the equation giving the variable D_L has been eliminated, this being $(3.28)_1$ associated with the constitute law between C' and D_L. (3.29) alone cannot therefore constitute a computation procedure. However, this is a necessary condition, which must be met by the linearized deformation D_L.

This principle will be called the "principle of virtual stresses"[8].

8. – <u>APPLICATION OF VARIATIONAL PRINCIPLES</u>

8.1. – <u>Stress functions</u>

Let us take a three-dimensional medium in equilibrium, under the sole action of surface densities F on Σ_F, which can be zero, namely satisfying equations (1.23) :

$$(3.30) \quad \left[\begin{array}{l} \text{div } C = 0, \quad C = \overline{C} \text{ in } \Omega \\[2ex] Cn = F \text{ on } \Sigma_F \end{array} \right.$$

We know that $(3.30)_1$ and $(3.30)_2$ constitute closure conditions for the stress C, such that, if C is of class C^1, and by application of Poincaré's theorem, there exists in a simply connected open set, an Hermitian endomorphism field B of class C^3, defined to within any symmetrical vector gradient[9], such that :

$$\left[\begin{array}{l} C = \text{rot } \overline{\text{rot } B} \\[2ex] \text{with :} \\[2ex] B = \overline{B} = B_1 + \text{grad } U_1 + \overline{\text{grad } U_1} \end{array} \right.$$

Application of the de Rham theorem also enables us to write the global conditions of existence, namely on the boundary, in the case of multiply connected media.

It is interesting to obtain the same results, using variational methods[8].

In fact (3.30) is equivalent to the principle of virtual displacements :

$$
\left[
\begin{array}{l}
\displaystyle\int_\Omega T_r\left(C\,\frac{\partial\delta U}{\partial M}\right)d\Omega - \int_{\Sigma_F} \overline{F}\delta U d\Sigma = 0, \ \ \forall\,\delta U \ \text{K.A} \\[2em]
C = \overline{C},
\end{array}
\right.
$$

or again :

$$
(3.31)\qquad
\left[
\begin{array}{l}
\displaystyle\int_\Omega T_r(C\delta D_L)d\Omega - \int_{\Sigma_F} \overline{F}\delta U d\Sigma = 0 \ \ \forall\,\delta U \ \text{C.A} \\[2em]
\text{with}\ \ \delta D_L = \frac{1}{2}\left[\frac{\partial\delta U}{\partial M} + \overline{\frac{\partial\delta U}{\partial M}}\right].
\end{array}
\right.
$$

Let us now consider D_L as an independent variable in Ω. In this case, D_L must meet constraint (1.19), namely :

$$(3.32)\qquad \text{rot}\ \overline{\text{rot}\ D_L} = 0 \ \text{in}\ \Omega,\ \text{with}\ D_L = \overline{D_L}\ .$$

We describe as a "kinematically admissible deformation field", a Hermitian field on value D_L and class C^2, verifying (3.32) in Ω, and $(3.31)_2$ on Σ.

We can then introduce constraint (3.32) in $(3.31)_1$, using a Lagrangian multiplier B, an Hermitian endomorphism of $\vec{E}_3$, which is taken to be non-independent [see (A.101)] , to simplify the writing ($\delta B = 0$), giving :

$$
(3.33)\qquad
\left[
\begin{array}{l}
\displaystyle\int_\Omega T_r(C\delta D_L)d\Omega - \int_\Omega T_r(B\ \text{rot}\ \overline{\text{rot}\ \delta D_L})d\Omega - \int_{\Sigma_F} \overline{F}\delta U d\Sigma = 0 \\[2em]
\forall\,\delta D_L,\ \delta U\ \text{K.A},\ \text{and with (3.32).}
\end{array}
\right.
$$

We will develop this calculation, as an exercise. For this purpose, we use a natural basis S at $M \in \overline{\Omega}$. We know [see (A.63)] that :

$$\text{rot } \delta D_L = i(\overline{^i S^{-1}}) \partial_i \delta D_L, \quad (i = 1, 2, 3)$$

The volume integral of (3.33) is then written, in succession :

$$\int_\Omega T_r (C \delta D_L - B i(\overline{^i S^{-1}}) \partial_i \overline{\text{rot } \delta D_L}) d\Omega$$

$$= \int_\Omega T_r (C \delta D_L + i(\overline{^i S^{-1}}) B \partial_i \text{rot } \delta D_L) d\Omega$$

$$= \int_\Omega T_r (C \delta D_L + i(\overline{^i S^{-1}}) \partial_i \left[B \text{ rot } \delta D_L \right] - \text{rot } B \text{ rot } \delta D_L) d\Omega$$

$$= \int_\Omega T_r (C \delta D_L + \text{rot} \left[B \text{ rot } \delta D_L \right] - \text{rot } B \, i(\overline{^i S^{-1}}) \partial_i \delta D_L) \, d\Omega$$

$$= \int_\Omega T_r (C \delta D_L + \text{rot} \left[B \text{ rot } \delta D_L \right] + i(\overline{^i S^{-1}}) \overline{\text{rot } B} \, \partial_i \delta D_L) d\Omega$$

$$= \int_\Omega T_r (C \delta D_L + \text{rot} \left[B \text{ rot } \delta D_L \right] + \text{rot} \left[\overline{\text{rot } B} \delta D_L \right] - \text{rot } \overline{\text{rot } B} \delta D_L) d\Omega$$

$$= \int_\Omega T_r (\left[C - \text{rot } \overline{\text{rot } B} \right] \delta D_L + \text{rot} \left[B \text{ rot } \delta D_L + \overline{\text{rot } B} \, \delta D_L \right]) d\Omega.$$

The second term provides a surface integral, by application of the Stokes formula. Thus :

$$\int_\Omega T_r (\text{rot} \left[B \text{ rot } \delta D_L + \overline{\text{rot } B} \, \delta D_L \right]) d\Omega$$

can be calculated by expressing the trace in a natural basis S, assumed to be constant, as it is an invariant [see (A.8)] :

$$\int_\Omega T_r \left(\mathrm{rot} \left[B \ \mathrm{rot} \ \delta D_L + \overline{\mathrm{rot} \ B} \ \delta D_L \right] \right) d\Omega$$

$$= \int_\Omega {}^i s^{-1} \ \mathrm{rot} \left[B \ \mathrm{rot} \ \delta D_L + \overline{\mathrm{rot} \ B} \ \delta D_L \right] S_i \ d\Omega \ , \quad i = 1,2, \ 3;$$

$$= \int_\Omega \mathrm{rot} \overline{\left[\left[B \ \mathrm{rot} \ \delta D_L + \overline{\mathrm{rot} \ B} \ \delta D_L \right] S_i \right] \ {}^i s^{-1}} \ d\Omega \ .$$

We can then apply formula (A.56) :

$$\overline{\mathrm{rot} \ V} = \mathrm{div} \left[i(V) \right]$$

followed by the Stokes formula. Thus :

$$\int_\Omega T_r \left(\mathrm{rot} \left[B \ \mathrm{rot} \ \delta D_L + \overline{\mathrm{rot} \ B} \ \delta D_L \right] \right) d\Omega$$

$$= \int_\Omega \mathrm{div} \left[i \left(\left[B \ \mathrm{rot} \ \delta D_L + \overline{\mathrm{rot} \ B} \ \delta D_L \right] S_i \right) \left(\overline{{}^i s^{-1}} \right) \right] d\Omega$$

$$= \int_\Sigma \bar{n} \ i \left(\left[B \ \mathrm{rot} \ \delta D_L + \overline{\mathrm{rot} \ B} \ \delta D_L \right] S_i \right) \left(\overline{{}^i s^{-1}} \right) d\Sigma$$

$$= \int_\Sigma {}^i s^{-1} \ i(n) \cdot \left[B \ \mathrm{rot} \ \delta D_L + \overline{\mathrm{rot} \ B} \ \delta D_L \right] S_i \ d\Sigma$$

$$= \int_\Sigma T_r \left(i(n) \cdot \left[B \ \mathrm{rot} \ \delta D_L + \overline{\mathrm{rot} \ B} \ \delta D_L \right] \right) d\Sigma \ .$$

But $(3.31)_2$ is assumed to hold on Σ (the integrability of D_L being ensured), so that the surface term therefore becomes :

$$\int_\Sigma T_r(i(n)\left[\frac{B}{2}\,\text{rot}\,\frac{\partial\delta U}{\partial M} + \overline{\text{rot}\,B}\left[\frac{\partial\delta U}{\partial M} - \frac{1}{2}\left[\frac{\partial\delta U}{\partial M} - \overline{\frac{\partial\delta U}{\partial M}}\right]\right]\right])d\Sigma$$

$$=\int_\Sigma T_r\left(i(n)\left[\frac{B}{2}\,\text{rot}\,\frac{\partial\delta U}{\partial M} + \overline{\text{rot}\,B}\left[\frac{\partial\delta U}{\partial M} - i\left(\frac{\text{rot}\,\delta U}{2}\right)\right]\right]\right)d\Sigma$$

$$=\int_\Sigma\left[T_r\left(i(n)\left[\frac{B}{2}\,\text{rot}\,\frac{\partial\delta U}{\partial M} + \overline{\text{rot}\,B}\,\frac{\partial\delta U}{\partial M}\right]\right)\right.$$

$$\left. - \frac{1}{2}\,T_r(i(n)\overline{\text{rot}\,B}\,i(\text{rot}\,\delta U))\right]d\Sigma$$

where we apply the double vectorial product formula, and take account of
$T_r(\text{rot}\,B) = 0$ when $B = \overline{B}$, [see (A.21), (A.5) and (A.11)$_3$].

Again expressing the trace in the natural basis, assumed to be constant,
we obtain for this surface integral :

$$\int_\Sigma\left[{}^i S^{-1}\,i(n)\left[\frac{B}{2}\,\text{rot}\,\partial_i\delta U + \overline{\text{rot}\,B}\,\partial_i\delta U\right] - \frac{n}{2}\,\text{rot}\,B\,\text{rot}\,\delta U\right]d\Sigma$$

$$=\int_\Sigma\left[-\bar{n}\,i(\overline{{}^i S^{-1}})\left[\partial_i\left[\frac{B}{2}\,\text{rot}\,\delta U\right] - \frac{\partial_i B}{2}\,\text{rot}\,\delta U + \partial_i\left[\overline{\text{rot}\,B}\,\delta U\right] - \right.\right.$$

$$\left.\left. - \partial_i\,\overline{\text{rot}\,B}\,\delta U\right] - \frac{n}{2}\,\text{rot}\,B\,\text{rot}\,\delta U\right]d\Sigma$$

$$=\int_\Sigma\left[-\bar{n}\,\text{rot}\left[\frac{B}{2}\,\text{rot}\,\delta U\right] + \frac{\bar{n}}{2}\,\text{rot}\,B\,\text{rot}\,\delta U - \bar{n}\,\text{rot}\left[\overline{\text{rot}\,B}\,\delta U\right]\right.$$

$$\left. + \bar{n}\,\text{rot}\,\overline{\text{rot}\,B}\,\delta U - \frac{n}{2}\,\text{rot}\,B\,\text{rot}\,\delta U\right]d\Sigma.$$

Assuming that the edge of Σ is empty, for sake of simplification, namely Σ is not a pseudo-manifold, the flux of a rotational being zero, we are then left for this same integral with :

$$\int_{\Sigma} \bar{n} \ \mathrm{rot} \ \overline{\mathrm{rot} \ B} \ \delta U \ d\Sigma.$$

(This simplification hypothesis can also be abandoned easily, if singular lines exist on Σ).

Finally (3.33) is written :

$$\left[\int_{\Omega} T_r \left(\left[C - \mathrm{rot} \ \overline{\mathrm{rot} \ B} \right] \delta D_L \right) d\Omega + \int_{\Sigma} \bar{n} \ \mathrm{rot} \ \overline{\mathrm{rot} \ B} \ \delta U d\Sigma \right.$$

$$- \int_{\Sigma_F} \bar{F} \ \delta U d\Sigma = 0$$

$$\forall \delta D_L \text{ and } \delta U \ \text{K.A.}$$

The Euler equations for the principle are then written :

$$(3.34) \quad \left[\begin{array}{l} C = \mathrm{rot} \ \overline{\mathrm{rot} \ B}, \ \forall B = \bar{B} \ \text{ in } \ \Omega \\[2ex] \mathrm{rot} \ \overline{\mathrm{rot} \ B} \ n = F \qquad \text{on } \ \Sigma_F \ . \end{array} \right.$$

We can verify immediately that this method also gives the conditions for global closure, in the case of multiply connected media, from :

$$\int_{\Sigma_F} \left[\bar{n} \ \mathrm{rot} \ \overline{\mathrm{rot} \ B} - \bar{F} \right] \delta U d\Sigma = 0 \ \ \forall \delta U \ \text{K.A.}$$

It is merely necessary to take a rigid body displacement field δU, having the form :

$$\delta U = \delta_o U + i(\delta_o \theta)(M), \quad \forall \delta_o U, \ \delta_o \theta \ \text{ constant on each cycle non-}$$

homotopic with zero in Ω.

We find again Gurtin's theorem in particular for holes homotopic with spheres, according to which a stress function $B = \bar{B}$ gives a univoque stress field, if surface densities F are self-equilibrated on each non-connected edge[8].

8.2. - <u>Linearized compatibility condition</u>

We can also use a variational principle to find the condition of compatibility (3.32)[8]. For this purpose, it is sufficient to use the principle of virtual stresses (see paragraph 7), according to which the linearized deformation D_L must satisfy the principle :

$$(3.35) \quad \left[\int_\Omega T_r(\delta C\ D_L)d\Omega - \int_{\Sigma_U} \overline{\delta Cn}\ U_d\ d\Sigma = 0 \right.$$

$$\left. \forall \delta C \ \text{ S.A. homogeneous, with } D_L = \overline{D_L}. \right.$$

Now according to the preceding result, we need merely take :

$$\delta C = \text{rot } \overline{\text{rot } \delta B} \ \text{ in } \Omega \ , \ \forall \delta B = \overline{\delta B}.$$

The boundary closure conditions are trivially verified for zero boundary loads. (3.35) is then written :

$$\left[\int_\Omega T_r(\text{rot } \overline{\text{rot } \delta B}\ D_L)d\Omega - \int_{\Sigma_U} \overline{\text{rot } \overline{\text{rot } \delta B}\ n.\ U_d}d\Sigma = 0 \right.$$

$$\forall \delta B = \overline{\delta B} \ \text{ verifying}$$

$$\text{rot } \overline{\text{rot } \delta B}\ n = 0 \ \text{ on } \Sigma_F.$$

By a calculation in all points similar with that shown in paragraph 8.1, we find :

$$\int_\Omega T_r(\text{rot } \overline{\text{rot } D_L}\ \delta B)d\Omega - \int_\Sigma T_r(i(n)\left[D_L \text{ rot } \delta B + \overline{\text{rot } D_L}\ \delta B \right]) \ d\Sigma$$

$$- \int_{\Sigma_U} \overline{\text{rot } \overline{\text{rot } \delta B}\ n.\ U_d}d\Sigma = 0$$

$$\forall \delta B = \overline{\delta B}, \text{ with rot } \overline{\text{rot } \delta B}\ n = 0 \text{ on } \Sigma_F.$$

The first integral gives the necessary condition :

$$(3.36) \qquad \text{rot } \overline{\text{rot } D_L} = 0 \ \text{ in } \Omega \ \text{ with } D_L = \overline{D_L}.$$

If D_L satisfies the conditions of global closure, there exists a displacement field U, such that :

$$D_L = \frac{1}{2}\left[\frac{\partial U}{\partial M} + \overline{\frac{\partial U}{\partial M}}\right] \quad \text{in} \quad \Omega.$$

The surface integral is now written :

$$\int_{\Sigma_U} \bar{n} \; \text{rot} \; \overline{\text{rot} \; \delta B} \; U d\Sigma - \int_{\Sigma_U} \bar{n} \; \text{rot} \; \overline{\text{rot} \; \delta B} \; U_d \; d\Sigma = 0$$

which gives us again :

$$U = U_d \quad \text{on} \quad \Sigma_U.$$

To find the conditions of global closure in the case of non-simply connected media, it is simpler to use (3.36), which gives directly, in the case where circuits C_k non homotopic with zero exist (de Rham's theorem) :

$$(3.37) \quad \left[\begin{array}{l} \displaystyle\int_{C_k} \text{rot} \; D_L \; dM = 0 \\[2em] \displaystyle\int_{C_k} \left[D + i(\Omega)\right] dM = 0 \, . \end{array} \right.$$

In fact (3.36) gives the existence of a vector field Ω in any simply-connected open set, such that [see (A.41) and (A.60)] :

$$(3.38) \quad \text{rot} \; D_L = \frac{\partial \Omega}{\partial M}$$

which gives the condition $(3.37)_1$ on the circuits non-homotopic with zero. In addition, we know that $D_L + i(\Omega)$ is a closed endomorphism, hence $(3.37)_2$, which is also written :

$$\int_{C_k} \left[D_L + i(\Omega) \right] dM = \int_{C_k} D_L dM - \int_{C_k} i(dM)(\Omega)$$

$$= \int_{C_k} D_L dM - \int_{C_k} d\left[i(M)(\Omega) \right] + \int_{C_k} i(M) \frac{\partial \Omega}{\partial M} dM$$

$$= \int_{C_k} \left[D_L + i(M).\mathrm{rot}\ D_L \right] dM,$$

using (3.38) and the fact that the edge of C_k is empty.

Finally, the conditions of global closure are written :

$$(3.39) \quad \begin{bmatrix} \displaystyle\int_{C_k} \mathrm{rot}\ D_L\, dM = 0 \\[4ex] \displaystyle\int_{C_k} \left[D_L + i(M).\mathrm{rot}\ D_L \right] dM = 0\,. \end{bmatrix} \qquad \text{with}\quad D_L = \overline{D_L}$$

These conditions have a simple kinematic interpretation, based on the
unique   existence of rotation and displacement, at any point of a circuit.

8.3. – <u>Application of the finite element method</u>

The variational principles described above can all be discretized using
the finite element method. Clearly, each of these principles presents both
advantages and disadvantages.

Principles with several fields have the disadvantage of increasing the
number of unknown fields, and consequently the number of degrees of freedom.
On the other hand, they offer the advantage of introducing constraints, or
in general equations, in weak form, therefore only imposing their verificat-
ion in the mean in the medium, which equates precisely to greater freedom
in the search for solutions, by broadening the class of the latter. Approximate solutions then belonging to wider classes, can generally be adapted
more easily to the conditions of a problem.

We also note in general, that in a principle with several fields, not all the unknown fields are necessarily taken in the class of continuous functions. In fact it is merely necessary to consider the weak forms used, to know which are the widest classes to which these fields should belong respectively (and in which it is appropriate to look for theorems of existence and uniqueness). For example, we observe that distributed multiplier fields can be selected very frequently in the class of square summable functions, and in principle do not have to be continuous at the interfaces.

Other types of constraints can be introduced in weak form in a principle, according to the particular type of problem set, using Lagrangian multipliers, the mechanical interpretation of which is always easy. These multipliers always constitute unknown fields, as mentioned above, and therefore represent supplementary unknowns. These constraints can concern the constitutive law, for example in the case of an incompressible medium, or the type of solution displacement, for example in the case of isochoric movements (with no volume variation), which is different. Redundant constraints must naturally be avoided.

Assembly, which is one of the most important operations of the method, must be made at the interfaces on the unknown fields, provided that continuity is required or deliberately preferred. However, nodal assembly at the interfaces is frequently considered sufficient, with no requirement for complete continuity. If continuity is not selected, not even locally, it is essential to introduce it in the mean, using interface multipliers, by replacing the assembly of the primary unknowns by that of the discretized multipliers.

To illustrate certain of the preceding remarks, we shall take as an example, the <u>application of the Hellinger-Reissner principle</u>.

Principle (3.13) can thus be written in a modified form, with integration of the first integral parts :

$$\delta\left[\int\!\!\int_{\Omega_o}\left[-[\operatorname{div}\ t + \overline{f}]U - \frac{1}{2}\,\beta(t)(t)\right]\,d\Omega_o + \int_{\Sigma_{oF}}\left[\overline{tn} - \overline{F}\right]Ud\Sigma_o + \int_{\Sigma_{oU}}\overline{tn}U_d\right] = 0,\forall t,U,$$

and can be applied in this form. Its discretization leads to the adoption of a square summable stress class, the divergence of which is also of the square summable type [space H (div., Ω_o)[24]]. The displacement field U in Ω_o, this will be also taken in the space of the square summable vectorial functions defined in Ω_o, not necessarily continuous at the interface of the finite elements. The displacement field U on Σ_{oF} must belong to a class of square summable vectorial functions defined on Σ_{oF}. This latter field then constitutes a field independent from the preceding one (Ω_o is still the whole domain and Σ_o its boundary).

We should add that assembly of an element is achieved by writing that the stress tn is continuous on its boundary, eliminating the internal edge integrals, or by adopting an independent displacement field defined on its boundary, thus achieving continuity of these stresses in the mean (dual hybrid principle).

On the other hand, in principle (3.13) in its unmodified form, the adoption of a non-continuous square summable stress field in Ω_o, must be accompanied by the choice of an independent stress field tn, defined on Σ_{oU}. Using an argument similar to that above, but of dual character, the assembly of each element is achieved by writing that the displacement is continuous on its boundary, eliminating the internal edge integrals ; or by adopting an independent unknown field tn on the boundary of the element, thus giving continuity of these displacements in the mean (primal hybrid principle).

Another illustration, for which we shall give a certain number of details, will be the <u>application of the principle of complementary energy</u>.

This is a principle with one unknown field, the stress field, in other words a variable of the static type. This leads to finite elements of the <u>equilibrium type</u>[5,10] (as opposed to the displacement model). If we choose to interpolate this stress field, so that it is continuous on the interfaces, we obtain a "codiffusive" finite element model (namely conforming as to the stresses). The assembly is of the interface type. If the interfaces are linear (straight lines and plane), each is defined by its normal. For codiffusive elements, assembly is then easy, and the stress field satisfies the equilibrium conditions for the complete structure, after assembly.

If on the other hand, we admit stress fields which are totally discontinuous on the interfaces, "assembly" is nevertheless required for the structure to be in equilibrium as a whole. We must then ensure continuity of the stress field in the mean. For this purpose, we introduce the interface stress discontinuity by means of a multiplier, which incidently is a displacement. In fact let us take the stationarity equation written on any domain of the medium, such that $\Sigma_U = \emptyset$

The equilibrium solution satisfies :

$$(3.40) \qquad \delta \int_\Omega \frac{1}{2} \beta\,(C)\,(C)\,d\Omega = 0\ \forall C\ \text{S.A.}$$

Let us now separate Ω arbitrarily, by means of an interface Σ^*, into two domains, Ω_1 and Ω_2. For Ω_1, the external normal to Σ^* will be n_1, and $n_2 = -n_1$ for Ω_2. Equilibrium on Σ^* is given by :

$$(3.41) \qquad \left[C_1 - C_2\right] n_1 = 0 \text{ on } \Sigma^* \text{ (as C must be S.A. in } \Omega).$$

Clearly, if :

$$C_1 n_1 \neq C_2 n_1$$

this condition (3.41) must be introduced in the principle, using a multiplier U. Equation (3.40) then becomes :

$$\delta \left[\int_{\Omega_1} \frac{1}{2} \beta(C)(C)\,d\Omega - \int_{\Sigma^*} \left[\overline{C_1 n_1} - \overline{C_2 n_1}\right] Ud\Sigma \right.$$

$$\left. + \int_{\Omega_2} \frac{1}{2} \beta(C)(C)\,d\Omega \right] = 0,\ \forall\ \text{C.S.A. in } \overline{\Omega}$$
$$\forall\, U \qquad \text{on } \Sigma^*.$$

For example, for the finite elements occupying domain Ω_1 with boundary Σ_1, we must calculate the quantities :

$$\delta \left[\int_{\Omega_1} \frac{1}{2} \beta(C)(C)\,d\Omega - \int_{\Sigma_1} \overline{Cn}\ Ud\Sigma \right].$$

The equation for the complete structure will be written by making the sum of these quantities equal to zero, after assembly of the discretized displacements U. (If necessary, we complete with the term :

$$- \int_{\Sigma_U} \overline{Cn}\ U_d\ d\Sigma$$

if boundary Σ_U of the structure is non-empty).

Furthermore, it should be noted that the maximization of the complementary energy should be made in the class of all <u>statically admissible</u> stresses. If the volume force density is non-zero, we can decompose C into a sum :

$$C = C_1 + C_2$$

where C_1 is selected in a class satisfying :

(3.42)
$$\begin{cases} \text{div } C_1 + \overline{f} = 0 \quad \text{in } \Omega \\ C_1 n = F \qquad \text{on } \Sigma_F. \end{cases}$$

and C_2 in a class such that :

(3.43)
$$\begin{cases} \text{div } C_2 = 0 \quad \text{in } \Omega \\ C_2 n = 0 \quad \text{on } \Sigma_F. \end{cases}$$

The stress C_1 giving non-homogeneous equilibrium, is interpolated separately to satisfy equation (3.42). In fact an interpolation of the type :

$$C_1 = \underset{\sim}{T}_1(M) \cdot {}^1\alpha , \quad {}^1\alpha \in \mathbb{R}^{m_1}$$

being selected, (3.42) will only be satisfied strictly for densities f and F belonging to the class defined by $\underset{\sim}{T}_1(M)$.

The stress C_2 will also be interpolated by :

$$C_2 = T_2(M) \cdot {}^2\alpha , \quad {}^2\alpha \in \mathbb{R}^{m_2},$$

so as to satisfy (3.43). For example, we can derive C_2 from a stress function B, the components of which being in fact interpolated.

Stress C supplies interface stresses Cn, for each element, these being the unknown generalized forces Q of the discretized problem. To simplify, we will assume that C is continuous at the interfaces. For C to be statically admissible, all interface stresses must be equal, on all interfaces introduced by the finite element modelling, in other words equilibrium must be achieved on each interface, providing the conditions for <u>assembly</u> of all the elements. If the interfaces are linear, they are defined by their normals n.

On each interface of element e, with normal n_k, we must define as many generalized forces ^eQ_k as are necessary to achieve assembly, and as there are α coefficients in the interpolation (in the same way as a sufficient number of nodes must be introduced in the displacement method). For example, for an internal interface where density F equals zero, the stress at point M will be :

$$(3.44) \qquad \left[C_2 \; n_k \right](M) = \underset{\sim}{T}_2(M) n_k . ^2\alpha .$$

We must choose as many points M on the interface, and therefore as many generalized forces ^eQ_k per interface, as are necessary taking account of the order of column $^2\alpha$. We can also define these generalized forces by mean values, or stress resultants on the interface. If we write (3.44) for all interfaces of element e, we obtain a connection matrix $^e\mathbf{C}_2$, such that :

$$^eQ = {}^e\mathbf{C}_2 . ^2\alpha,$$

where ^eQ is the column of generalized forces for element e.

It should be noted that this connection matrix is rectangular, as the value of ^eQ in the solution of a static problem must satisfy the equilibrium condition for element e. The components of ^eQ are therefore not independent, and $^e\mathbf{C}_2$ is rectangular.

If the volume density of given forces in element e is now non-zero, the interpolation of C must meet $(3.42)_1$, therefore :

$$(3.45) \qquad C = \underset{\sim}{T}_1(M) . ^1\alpha + \underset{\sim}{T}_2(M) . ^2\alpha \quad \text{in} \quad \Omega_e,$$

and for the interface defined by n_k, we have :

$$\left[C \; n_k \right](M) = \left[\underset{\sim}{T}_1(M) . ^1\alpha + \underset{\sim}{T}_2(M) . ^2\alpha \right] n_K .$$

For any element e, we then obtain :

$$\left[{}^e\mathbf{C}_1 \quad {}^e\mathbf{C}_2 \right] \left[\begin{array}{c} ^1\alpha \\ ^2\alpha \end{array} \right] = {}^eQ .$$

If in addition, element e has a common boundary with $\Sigma_F(\Sigma_{eF} \neq \emptyset)$, the surface density F must also be balanced by the field C_1 (in fact F must belong to the class defined by the interpolation of C_1). Thus for the interfaces concerned, with normal n_ℓ :

$$C_1 \, n_\ell = \underset{\sim}{T}_1(M).n_\ell \, . \, {}^1\alpha = F_\ell(M).$$

For interfaces of this type for element e, we have :

$$^e C_F . \, {}^1\alpha = {}^e Q_d,$$

where ${}^e Q_d$ are the given generalized external surface forces. For any element e, we therefore have :

$$(3.46) \qquad e C_\alpha = {}^e Q \quad ou \quad \begin{bmatrix} {}^e C_1 & {}^e C_2 \\ {}^e C_F & 0 \end{bmatrix} \begin{bmatrix} {}^1\alpha \\ {}^2\alpha \end{bmatrix} = \begin{bmatrix} {}^e Q \\ {}^e Q_d \end{bmatrix}$$

Finally, if we still call α the set of the α parameters relative to all elements placed in certain order, we obtain a general connection matrix $\mathcal{C}$, such that :

$$(3.47) \qquad \mathcal{C}\alpha = Q \quad or \quad \begin{bmatrix} {}^1\mathcal{C}_1 & {}^1\mathcal{C}_2 \\ {}^2\mathcal{C}_1 & 0 \end{bmatrix} \begin{bmatrix} {}^1\alpha \\ {}^2\alpha \end{bmatrix} = \begin{bmatrix} {}^1Q \\ {}^2Q \end{bmatrix}$$

with ${}^2\mathcal{C}_2 = 0$, and ${}^2Q$ as the column of given generalized surface forces.

Assembly is carried out after changes of reference system, for the representation of all forces in a common reference system, where appropriate, namely :

$$Q = S' \, Q' \quad or \quad {}^e Q = {}^e S' Q'$$

The assembly then gives the independent forces Q", such that :

$$Q' = S''Q''.$$

Whence :

$$(3.48) \qquad \mathcal{C}''\alpha \; = S'S''Q''$$

The principle of complementary energy gives the equation :

$$(3.49) \qquad \delta \left[\int_{\Omega} \frac{1}{2} \underset{\sim}{\beta}(C)(C)d\Omega - \int_{\Sigma_U} \overline{Cn} \; U_d \; d\Sigma \right] = 0, \forall \, C \; \text{S.A.}$$

Putting (3.45) in (3.49), we find a discretized equation having the form :

$$(3.50) \qquad \left[\begin{array}{l} \delta\left[\dfrac{1}{2} \bar{\alpha} \; H\alpha - \overline{\mathcal{C}\alpha} \; q_d \right] = 0, \; \text{with } H = \bar{H} \\[2ex] \alpha \text{ satisfying } {}^{2}C_1 . {}^{1}\alpha = {}^{2}Q. \end{array} \right.$$

It is then natural to introduce $(3.50)_2$ by means of a Lagrangian multiplier, which we shall call ${}^{2}q$, the conjugated displacement of given generalized surface forces. We then obtain :

$$\delta \left[\frac{1}{2} \bar{\alpha} \; H\alpha - \bar{\alpha} \; \overline{\mathcal{C}} \; q_d - \overline{{}^{2}q} \left[{}^{2}C_{,}\alpha - {}^{2}Q \right] \right] = 0, \forall \, \alpha, \; {}^{2}q.$$

Hence the equation :

$$(3.51) \qquad \left[\begin{array}{cc} H & -\overline{{}^{2}\mathcal{C}} \\[2ex] -{}^{2}\mathcal{C} & 0 \end{array} \right] \left[\begin{array}{c} \alpha \\[2ex] {}^{2}q \end{array} \right] = \left[\begin{array}{c} \overline{\mathcal{C}} \; q_d \\[2ex] {}^{2}Q \end{array} \right].$$

The solution is then written :

$$\alpha = H^{-1} \left[\overline{{}^{2}\mathcal{C}} \; {}^{2}q + \overline{\mathcal{C}} \; q_d \right]$$

$$\; {}^{2}C\alpha = {}^{2}C \, H^{-1} \; \overline{{}^{2}\mathcal{C}} \; {}^{2}q + {}^{2}C \, H^{-1} \; \overline{\mathcal{C}} \; q_d = -{}^{2}Q$$

$$\; {}^{2}q = \left[{}^{2}C \, H^{-1} \; \overline{{}^{2}\mathcal{C}} \right]^{-1} \left[{}^{2}Q - {}^{2}C \, H^{-1} \; \overline{\mathcal{C}} \; q_d \right]$$

$$\alpha = H^{-1}\left[\overline{{}^{2}C}\left[{}^{2}C\,H^{-1}\,\overline{{}^{2}C}\right]^{-1}\left[-{}^{2}Q-{}^{2}C\,H^{-1}\,\overline{C}\,q_{d}\right]+\overline{C}\,q_{d}\right]$$

$${}^{1}Q = {}^{1}C\,H^{-1}\left[\overline{{}^{2}C}\left[{}^{2}C\,H^{-1}\,\overline{{}^{2}C}\right]^{-1}\left[-{}^{2}Q-{}^{2}C\,H^{-1}\,\overline{C}\,q_{d}\right]+\overline{C}\,q_{d}\right].$$

This method requires the inversion of a high order matrix. It can be modified in such a way that this disadvantage is eliminated, by first introducing generalized self-balanced stress modes Q, a linear combination of which, increased by one statically admissible mode, represents the solution. These self-balanced modes are generally few in number for the complete structure, and the inversion is then that of a low order matrix. The drawback is then the search for self-balanced stress modes[11,12].

Another application consists in considering multipliers of the preceding type at the element level, for example assuming all generalized forces ${}^{e}Q$ to be known, with displacements not given on Σ_{e}. The equilibrium of the element is then given by equation (3.50), written out in this case, to yield :

$$^{e}Q = {}^{e}C\cdot H_{e}^{-1}\cdot {}^{e}C\cdot {}^{e}q = {}^{e}K_{e}\cdot {}^{e}q.$$

This gives a stiffness matrix as in the displacement models[5,10], which can be used in the same manner. However, this matrix is always singular by reason of the relations expressing the global equilibrium of the element. Furthermore, utilization of this matrix can be delicate.

It should also be noted that the generalized displacements q, resulting from application of the principle of complementary energy, are not continuous. In fact this principle, as we have seen in its particular form of the principle of virtual stresses, only provides compatible deformation in the mean, and not locally.

Another method of using this principle could be to adopt stress fields statically admissible in the mean, corresponding to the adoption of a class satisfying the principle of virtual displacements, finally taking us back to the Hellinger-Reissner two field principle.

Finally, it should be noted that the stiffnesses found by application of the complementary energy principle, are underestimated by comparison with the exact stiffnesses, contrary to those supplied by the displacement method, and this for "symmetrical" reasons. This argument will be used in the following paragraph.

8.4. – Upper bound for the error of an approximate solution

Simultaneous utilization of the two, primal and dual, methods for the same problem, allows to find an upper bound for the error in the solution, in certain commonly encountered cases[13,5].

In fact according to the primal principle, the exact solution U minimizes the total potential $\mathcal{J}(\underset{\sim}{V})$, while according to the dual principle, the exact solution C maximizes the complementary energy $\mathcal{D}(\underset{\sim}{t})$, the fields V and t being respectively kinematically admissible and statically admissible. Thus in addition :

$$\mathcal{J}(\underset{\sim}{U}) = \mathcal{D}(\underset{\sim}{C}).$$

In fact if we take in equation (3.2), the exact solution as a virtual field δU, we obtain :

$$\int_{\Omega_o} \underset{\sim}{\alpha}(D_L)(D_L)d\Omega_o - \int_{\Omega_o} \bar{f}\, U\, d\Omega_o - \int_{\Sigma_{oF}} \bar{F}\, U\, d\Sigma_o = 0.$$

Therefore :

$$\mathcal{J}(\underset{\sim}{U}) = -\int_{\Omega_o} \frac{1}{2}\underset{\sim}{\alpha}(D_L)(D_L)d\Omega_o \;,\text{with}\; D_L = \frac{1}{2}\left[\frac{\partial U}{\partial M_o} + \frac{\overline{\partial U}}{\partial M_o}\right],$$

or :

$$\mathcal{J}(\underset{\sim}{U}) = -\int_{\Omega_o} \frac{1}{2}T_r(C\,D_L)d\Omega_o.$$

As regards $\mathcal{D}(\underset{\sim}{C})$, we have for $U_d = 0$, or $\Sigma_{oU} = \emptyset$

$$\mathcal{D}(\underset{\sim}{C}) = -\int_{\Omega_o} \frac{1}{2}\underset{\sim}{\beta}(C)(C)d\Omega_o = -\int_{\Omega_o} \frac{1}{2}T_r(C\,D_L)d\Omega_o. \quad \text{Q.E.D.}$$

We can then write :

$$\mathcal{D}(\underset{\sim}{t}) \leqslant \mathcal{D}(\underset{\sim}{C}) = \mathcal{F}(\underset{\sim}{U}) \leqslant \mathcal{F}(\underset{\sim}{V}).$$

This gives an overestimate in the energy error, equal to :

$$\mathcal{F}(\underset{\sim}{V}) - \mathcal{D}(\underset{\sim}{t})$$

for the two approximate solutions.

To have an error as a function of the idealization, calculations, made with the equilibrium and displacement models, must naturally relate to common modellings.

8.5. - <u>Non-conforming elements - Patch test</u>

To calculate finite elements of the displacement type, conforming or non-conforming, we can use variational principles in the following convenient manner. We consider the principle of potential energy, where we assume the volume density f of external forces to be zero, and increased by one term, integrated on boundary Σ_U, representing a mean value displacement continuity by means of a Lagrange multiplier T. Thus for an element occupying a domain Ω_e with boundary Σ_e, and using the previous notations, we have :

$$(3.52) \qquad \delta \left[\int_{\Omega_e} \frac{1}{2} \underset{\sim}{\alpha}(D_L)(D_L)\,d\Omega - \int_{\Sigma_{eF}} \bar{F}\,U d\Sigma - \int_{\Sigma_{eU}} \bar{T}\Big[U-U_d\Big]d\Sigma \right] = 0 .$$

If we desire a pure displacement model, with a continuous displacement U, it is merely necessary to make $U = U_d$ in (3.52) and $\Sigma_{eF} = \Sigma_e$. Therefore imposing a force density F on Σ_e, or nodal forces Q, the formula exhibits the classical element stiffness matrix K_e by the usual method. Hence :

$$K_e\ q = Q$$

Assembly is made on the actual displacement q.

On the other hand, if we wish to calculate a non-conforming model[14], we develop the displacement U in the usual form :

$$U = \underset{\sim U}{T}(M).\alpha ,$$

and the multiplier T on interfaces Σ_{eU} in the form :

$$T = \underset{\sim T}{T}(M) . Q \, , \, (M \in \Sigma_{eU}) .$$

If we relax the continuity of U over the whole of the element boundary Σ_e, we make $\Sigma_{eU} = \Sigma_e$ and $\Sigma_{eF} = \emptyset$. We then use the equation :

$$(3.53) \quad \left[\begin{array}{l} \delta \left[\int_{\Omega_e} \frac{1}{2} \underset{\sim}{\alpha}(D_L)(D_L) d\Omega - \int_{\Sigma_e} \bar{T} \left[U - U_d \right] d\Sigma \right] = 0 \\[2em] \forall \alpha, \, Q, \text{ with :} \\[1em] D_L = \frac{1}{2} \left[\frac{\partial U}{\partial M} + \overline{\frac{\partial U}{\partial M}} \right] \\[1em] U = \underset{\sim U}{T}(M) \alpha \, , \, T = \underset{\sim T}{T}(M) . Q, \, (M \in \Sigma_e) . \end{array} \right.$$

This equation expresses the equilibrium of element e under the effect of an imposed displacement U_d on Σ_e, in fact represented by a conjugated interface displacement q' of Q. (3.53) then gives us, using obvious notations :

$$\bar{\alpha} \, I \, \delta\alpha - \bar{Q} \, R \, \delta\alpha - \overline{\delta Q} \, R\alpha + \overline{\delta Q} \, q' = 0, \, \forall \delta\alpha \, , \, \delta Q .$$

Hence :

$$\begin{bmatrix} I & -\bar{R} \\ -R & 0 \end{bmatrix} \begin{bmatrix} \alpha \\ Q \end{bmatrix} = \begin{bmatrix} 0 \\ -q' \end{bmatrix}$$

$$\alpha = I^{-1} \, \bar{R} \, Q$$

$$R \, I^{-1} \, \bar{R} \, Q = q' .$$

This gives us an element stiffness matrix K_e :

$$Q = K_e \, q' \text{ with } K_e = \left[R \, I^{-1} \, \bar{R} \right]^{-1 \, (*)} .$$

(*) As a general rule, for $RI^{-1}\bar{R}$ to be invertible, we must choose an interpolation $T = T(M)Q$ such that $\bar{R}$ be regular, otherwise $\exists Q \neq 0$, such that $RQ = 0$, compromising invertibility[24].

If R is invertible, obviously we have :

$$K_e = \overline{R}^{-1} \, I \, R^{-1}.$$

We can also devise mixed models without difficulty, in which continuity is only relaxed for certain components, or for certain displacement modes, the others remaining continuous[14]. The boundary Σ_e is then mixed, associating types Σ_{eF} and Σ_{eU}.

If we call q the continuous modal or nodal discretized displacement variables, there exists a rectangular connection matrix $\mathbf{C}$ between q and α, (q assumed to be given by $\Omega - \Omega_e$, as U_d) :

$$q = \mathbf{C}\alpha \, .$$

We then introduce this constraint in (3.53), using the conjugated Lagrangian multiplier Q' of the variable q, which translates the generalized continuous displacements. With q assumed to be given :

$$\left[\begin{array}{l} \delta \left[\int_{\Omega_e} \frac{1}{2} \underset{\sim}{\alpha}(D_L)(D_L) - \int_{\Sigma_e} \overline{T}\left[U - U_d\right] d\Sigma - \overline{Q'}\left[q - \mathbf{C}\alpha\right] \right] = 0 \\[4mm] \forall \, \alpha, \, Q', \, Q \text{ with :} \\[4mm] U = \underset{\sim}{T}_U(M) . \alpha \, , \, T = \underset{\sim}{T}_T(M) . Q \; (M \in \Sigma_e). \end{array} \right.$$

Hence :

$$\overline{\alpha} \, I \, \delta\alpha - \overline{Q} \, R \, \delta\alpha - \overline{\delta Q} \, R\alpha + \overline{\delta Q} \, q' - \overline{\delta Q'}[q - \mathbf{C}\alpha] + \overline{Q'}\mathbf{C}\delta\alpha = 0, \; \forall\alpha, \, Q', Q$$

$$\begin{bmatrix} I & -\overline{R} & -\overline{\mathbf{C}} \\ -R & 0 & 0 \\ -\mathbf{C} & 0 & 0 \end{bmatrix} \begin{bmatrix} \alpha \\ Q \\ Q' \end{bmatrix} = \begin{bmatrix} 0 \\ -q' \\ -q \end{bmatrix}$$

$$\alpha = I^{-1}\left[R \, Q - \mathbf{C}Q'\right]$$

$$\mathbf{C}\alpha = q$$

$$R\alpha = -q'.$$

We now put :

$$\begin{bmatrix} -C \\ R \end{bmatrix} = {}^e\mathfrak{D}, \quad \begin{bmatrix} q \\ q' \end{bmatrix} = {}^e q, \quad \begin{bmatrix} Q' \\ Q \end{bmatrix} = {}^e Q$$

and obtain :

$$\alpha = I^{-1} \, \overline{{}^e\mathfrak{D}}{}^e Q$$

$$^e\mathfrak{D}\alpha = {}^e\mathfrak{D}I^{-1} \, \overline{{}^e\mathfrak{D}}{}^e Q = {}^e q$$

This gives a new element stiffness matrix K_e, such that :

$$^e Q = {}^e K_e \cdot {}^e q \quad \text{with} \quad {}^e K_e = \left[{}^e\mathfrak{D}I^{-1} \, \overline{{}^e\mathfrak{D}} \right]^{-1},$$

between the set of the generalized continuous and discontinuous displacement variables, and the conjugated generalized forces.

<u>Patch test</u>[15,20]

We can use the principle of virtual stresses (see paragraph 7),to give a
simple interpretation of the Patch test mentioned in chapter II (paragraph
1,2, remark 11), this interpretation demonstrating that this criterion is
a necessary condition of convergence for non-conforming finite elements of
the displacement type.

Let us consider in fact a domain Ω_e with boundary Σ_e, internal to domain
Ω with boundary Σ, occupied by the medium in any state.   With the previous notations, if Σ_e has no common boundary with Σ_U, and in the case of linearized deformation, as we saw in paragraph 7, any compatible deformation D_L
must meet the principle :

$$(3.54) \qquad \int_{\Omega_e} \delta\tilde{C}\, D_L \, d\Omega = 0 \quad \forall \delta\tilde{C} \text{ S.A.H.}$$

In this case, D_L satisfies the linearized conditions of compatibility :

$$\text{rot } \overline{\text{rot } D_L} = 0 \qquad [\text{see } 1.18].$$

This is the Euler equation for principle (3.54).

We know that there then exists a unique displacement field U in a simply connected domain Ω_e, such that :

$$D_L = \frac{1}{2}\left[\frac{\partial U}{\partial M} + \overline{\frac{\partial U}{\partial M}}\right].$$

If we partition Ω_e by means of an internal boundary Σ_e^*, into two domains Ω_{e1} and Ω_{e2}, (3.34) becomes :

$$\int_{\Omega_{e1}\cup\Omega_{e2}} T_r(\delta C\,\frac{\partial U}{\partial M})\,d\Omega = 0,\ \forall\delta C\ \ S.A.H,$$

namely :

(3.55)
$$\int_{\Sigma_e^*} \overline{\delta C\ n_1}\left[U_1 - U_2\right] d\Sigma = 0,\ \forall\delta C\ \ S.A.H.$$

or also $\delta C\ n_1$ on Σ_e^*, as we give no external force density on Σ_e^*. In this formula, n_1 is the unit normal to Σ_e^*, external to Ω_{e1}, and U_1 and U_2 are the values of U on the boundaries of Ω_{e1} and Ω_{e2} respectively, restricted to Σ_e^*, (the integral on boundary Σ_e is zero, since δC is homogeneous and statically admissible, and $Cn = 0$ on Σ_e).

Equation (3.55) gives, theoretically :

$$U_1 = U_2 \text{ on } \Sigma_e^*.$$

A necessary condition for continuity of the displacement field U in Ω_e, is (3.54) for example, with a constant stress field $\delta C = C_o$, in Ω_e. C_o is statically admissible for Ω_e, as firstly :

$$\text{div } C_o = 0 \text{ in } \Omega_e,$$

and secondly we give no external force on Σ_e. (Furthermore, we know that this is theoretically impossible in a first gradient theory : see chapter II, paragraph 1.2, remark 10). Therefore $\Sigma_{eF} = \emptyset$. Hence the necessary condition :

(3.56)
$$\int_{\Omega_e} T_r(C_o\,\frac{\partial U}{\partial M})\,d\Omega = 0\ ,\ \ \forall\ C_o \text{ constant.}$$

(Furthermore, if $\Sigma_{eF} \neq \emptyset$, continuity is no longer ensured on this part of the boundary).

Now domains Ω_{e1} and Ω_{e2} represent, for example, two finite elements adjacent along interface Σ_e^*. The necessary condition (3.56) is therefore replaced by :

$$(3.57) \qquad \int_{\Sigma_e^*} \overline{C_o n_1} \left[U_1 - U_2 \right] d\Sigma = 0.$$

But we can consider all the finite elements adjacent to Ω_{e1}, of the boundary Σ_{e1}. (3.56) and (3.57) can now be written, for this element :

$$(3.58) \qquad \int_{\Omega_{e1}} T_r (C_o \frac{\partial U_1}{\partial M}) d\Omega = 0 \qquad , \qquad \forall\, C_o \text{ constant,}$$

or :

$$(3.59) \qquad \int_{\Sigma_{e1}} \overline{C_o n_1} \left[U_1 - U_2 \right] d\,\Sigma = 0,$$

U_1 is the deplacement in Ω_{e1}, and on its boundary Σ_{e1}, U_2 is the displacement in the adjacent domains.

Now for a given type of non-conforming element, (3.59), and consequently (3.58), represent an equation of continuity in the mean. For such a type of element :

$$U = \underset{\sim}{T}(M).\alpha = \underset{\sim}{T_i}(M).{}^i\alpha \quad , \quad \alpha \in \mathbf{R}^n.$$

The solution is given by :

$$U_1 = \underset{\sim}{T}(M).\alpha_1 \quad \text{for element } e_1$$

$$U_2 = \underset{\sim}{T}(M).\alpha_2 \quad \text{for the adjacent elements.}$$

Whence :

$$U_1 - U_2 = \underset{\sim}{T}(M) \left[\alpha_1 - \alpha_2 \right].$$

And for any problem (3.59) becomes :

$$(3.60) \qquad \int_{\Sigma_{e_1}} \overline{C_o n_1} \, \underset{\sim}{T}(M)\alpha \; d\Sigma \; = 0, \; \forall \alpha, \; (M \in \Sigma_{e_1}), \forall C_o \, C^t.$$

We can also use this formula in the following form, constituting the patch test for element e :

$$(3.61) \qquad \int_{\Omega_e} T_r(C_o \, \frac{\partial \underset{\sim}{T}_i(M)}{\partial M}).^i\alpha \, d\Omega \; = 0, \forall \alpha, \forall C_o C^t.$$

If the boundaries of e are linear (straight lines, planes), (3.60) is written :

$$(3.62) \qquad \underset{k}{\Sigma} \, \overline{C_o n_k} \int_{\Sigma_{e_k}} \underset{\sim}{T}(M) d\Sigma \; = 0 \; , \; \forall C_o \, C^t,$$

where the boundaries Σ_{ek} of the element have n_k as their external unit normals.

In general, the displacement field U for element e is interpolated by an expansion,   one part of which can be continuous after assembly, and one part discontinuous. We therefore apply the patch test to the discontinuous part at the interfaces. If C is the connection matrix between the generalized nodal displacements q, and the variables $^1\alpha$, the development is written:

$$U = \underset{\sim}{T}_1(M)\,^1\alpha + T_2(M).\,^2\alpha \;\; \text{with} \;\; C.\,^1\alpha = q$$

$$U = \underset{\sim}{T}_1(M) \, C^{-1} q + \underset{\sim}{T}_2(M).\,^2\alpha \; .$$

We apply criterion (3.62) to the component T_2 of the development.

We note that when the number of elements tends towards infinity, their dimensions tending towards zero, the stress on point M of Ω_e tends towards a constant, namely C_o, when the approximate solution tends towards the exact solution. The first member of (3.57) represents the work of the stresses due to discontinuities of the solution at the interfaces. The sum of these

elementary works tends, by (3.61), towards the deformation energy throughout
the medium due to these discontinuities, in the form of a sum of which each
term tends towards zero if the patch test is checked. If this sum tends
towards zero, the solution converges in the sense of a norm defined by
the deformation energy.

8.6. – Hellinger-Reissner principle and saddle-point

The solution satisfying the Hellinger-Reissner principle is a saddle-
point of the two field function defined by this principle. This function[21]
is written [see (3.13)] :

$$(3.63) \qquad \underset{\sim}{R}(\underset{\sim}{C},\underset{\sim}{U}) = \int_{\Omega_o} \left[\frac{1}{2} \underset{\sim}{\tilde{C}} \left[\frac{\partial U}{\partial M_o} + \overline{\frac{\partial U}{\partial M_o}} \right] - \frac{1}{2} \underset{\sim}{\beta}(C)(C) - \overline{f}\, U \right] d\Omega_o$$

$$- \int_{\Sigma_{oF}} \overline{F}\, U\, d\Sigma_o + \int_{\Sigma_{oU}} \overline{Cn} \left[U_d - U \right] d\Sigma_o.$$

Taking C_o and U_o as the exact solution of the equation :

$$\delta R = 0, \; \forall C \text{ and } U,$$

we put :

$$\left. \begin{array}{l} C = C_o + C_1 \\[2em] U = U_o + U_1 \end{array} \right\} \qquad \forall C_1 \text{ and } U_1.$$

If we execute a Taylor-MacLaurin expansion of R from C_o and U_o, we obtain
[see (1.38)] :

$$(3.64) \qquad \underset{\sim}{R} \begin{pmatrix} \underset{\sim}{C}_o + \underset{\sim}{C}_1 \\ \underset{\sim}{U}_o + \underset{\sim}{U}_1 \end{pmatrix} - \underset{\sim}{R} \begin{pmatrix} \underset{\sim}{C}_o \\ \underset{\sim}{U}_o \end{pmatrix} = \underset{\sim}{R}' \begin{pmatrix} \underset{\sim}{C}_o \\ \underset{\sim}{U}_o \end{pmatrix} \begin{pmatrix} \underset{\sim}{C}_1 \\ \underset{\sim}{U}_1 \end{pmatrix} + \frac{1}{2} \underset{\sim}{R}'' \begin{pmatrix} \underset{\sim}{C}_o \\ \underset{\sim}{U}_o \end{pmatrix} \begin{pmatrix} \underset{\sim}{C}_1 \\ \underset{\sim}{U}_1 \end{pmatrix} \begin{pmatrix} \underset{\sim}{C}_1 \\ \underset{\sim}{U}_1 \end{pmatrix} = \Delta R.$$

The expansion stops at the second order as shown by (3.63). Now the
solution satisfies the condition of stationarity $\delta R = 0$, namely :

$$\underset{\sim}{R}{}' \left(\begin{array}{c} \underset{\sim}{C}_o \\[2mm] \underset{\sim}{U}_o \end{array} \right) = 0 .$$

It therefore remains in (3.64) :

$$\Delta R = \frac{1}{2} \, \underset{\sim}{R}{}'' \left(\begin{array}{c} \underset{\sim}{C}_o \\[2mm] \underset{\sim}{U}_o \end{array} \right) \left(\begin{array}{c} \underset{\sim}{C}_1 \\[2mm] \underset{\sim}{U}_1 \end{array} \right) \left(\begin{array}{c} \underset{\sim}{C}_1 \\[2mm] \underset{\sim}{U}_1 \end{array} \right) .$$

We immediately find :

$$\Delta R = \int_{\Omega_o} \left[\frac{1}{2} \, \tilde{C}_1 \left[\frac{\partial U_1}{\partial M_o} + \overline{\frac{\partial U_1}{\partial M_o}} \right] - \frac{1}{2} \, \beta(C_1)(C_1) \right] d\Omega_o - \int_{\Sigma_{oU}} \overline{C_1 n} \, U_1 \, d\Sigma .$$

Now :

$$\underset{\sim}{C}_1 = 0 \Longrightarrow \Delta R = 0 \quad \forall \, \underset{\sim}{U}_1 .$$

Therefore :

$$(3.65) \qquad \underset{\sim}{R}(\underset{\sim}{C}_o , \underset{\sim}{U}) = \underset{\sim}{R}(\underset{\sim}{C}_o , \underset{\sim}{U}_o) .$$

Then :

$$U_1 = 0 \Longrightarrow \Delta R = - \int_{\Omega_o} \frac{1}{2} \, \beta(C_1)(C_1) d\Omega_o \leqslant 0 \quad \forall \, C_1 .$$

Therefore :

$$(3.66) \qquad \underset{\sim}{R}(\underset{\sim}{C} , \underset{\sim}{U}_o) \leqslant \underset{\sim}{R}(\underset{\sim}{C}_o , \underset{\sim}{U}_o) .$$

Together the two relations (3.65) and (3.66) give :

$$\underset{\sim}{R}(\underset{\sim}{C} , \underset{\sim}{U}_o) \leqslant \underset{\sim}{R}(\underset{\sim}{C}_o , \underset{\sim}{U}_o) = \underset{\sim}{R}(\underset{\sim}{C}_o , \underset{\sim}{U}) \qquad Q.E.D.$$

Furthermore, we immediately see that the saddle point $\underset{\sim}{C}_o$, $\underset{\sim}{U}_o$ is the solution for the dual and primal problems respectively, and reciprocally. Uniqueness can also be demonstrated.

The preceding theorem enables us to find the solution of a static problem, using a search algorithm for the saddle point of function R^{22}.

8.7. – Legendre transformation

We consider the volume density of the deformation energy α at a point M of an arbitrary **hyperelastic** medium, α being differentiable :

$$(3.67) \qquad \alpha = \underset{\sim}{\alpha}(D)$$

This energy gives the stress C at M, by :

$$(3.68) \qquad \underset{\sim}{C} = \frac{\partial \alpha}{\partial D} \,,$$

Through conditions of continuity, differentiability and regularity, this enables us to calculate, at point M :

$$D = \underset{\sim}{\gamma}(C).$$

If we put :

$$(3.69) \qquad \beta = \underset{\sim}{\beta}(C) = \underset{\sim}{C}\, D - \underset{\sim}{\alpha}(D), \quad \text{with } D = \underset{\sim}{\gamma}(C),$$

we have, $\forall$ differential fields δD and δC, dependent or not ;

$$\delta\beta = \frac{\partial\beta}{\partial C}\,\delta C + \frac{\partial\beta}{\partial D}\,\delta D = \underset{\sim}{\delta C}\, D + \underset{\sim}{C}\,\delta D - \frac{\partial\alpha}{\partial D}\,\delta D.$$

But with (3.68), we find :

$$\delta\beta = \underset{\sim}{\delta C}D = \underset{\sim}{D}\delta C.$$

Whence :

$$(3.70) \qquad \begin{cases} \dfrac{\partial\beta}{\partial C} = \underset{\sim}{D} \\[2ex] \dfrac{\partial\alpha}{\partial D} = \underset{\sim}{C}. \end{cases}$$

The β function is the complementary energy. Furthermore, we verify, if α is a quadratic form of D :

$$\alpha = \frac{1}{2}\left[\underset{\sim}{\alpha}(D)\right](D)$$

$$\underset{\sim}{C} = \frac{\partial \alpha}{\partial D} = \left[\underset{\sim}{\alpha}(D)\right]$$

$$\alpha = \frac{1}{2}\ \underset{\sim}{C}\ D$$

$$\beta = \underset{\sim}{\beta}(C) = \underset{\sim}{C}\ D - \frac{1}{2}\underset{\sim}{C}\ D = \frac{1}{2}\underset{\sim}{C}\ D = \alpha$$

The Legendre transformation is a transformation by duality. It consists in adopting here, as an independent variable, not the deformation D but the stress C, a dual variable.

All that is necessary is to introduce the variable D in a principle, as an independent variable, if it is not so already.

Let us then consider the primal principle for a hyperelastic medium (<u>linear or not</u>). This gives the equation :

$$\left[\delta\left[\int_{\Omega_o}\alpha d\Omega_o - \int_{\Omega_o}\overline{f}Ud\Omega_o - \int_{\Sigma_o}\overline{F}Ud\Sigma_o\right] = 0 \quad , \quad \forall U \text{ K.A.} \atop \text{with}\quad \alpha = \underset{\sim}{\alpha}(D)\ , \quad D = \underset{\sim}{D}(U) = \overline{D}\ .\right.$$

Introducing the independent variable D by means of a multiplier $t = \overline{t}$, we then obtain, as before :

$$\delta\left[\int_{\Omega_o}\left[\alpha - \underset{\sim}{t}\left[D - \underset{\sim}{D}(U)\right]\right]d\Omega_o - \int_{\Omega_o}\overline{f}Ud\Omega_o - \int_{\Sigma_{oF}}\overline{F}Ud\Sigma_o\right] = 0, \forall D, t, \text{ and } U \quad \text{K.A.}$$

or

(3.71)
$$\left[\int_{\Omega_o}\left[\left[\frac{\delta\alpha}{\delta D} - \underset{\sim}{t}\right]\delta D + \underset{\sim}{t}\delta\left[\underset{\sim}{D}(U)\right] - \underset{\sim}{\delta t}\left[D - \underset{\sim}{D}(U)\right]\right]d\Omega_o - \int_{\Omega_o}\overline{f}\delta Ud\Omega_o - \int_{\Sigma_{oF}}\overline{F}\delta Ud\Sigma_o = 0\right.$$

$$\forall \delta D, \ \delta t, \ \delta U \quad \text{K.A.}$$

If we identify t with the stress C, in verifying the equation :

$$\frac{\partial \alpha}{\partial D} = \underset{\sim}{t} = \underset{\sim}{C},$$

the deformation D must be eliminated by means of this same equation, namely:

$$D = \frac{\widetilde{\partial \beta}}{\partial C} \quad \left(= \underset{\sim}{\gamma}(C) \right)$$

Equation (3.71) now becomes :

$$(3.72) \qquad \delta \left[\int_{\Omega_o} \left[\tilde{C} \, \underset{\sim}{D}(U) - \beta(C) \right] d\Omega_o - \int_{\Omega_o} \overline{f} U d\Omega_o - \int_{\Sigma_{oF}} \overline{F} U d\Sigma_o \right] = 0, \; \forall U \;\; K.A., \;\; \forall C.$$

Hence the rule.

Principle (3.72) constitutes incidentally the Hellinger-Reissner principle with two fields, for a non-linear hyperelastic medium.

Formula (3.69) is used classically in the energy principles, to express internal energy as a function of the dual variables. But its utilization is also extremely convenient for expressing the energy, a function of state variables, as a function of certain dual variables only, for example certain stress components, on which we wish to make special hypotheses.

For example, let $^i\varepsilon$ be certain components of D, and $^i\sigma$ the dual components retained, (3.69) becomes :

$$\underset{\sim}{\beta} = \beta(^i\sigma, ^j\varepsilon) = {}^i\sigma \, {}^i\varepsilon - \underset{\sim}{\alpha}(\varepsilon) \quad \text{for} \quad j \neq i.$$

Under that conditions (3.72) becomes :

$$(3.73) \qquad \left[\delta \left[\int_{\Omega_o} \left[{}^i\sigma, {}^i\underset{\sim}{\varepsilon}(U) - \beta \right] d\Omega_o - \int_{\Omega_o} \overline{f} U d\Omega_o - \int_{\Sigma_{oF}} \overline{F} U d\Sigma_o \right] = 0, \forall U \;\; K.A., \;\; \forall {}^i\sigma \right.$$

with :

$$\underset{\sim}{\beta} = \beta(^i\sigma, ^j\varepsilon) = {}^i\sigma \, {}^i\varepsilon - \underset{\sim}{\alpha}\left({}^i\varepsilon, {}^j\underset{\sim}{\varepsilon}(U) \right), \text{ where } {}^i\varepsilon = {}^i\underset{\sim}{\varepsilon}\left({}^i\sigma, {}^j\underset{\sim}{\varepsilon}(U) \right),$$

and $j \neq i$.

The Legendre transformation (3.69) is generalized in the theory of duality, by the Fenchel transformation[22] in the case of non-differentiable functionals.

CHAPTER IV

VIBRATION OF LINEAR STRUCTURES

1. - <u>INTRODUCTION</u>

All mechanical structures subjected to variable external forces, are
liable to vibrate. These excitation forces, depending on time, are periodic,
transient or random.

Vibrations in mechanical systems usually have detrimental effects, in
practice of three types : resonances, producing excessive noise or ruptures
by excessive deformation amplitude ; fatigue, resulting in the rupture of
materials or damage to equipments ; and finally instabilities of all kinds
(combustion, handling, servo-control, aeroelastic, instabilities resulting
from dynamic buckling, etc.).

Resonance phenomena occur in passive systems, while instabilities are
encountered in active systems, namely those associated with energy sources.

In a large number of cases, linear analyses relating to small movements
are considered sufficient to describe the dynamic behaviour of the structures, and only linear, or linearized structures will be dealt with in this
chapter. However, it should not be forgotten that non-linear analysis is
indispensable for large variations of the state variables. This will be
considered in the following chapter.

Linear structures are covered by three types of classical calculation,
designed to forecast their vibratory behaviour :

- Calculation of the free vibration of conservative linear structures.
Damping forces, usually inaccessible to calculation, are generally estimated
or measured in their generalized modal aspect, and introduced afterwards.

- Calculation of the response of passive dissipative structures to excitations of a periodic, transient or random type.

- Calculation of active structures, in order to determine the domains of
instability of the parameters defining the structures and the external
actions.

In the following pages, we shall commence by describing the theory of
discretized structures. We shall then give certain indications concerning
continuous structures, and also some useful theorems. We shall conclude with

a selection of classical formulae, giving the response of a linear structure to forced excitations, with particular reference to modal methods.

2. - <u>DISCRETIZED STRUCTURES</u>

Let us consider a conservative linear elastic structure, occupying a domain Ω_o with boundary Σ_o in the natural state, and let us assume the structure to be subjected to a given volume density of loading f, and a given surface density of loading F on Σ_o. If the loads depend on time, resultant displacement U at point M_o of $\overline{\Omega_o}$ will depend on M_o and time t :

$$U = \underset{\sim}{U}(M_o,t), \quad f = \underset{\sim}{f}(M_o,t), \quad F = \underset{\sim}{F}(M_o,t),$$

where M_o and t are Lagrange variables.

The structure is then subjected to a volume density of inertial forces, and if we call $\rho = \underset{\sim}{\rho}(M_o)$ the density at point M_o, the volume density of inertial forces is written $-\rho\ddot{U}$, at any instant t, where $\ddot{U} = \dfrac{\partial^2 U}{\partial t^2}$.

Discretization, for example using the displacement finite element method, is obtained in each element e by interpolation :

$$U = \underset{\sim}{T}(M_o).\underset{\sim}{q}(t),$$

where $\underset{\sim}{q}(t)$ is the column of (generalized) nodal displacement.

The volume density of inertial forces is therefore written :

$$- \rho\ddot{U} = - \underset{\sim}{\rho}(M_o)\ \underset{\sim}{T}(M_o)\underset{\sim}{\ddot{q}}(t),$$

and gives a virtual work :

$$\int_{\Omega_e} - \rho\ddot{U}\delta U d\Omega = - \overline{\delta q}\left[\int_{\Omega_e} \underset{\sim}{\rho}(M_o)\ \overline{\underset{\sim}{T(M_o)}}\ \underset{\sim}{T}(M_o)d\Omega\right]\underset{\sim}{\ddot{q}}(t)$$

$$= - \overline{\delta q}\ M_e\ \underset{\sim}{\ddot{q}}(t), \quad M_e = \overline{M_e}.$$

M_e is the mass matrix for element e, related to the nodal displacement q.

Finally, after assembly, the discretized equation for the structure is written :

$$(4.1) \qquad \begin{cases} M\,\ddot{\underset{\sim}{q}}(t) + K\,\underset{\sim}{q}(t) = \underset{\sim}{Q}(t) \qquad q,\, Q \in R^n \\[2ex] \text{with} \quad M = \bar{M}, \quad K = \bar{K}. \end{cases}$$

Matrices M and K are the mass and stiffness matrices for the discretized global structure.

Naturally, we reach a similar result using a global Ritz method.

Clearly, this result can be obtained from Hamilton's principle :

After discretization, the deformation energy of the system is written :

$$W = \frac{1}{2}\,\overline{\underset{\sim}{q}(t)}\,K\,\underset{\sim}{q}(t)$$

and the kinetic energy :

$$E = \frac{1}{2}\,\overline{\dot{\underset{\sim}{q}}(t)}\,M\,\dot{\underset{\sim}{q}}(t)$$

The two quadratic forms, W and E are semi-positive in the general case.

The Hamilton principle classically gives us the Lagrangian equation :

$$\frac{d}{dt}\,\frac{\partial E}{\partial \dot{q}} + \frac{\partial W}{\partial q} = \bar{Q},$$

which takes us back to (4.1)

If the displacements U_d are now cancelled on a part $\Sigma_{oU} \neq \emptyset$ of Σ_o, they provide given generalized displacements ^1q, and unknown reaction forces ^1Q, which must be eliminated (see paragraph 1.2, note 9). We then obtain an equation of the same type as (4.1). We have also seen that we can take account of given displacements, by means of a Lagrangian multiplier, which proves to be ^1Q. If $\Sigma_U = \emptyset$, the structure is completely free, and the stiffness matrix K is singular, such that :

$$\exists q_s \neq 0 \mid K\,q_s = 0$$

The generalized displacements q_s, the number of which is 6, are rigid body modes.

This situation also occurs if $\Sigma_U \neq \emptyset$, but only if the given displacements are such that certain degrees of freedom still allow rigid body displacements.

If we are interested in the free movements of the structure, $\underset{\sim}{Q}(t) = 0$, and equation (4.1), after elimination of support reactions, becomes :

$$(4.2) \qquad M \; \underset{\sim}{\ddot{q}}(t) + K \; \underset{\sim}{q}(t) = 0,$$

of which the non-zero solutions are harmonic, of the type :

$$\underset{\sim}{q}(t) = q \; e^{j\omega t}.$$

Equation (4.2) then becomes the classical equation for a discretized conservative linear system, in free vibration :

$$(4.3) \qquad [- \omega^2 \; M + K] \; q = 0$$

When the rigid body displacements have been eliminated, and if the displacements q are independent [*], the two matrices M and K are positive definite, as also the corresponding quadratic forms E and W. We can then state the following theorem :

> Theorem :
>
> The mass and stiffness matrices M and K, positive definite, are represented by real diagonal matrices in a common unitary basis, this being the eigenmode basis of the matrix $M^{-1}K$. The eigenvalues of $M^{-1}K$ are real and positive, and the associated eigenvectors, or eigen subspaces, are real and orthogonal in the sense of the scalar product defined by M.

In fact let us consider the kinetic energy $E = \frac{1}{2} \; \overline{\dot{q}} \; M\dot{q}$, it defines a symmetrical bilinear form on any two columns, q_1 and q_2, of $\mathbb{R}^n$ by :

$$\underset{\sim}{E}(q_1)(q_2) = \overline{q_1} \; M \; q_2, \quad M = \overline{M},$$

where $\overline{q}$ is the transpose of q, in the sense of the standard scalar product of $\mathbb{R}^n$.

---

[*] In a linear system, these displacements may have been badly chosen.

The quadratic form $\frac{1}{2} \underset{\sim}{E}(q)(q)$ being positive definite, we can define a new scalar product on $\mathbb{R}^n$, such that [see (A.10)] :

(4.4) $\qquad \underset{\sim}{E}(q_1)(q_2) = \tilde{q}_1 \, q_2 = \overline{q_1} \, M \, q_2$

where $\tilde{q}$ designates the transpose of q, in the sense of this new scalar product.

But the potential energy W also enables us to define a symmetrical bilinear form on $\mathbb{R}^n$, such that :

$$\underset{\sim}{W}(q_1)(q_2) = \overline{q_1} \, K \, q_2$$

Under these conditions, there exists a unique matrix A, Hermitian in the sense of the new scalar product, such that [1] :

(4.5) $\qquad \underset{\sim}{W}(q_1)(q_2) = \tilde{q}_1 \, A \, q_2 = \overline{q_1} \, K \, q_2$, with $A = \tilde{A}$

and A is obviously positive definite.

A then has a simple spectral decomposition : its eigenvalues are real and positive, single or multiple. A is represented by a real diagonal matrix in unitary basis S which is the basis of its eigenvectors, with possible repetition of the diagonal terms if the eigenvalues are multiple, in which case they are associated with eigen subspaces of dimensions larger than 1.

But (4.4) gives :

$$\tilde{q}_1 = \overline{q}_1 \, M, \quad \forall q_1,$$

whence :

$$\overline{q_1} = \tilde{q}_1 \, M^{-1},$$

and by (4.5) :

$$\tilde{q}_1 \, A \, q_2 = \overline{q_1} \, K \, q_2 = \tilde{q}_1 \, M^{-1} K \, q_2, \quad \forall q_1, \, q_2.$$

Therefore :

(4.6) $\qquad A = M^{-1} K$

A is represented by a real diagonal matrix Λ in the basis S of the orthogonal eigenmodes, made unitary (as $\mathbb{R}^n$ is positive) by normalization. Thus :

$$(4.7) \qquad \tilde{S} S = 1_{\mathbb{R}^n} = \bar{S} M S$$

and :

$$(4.8) \qquad \tilde{S} A S = \Lambda \text{ positive real diagonal } \text{(as A is positive definite).}$$

It should also be noted that if we call $|_i$ the basis columns of $\mathbb{R}^n$, and $^i|$ the basis lines of $\mathbb{R}^n$, we have :

$$\overline{|}_i = {}^i| \quad , \quad \widetilde{|}_i = {}^i|$$

$$\overline{{}^i|} = |_i \quad , \quad \widetilde{{}^i|} = |_i .$$

In fact let S be a basis and q_1, q_2 any two columns :

$$q_1 = SX \ , \ q_2 = SY$$

$$\tilde{q}_1 q_2 = \widetilde{SX}\ SY = \tilde{S}\ \tilde{X}\ SY = \tilde{X}\ G'Y,$$

where $G' = \tilde{S} S$ is the Gram matrix of S "in this scalar product".

$$\overline{q_1}\ q_2 = \overline{SX}\ SY = \bar{X}\ \bar{S}\ S\ Y = \bar{X}\ G\ Y ,$$

with $G = \bar{S} S$. G and G' are regular.

Let $g'_{\mathbb{R}^n}$ be the fundamental metric tensor of $\mathbb{R}^n$, such that :

$$g'_{\mathbb{R}^n}(q_1)(q_2) = \tilde{q}_1 q_2 = g'_{\mathbb{R}^n}(SX)(SY) = g_{\mathbb{R}^n}(X)(\tilde{S} S Y),$$

with $g_{\mathbb{R}^n}$ the usual metric tensor of $\mathbb{R}^n$, such that :

$$g_{\mathbb{R}^n}(X)(Y) = \sum_j {}^jX\ {}^jY .$$

Let us put $X = |_i$:

$$g_{\mathbb{R}^n}(|_i)(\tilde{S} S Y) = \sum_j {}^j|_i\ {}^j\widetilde{SSY} = {}^i\widetilde{SSY} = {}^i|\widetilde{SSY} = |_i \widetilde{SSY}.$$

Thus :

$$\tilde{|}_i G'Y = {}^i| G'Y, \ \forall G'Y \text{ as } G' \text{ is regular.}$$

Hence $\tilde{\mathbb{1}}_i = {}^i\mathbb{1}$, independently of the metric tensor g'_{R^n} used. Q.E.D.

Under these conditions, (4.7) gives :

$$(4.9)\quad \begin{cases} \tilde{S}\,S = \mathbb{1}_{R^n} = \bar{S}\,MS \\[2ex] {}^i\tilde{S}\,S_j = \tilde{S}_i\,S_j = \bar{S}_i\,M\,S_j = \begin{cases} 0, & i \neq j \\ 1, & i = j \end{cases} \end{cases}$$

While (4.8) gives :

$$(4.10)\quad \begin{cases} \tilde{S}\,A\,S = \Lambda \text{ , diagonal matrix of real eigenvalues.} \\[2ex] {}^i\tilde{S}\,A\,S_j = \tilde{S}_i\,A\,S_j = {}^i\Lambda_j = \begin{cases} 0, & i \neq j \\ {}^i\Lambda_i, & i = j \end{cases} \end{cases}$$

If we adopt a normalization other than (4.7) for the eigenmodes S_i of A, this equates to considering another modal basis S', deduced from S by :

$$(4.11)\quad S' = S\,\sqrt{\mu}$$

where $\sqrt{\mu}$ is a positive real diagonal, such that :

$$(4.12)\quad \begin{cases} \tilde{S}'_i\,S'_i = {}^i\mu_i > 0, \text{ (i fixed)} \\[2ex] \tilde{S}'\,S' = \bar{S'}\,M\,S' = \sqrt{\mu}\,\tilde{S}\,S\,\sqrt{\mu} = \mu \\[2ex] {}^i\mu_i = \bar{S'_i}\,M\,S'_i \, . \end{cases}$$

${}^i\mu_i$ is the generalized mass of mode S'_i. Naturally we can adopt the most convenient normalization for mode S'_i, from which its generalized mass will result.

(4.10) gives, with (4.11) :

$$\tilde{S}\,A\,S = \bar{S}\,K\,S = \Lambda$$

$$= \left[\sqrt{\mu}\right]^{-1}\bar{S'}\,K\,S'\left[\sqrt{\mu}\right]^{-1} = \Lambda\,,$$

or

$$\overline{S}{}' \, K \, S' = \Lambda \mu \; ; \quad \Lambda, \mu \quad \text{positive diagonals.}$$

We put :

$$(4.13) \qquad \overline{S}{}' \, K \, S' = \gamma = \Lambda \mu.$$

The positive diagonal matrix γ is the matrix of generalized modal stiffnesses.

(4.12) and (4.13) demonstrate the theorem.

It should be noted that the matrix $A = M^{-1}K$ is not Hermitian in the usual metric of $\mathbb{R}^n$, but is Hermitian in the mass metric. Furthermore, the demonstration has been made in the general case where the eigenvalues can be multiple : in this case, the algebraic theorem invoked provides the orthogonality of the eigen subspaces associated with the eigenvalues, in the mass metric.

Let us now consider a column $q \in R^n$, representing the discretized nodal displacements. If q' represents q in the basis S' :

$$q = S' \, q'.$$

In S', (4.3) becomes the equation :

$$(4.14) \qquad \left[- \omega^2 \mu + \gamma\right] q' = 0,$$

which gives the following uncoupled equations, taking account of (4.13) :

$$- \omega_i^2 \, {}^i\mu_i + {}^i\gamma_i = o \quad \text{with } q' = |_i \,(\text{i fixed}), \text{ and } \omega_i^2 = {}^i\Lambda_i.$$

We therefore see that in S', the modes are the basis columns $|_i$ of $\mathbb{R}^n$, and with (4.13) :

$$\omega_i^2 = {}^i\Lambda_i.$$

All these results form the basis of experimental methods for the determination of the eigenmodes and natural frequencies, and their modal characteristics : generalized masses and generalized damping[2].

3. - RIESZ' THEOREM

This theorem gives spectral results relating to linear continuous structures, in the most frequent cases.

We saw in Chapter I that the static equilibrium solution for a linear elastic continuous structure, made its total potential energy minimum in the class of kinematically admissible displacements.

Let us take a structure occupying domain Ω_o of E_3, of boundary Σ_o, subjected to external loads of volume density f, and zero imposed displacement on a part Σ_{oU} of Σ_o. An external load surface density F is given on the complementary part Σ_{oF} of Σ_{oU}.

The total potential energy is written :

$$(4.15) \qquad \mathcal{J}(v) = \frac{1}{2} \, w(v)(v) - L(v) .$$

where $w = \frac{1}{2} \, w(v)(v)$ is a quadratic form on vectorial functions v, and L is a covector on these functions. We look for a solution u in the following space V :

$$V = \{v \mid v \in C^o(\overline{\Omega}_o), \; \frac{\partial^i v}{\partial^j x} \in L^2(\Omega), \; v \mid \Sigma_{oU} = 0\}$$

for kinematically admissible vectorial functions v. The partial derivatives of $v(x)$ at point $x \in \Omega$ in any local base, are denoted $\dfrac{\partial^i v}{\partial^j x}$, in summary form.

Solution u therefore verifies :

$$(4.16) \qquad \mathcal{J}(u) = \underset{v \in V}{\text{Inf}} \, \mathcal{J}(v) .$$

Providing V with a scalar product, such that :

$$\forall v_1, v_2 \in V, \; (v_1, v_2)_V = \int_\Omega \overline{v_1(x)} \, v_2(x) \, d\Omega_o + \int_{\Omega_o} \overline{D_L(v_1)} D_L(v_2) d\Omega_o ,$$

with :

$$D_L(v) = \frac{1}{2} \left[\frac{\partial v(x)}{\partial x} + \overline{\frac{\partial v(x)}{\partial x}} \right] ,$$

it can be demonstrated that V, when completed, is a Hilbert space, and in
this case the Sobolev space :

$$\left[H_o^1(\Omega) \right]^3 = \overline{V} \subset \left[L^2(\Omega) \right]^3$$

It can then be demonstrated[3] that, for current linear elastic constitutive laws, the form w defines a mapping with continuous and coercive
symmetrical bilinear scalar value, and that if the load densities f and F are
square summable, the linear form L is continuous on this space.

Under these conditions, problem (4.16) admits a unique solution in $\overline{V}$
(Lax-Milgram theorem).

If $\Sigma_U = \emptyset$, it is appropriate to eliminate the rigid body displacements
from $\overline{V}$, so that w remains coercive, in a similar way to that observed for
discrete systems.

The form w enables us to define a mapping k $\in \mathcal{L}(\overline{V}, \overline{V})$, such that :

$$\forall v_1, v_2 \in \overline{V}, \quad w = \underset{\sim}{w}(v_1)(v_2) = (v_1, k\, v_2)_{\overline{V}} = (k\, v_1, v_2)_{\overline{V}}$$

with k Hermitian, continuous and coercive (with $(v_1, v_2)_{\overline{V}}$ as scalar product).

In addition, $\forall f \in L^2$, $\exists L \in \overline{V}'$ such that (Riesz' theorem) :

$$\forall v \in \overline{V}, \quad (f, v)_{L^2} = L(v),$$

where V' is the continuous dual of $\overline{V}$. Now $L \in \overline{V}'$, $\exists v_f \in \overline{V}$, such that :

$$\forall v \in \overline{V}, \quad L(v) = (v_f, v)_{\overline{V}}.$$

Now problem (4.16), or more precisely :

$$\underset{\sim}{\mathcal{J}}(u) = \underset{v \in \overline{V}}{\mathrm{Inf}}\ \underset{\sim}{\mathcal{J}}(u)$$

can also be written in equivalent manner :

Find u such that :

$$\forall v \in \overline{V}, \quad (v, ku)_{\overline{V}} = (f, v)_{L^2} = (v_f, v)_{\overline{V}}.$$

Whence :

$$(4.17) \qquad ku = v_f \Rightarrow u = k^{-1} v_f$$

Furthermore, this is true $\forall v_f \in \overline{V}$ (Lax-Milgram theorem), but in addition we see that $\forall f \in L^2, \exists$ a continuous application $G \in \mathcal{L}(L^2, \overline{V})$, such that :

$$u = Gf, \quad \forall f \in L^2$$

Furthermore, the injection of $\overline{V}$ in L^2 is compact, and we deduce that G, which is continuous, is compact[*] from L^2 into L^2. Operator G corresponds to the Green function of the problem.

Let us now assume the surface load distribution F on Σ_{oF} to be <u>zero</u> ; and replace the external load volume density f by that for the inertial forces $- \underset{\sim}{\rho}(x) \dfrac{\partial^2 u(x,t)}{\partial t^2}$, where ρ is density on x, assumed to be continuous.

If we look for harmonic solutions to the problem, this volume density is written $- \omega^2 \rho o u$.

Clearly the vectorial function $-\omega^2 \rho o u \in \overline{V}$ if $u \in \overline{V}$. The problem is then as follows :

Find $u \in \overline{V}$, such that $u = - \omega^2 G \rho o u$,

or alternatively :

$$(4.18) \quad \left[\begin{array}{l} \text{Find } u \in \overline{V} \text{ such that } Au - \lambda u = 0 \text{ ;} \\ \text{where A is a mapping such that :} \\ A \in \mathcal{L}(\overline{V}, \overline{V}), \text{ where A is Hermitian and continuous.} \end{array} \right.$$

Now $\overline{V}$ is dense in L^2, so that A, which is continuous, can be extended on L^2 by continuity, and being compact in $\overline{V}$, <u>A is also compact in L^2</u>. Furthermore, A is positive definite under the preceding conditions. We can then apply the Riesz theorem, relative to the spectral decomposition of A, and

---

[*]   It should be remembered that, by definition, the compact application G
   from E into F (Hilbert spaces), is such that the image of any bounded
   set is relatively compact. Furthermore, G must be continuous.

which we will restrict to this particular case, to the compact operator A of L^2 in L^2.

Riesz' theorem[5,6]

The operator A from L^2 into L^2, linear compact and Hermitian, admits a spectrum Λ of eigenvalues λ real, positive and in a finite number, or which can be arranged in a sequence converging to zero, Λ including zero or not. We have $\lambda \leqslant |A|_{L^2}$. For each value of $\lambda \in \Lambda$, except perhaps for $\lambda = 0$, the eigen subspace E has a finite dimension. The E_λ are orthogonal one to another. L^2 is the direct Hilbertian sum of the E_λ, $\lambda \in \Lambda$. In the basis of the eigenmodes, A is orthodiagonal. Taking u_λ as an eigenmode, we have [*) :

$$Au_\lambda = \lambda u_\lambda .$$

Transposition of these results to the discrete case examined previously, is immediately obvious.

4. – VARIATIONAL PRINCIPLES[7,8]

In the following pages, we use the following formulations which facilitate the passage to the discrete case :

$$(v_1,v_2)_{L^2} = \overline{v_1}\, v_2 = \int_\Omega \overline{v_1(x)}\, v_2(x)d\Omega \quad \forall v_1,v_2 \in V \subset L^2(\Omega)$$

$$(v_1,k\,v_2)_{L^2} = (kv_1,v_2)_{L^2} = \overline{v_1}\, k\, v_2 = \int_\Omega \underset{\sim}{\alpha}(D_L(v_1))(D_L(v_2))d\Omega, \ k = \overline{k}$$

where k is the stiffness operator, or the stiffness matrix.

$$(v_1,mv_2)_{L^2} = (mv_1,v_2)_{L^2} = \overline{v_1}\, m\, v_2 = \int_\Omega \rho\overline{v_1}\, v_2\, d\Omega , \ m = \overline{m} .$$

Thus the deformation energy for a displacement u is written :

$$W = \frac{1}{2}\,\overline{u}\, k\, u , \ k = \overline{k} \quad \text{is positive definite,}$$

---

*) In this case, the left hand modes are the transposes of the right hand modes for the same eigenvalues.

while the kinetic energy is written :

$$E = \frac{1}{2}\,\bar{\dot{u}}\,m\,\dot{u}\,, \quad m = \bar{m}$$

In the sequel, we shall put : $E = \frac{1}{2}\,\bar{u}\,m\,u$

We can then state the following theorem :

> Theorem :
>
> a) The eigenmode u_1 of the vibration problem of a conservative linear elastic structure is the vector which minimizes the deformation energy in the class of the kinematically admissible displacements, with the condition $E = 1$. The corresponding eigenvalue is given by the value of this minimum.
>
> b) If in addition to the condition $E = 1$, we also impose :
>
> $$\bar{u}_1\,m\,u = 0,$$
>
> the solution for the new problem is the eigenvector u_2, which minimizes the deformation energy under these two constraints, still in the class V of the kinematically admissible displacements. The new minimum is equal to the eigenvalue $\lambda_2 \geqslant \lambda_1$. More generally, if we impose the conditions :
>
> $$\begin{cases} \bar{u_i}\,m\,u = 0\,, & (i = 1,2,\dots,n-1) \\[2mm] \dfrac{1}{2}\,\bar{u}\,m\,u = 1, \end{cases}$$
>
> the solution u_n, an eigenvector of rank n, minimizes the deformation energy in V, and the minimum is the eigenvalue λ_n. In addition :
>
> $$\lambda_n \geqslant \lambda_{n-1}$$

The proof is straightforward. Introducing a Lagrangian multiplier λ [see (1.102)] , u satisfies the necessary condition :

$$\delta\left[\frac{1}{2}\,\bar{u}\,ku - \lambda\left[\frac{1}{2}\,\bar{u}\,m\,u - 1\right]\right] = 0 \quad \forall \delta\,u \text{ K.A.}$$

namely :

(4.19) $[k - \lambda m]\,u = 0, \quad \text{with } \frac{1}{2}\,\bar{u}\,m\,u = 1.$

Therefore λ_1 and u_1 are, respectively, the eigenvalue and eigenmode of the solution to the spectral problem. In addition, λ_1 is the minimum of $\frac{1}{2}\,\bar{u}\,k\,u$, when u satisfies $\frac{1}{2}\,\bar{u}\,m\,u = 1$ because :

$$u = u_1 + \delta u \quad \forall \delta u \text{ C.A. } \left|\frac{1}{2}\,\bar{u}\,m\,u = 1,\right.$$

$$\frac{1}{2}\left[\bar{u}_1 + \overline{\delta u}\right]k\left[u_1 + \delta u\right] - \frac{1}{2}\bar{u}_1 k u_1 = \overline{u}_1\,k\,\delta u + \frac{1}{2}\,\overline{\delta u}\,k\,\delta u.$$

Now $\overline{u}_1\,k\delta u = 0$ when $\overline{u}_1\,m\delta u = 0$. This quantity is indeed positive.

Now taking the problem : find u kinematically admissible, which makes

$$\frac{1}{2}\,\bar{u}\,k\,u \text{ minimum when } \frac{1}{2}\,\bar{u}\,m\,u = 1 \quad \text{and} \quad \overline{u}_1\,m\,u = 0$$

We introduce a first multiplier λ, to take account of the normalization $\frac{1}{2}\,\bar{u}\,m\,u = 1$, which gives again (4.19). λ_2 and u_2 therefore verify the spectral problem. A second multiplier enables us to introduce the relation of orthogonality $\overline{u}_1\,m\,u = 0$, hence the equation :

$$\delta\left[\frac{1}{2}\,\bar{u}\,k\,u - \lambda\left[\frac{1}{2}\,\bar{u}\,m\,u - 1\right] - \mu\,\bar{u}_1\,m\,u\right] = 0 \quad \forall \delta u \text{ K.A.}$$

Therefore u_2 must satisfy :

$$\overline{u_2}\,k\,\delta u - \lambda\,\overline{u_2}\,m\,\delta u - \mu\,\overline{u_1}\,m\,\delta u = 0 \ \forall \delta u$$

with : $\quad \frac{1}{2}\,\overline{u_2}\,m\,u_2 = 1, \ \overline{u_2}\,m\,u_1 = 0.$

Taking $\delta u = u_2$, we obtain :

$$\lambda_2 = \frac{1}{2}\,\overline{u_2}\,k\,u_2.$$

λ_2 is effectively a minimum for the quantity $\frac{1}{2}\,\bar{u}\,k\,u$, in the class of the admissible increments, but this minimum is equal to or larger than λ_1, as the class of the admissible vectors is contained in the preceding class. Therefore :

$$\lambda_2 \geqslant \lambda_1 ,$$

and so on. Thus by this variational method, we find the orthogonal eigenvectors associated with the eigenvalues, in increasing order.

<u>Corollary</u>

Under the conditions of the preceding theorem, the eigenmodes for the vibration problem make the quotient :

$$\frac{\overline{v}\ k\ v}{\overline{v}\ m\ v}$$

minimum, this being called the <u>Rayleigh quotient</u>, with no normalization condition, the corresponding eigenvalues being respectively equal to these minima.

The proof is straightforward. We merely put :

$$E = \frac{1}{2}\,\overline{v}\ m\ v = \frac{1}{2}\,\tilde{v}\ v$$

where $\tilde{v}$ is the transpose of v in this new scalar product, and :

$$w = \frac{1}{2}\,\overline{v}\ k\ v = \frac{1}{2}\,\tilde{v}\ A\ v,\quad A = \tilde{A}\ \text{positive definite.}$$

A theorem of maximum/minimum

Theoretically we can find the nth eigenvalue and n^{th} eigenvector, without reference to the preceding $n-1$ eigenvectors.

Instead of imposing the conditions of orthogonality :

$$\overline{u_i}\ m\ u = 0,\quad i = 1,2,\ \ldots,\ n-1$$

with the eigenvectors u_i, we shall impose $n-1$ modified conditions :

$$\overline{v_i}\ m\ u = 0,\quad i = 1,2,\ldots,\ n-1$$

by means of $n-1$ admissible vectors v_i.

In this case $\frac{1}{2}\,\overline{u}\ k\ u$ has a lower bound which depends on the selected v_i, namely $d(v_1,v_2,\ldots,v_{n-1})$, with $\frac{1}{2}\,\overline{u}\ m\ u = 1$.

<u>Theorem</u>

Under these conditions :

$$\lambda_n = \underset{v_i \in V}{\text{Max.}}\ d(v_1,v_2,\ldots,v_{n-1}).$$

where V is the class of the admissible vectorial functions. This maximum is reached for $v_1 = u_1, \ldots\ldots, v_{n-1} = u_{n-1}$; $u_1, u_2, \ldots\ldots, u_{n-1}$ being the first n-1 eigenvectors.

In fact if :

$$v_1 = u_1, \ldots, v_{n-1} = u_{n-1},$$

$$d(u_1, u_2, \ldots, u_{n-1}) = \lambda_n,$$

as we have the conditions of the preceding theorem.

In addition :

$$d(v_1, v_2, \ldots, v_{n-1}) \leqslant \lambda_n.$$

In fact let us determine a particular vectorial function u, such that :

$$(4.20) \qquad \overline{v_i} \, m \, u = 0, \ i = 1, 2, \ldots, n-1$$

and such that :

$$(4.21) \qquad \frac{1}{2} \, \overline{u} \, m \, u = 1.$$

We look for a solution to this problem in the basis of the eigenmodes, truncated at rank n, limiting the class and increasing the minimum, which we will demonstrate that it reaches its maximum value λ_n.

Let us therefore take :

$$(4.22) \qquad u = \sum_{i=1}^{n} C_i \, u_i,$$

with coefficients C_i unknown. The n relations (4.20) and (4.21) enable us to calculate these coefficients. But (4.22), and the orthogonality of the u_i modes, give with (4.21) :

$$\sum_{i=1}^{n} C_i^2 = 1.$$

But :

$$\overline{u_i} \, m \, u_j = 0 \ \text{ for } i \neq j \text{ and } \frac{1}{2} \, \overline{u_i} \, k \, u_i = \lambda_i, \ \lambda_i \leqq \lambda_n, \ \forall i \leqq n$$

then gives :

$$\frac{1}{2}\,\bar{u}\,k\,u = \sum_{i=1}^{n} \lambda_i\,c_i^2 \leq \lambda_n \left[\sum_{i=1}^{n} c_i^2 \right] = \lambda_n.$$

We have therefore found a solution u, such that :

$$\frac{1}{2}\,\bar{u}\,k\,u \leq \lambda_n,$$

to the minimum problem, in the restricted class of the space subtended by the first n eigenmodes. The proof is completed by observing that this minimum, always less than λ_n in this class $\forall$ vectors v_i (i=1,2,...,n-1), reaches its maximum when we select the v_i equal to the eigenvectors u_i (i = 1,2, ..., n-1), Q.E.D.

The preceding theorem can also be stated as follows :

<u>Theorem of maximum/minimum</u>

Let R(u) be the Rayleigh quotient for the vibration problem, and P_{n-1} the plane (subspace of the kinematically admissible vectorial functions), subtended by n-1 vectors v_i (i = 1,2,..., n-1). The nth eigenvalue λ_n is such that :

$$\lambda_n = \text{Max.} \quad \text{Min. } R(u).$$
$$\forall P_{n-1},\ u \perp P_{n-1}$$

If $v_i = u_i$, v_i being a eigenvector of rank $\leq$ n-1 then we have :

$$\lambda_n = \text{Min. } R(u).$$
$$u \perp P_{n-1}(u_i)$$

In addition, we have the following dual theorem :

<u>Theorem of minimum/maximum</u>

Taking R(u) as the Rayleigh quotient for the vibration problem, and P_n the plane,(subspace of the kinematically admissible vectorial functions), subtended by n vectors v_i (i = 1,2, ..., n), the nth eigenvalue λ_n is such that :

$$\lambda_n = \text{Min. Max. } R(u).$$
$$\forall P_n, \quad u \in P_n$$

If $v_i = u_i$, u_i being an eigenvector of rank $\leqslant n$, we then have :

$$\lambda_n = \text{Max. } R(u).$$
$$u \in P_n(u_i)$$

In fact with any given plane P_n, and vector u such that :

$$u^* \quad P_n \text{ and } \overline{u}_i . u^* = 0, \quad (i = 1, 2, \ldots, n-1) \ (u_i \text{ eigenvectors of rank} \leqslant n-1).$$

This vector u^* exists, as it is defined by n-1 homogeneous equations for n unknowns.

Now we know that :

$$\lambda_n = \text{Min. } R(u)$$
$$u \perp P_{n-1}(u_i)$$

and the supplementary constraint $u^* \in P_n$ (any given P_n) $\Longrightarrow$

$$\lambda_n \leqslant R(u^*)$$

As $R(u^*) \leqslant \text{Max.} R(u)$, then $\lambda_n \leqslant \text{Max.} R(u)$
$\qquad u \in P_n \qquad\qquad\qquad u \in P_n$

Therefore $\lambda_n = \text{Min. Max. } R(u).$ $\qquad$ Q.E.D.
$$\forall P_n \quad u \in P_n$$

These theorems produce a certain number of general consequences : firstly, adding some constraints to a minimum problem, the minimum, namely the nth eigenvalue of the new problem, cannot decrease, and reciprocally.

Then given two minimum problems, having the same class of admissible functions, and such that the functionals satisfy :

$$\overline{u} k_1 u \leqslant \overline{u} k_2 u,$$

k_1 and k_2 being the operators relative to each of the problems, we have :

$$\text{min.} \overline{u} k_1 u \leqslant \text{Min.} \overline{u} k_2 u.$$

As a result, if Ω_1, Ω_2, ..., Ω_ℓ are the non-connected subdomains of Ω, on which admissible functions u are defined, we can consider the two following problems :

In the first :

$$u = 0 \text{ on } \partial\Omega.$$

In the second :

$$u = 0 \text{ on } \partial\Omega_1, \ \partial\Omega_2, \ ..., \ \partial\Omega_\ell.$$

Then let us take λ_n as the nth eigenvalue relative to the first problem, and λ_n^* the nth eigenvalue relative to the second and to the domain $\Omega^* = \Omega_1 \cup \Omega_2 \cup ... \cup \Omega_\ell$, taking account of their manifold order. Then :

$$\lambda_n^* \geqslant \lambda_n.$$

For conjugated boundary conditions (zero stresses on the boundary, for example $\frac{\partial u(M)}{\partial n} = 0$), we have the inverse situation. In fact we broaden the class of admissible functions in the second problem.

Taking another case, if the density is modified at each point of Ω and in the same direction, the nth eigenvalue is modified in the opposite direction (we need merely consider the Rayleigh quotient). If we change the medium, in such a way that quantity $\bar{u} \, k \, u$ increases (resp. decreases) at each point, the nth eigenvalue increases (resp. decreases) also.

In general in the case of a structure subjected to a linear kinematic constraint of the type :

$$\bar{w} \, m \, u = 0$$

where $u \in V$, the configuration space, and where $w \in V$ is given, we have the so called following theorem :

Theorem of separation

If λ_n and λ_{n+1} denote the nth and $n+1^{th}$ eigenvalue of the unlinked system, the nth eigenvalue λ_n^* of the linked system is such that :

$$\lambda_n \leqslant \lambda_n^* \leqslant \lambda_{n+1}.$$

In fact the first inequality has already been demonstrated, and let us call P the plane orthogonal to w, defined by the constraint, and P_n the plane subtended by the two eigenvectors u_n and u_{n+1}, corresponding to λ_n and λ_{n+1} respectively. Clearly dimension of $P_n \cap P \geqslant 1$.

Therefore :

$$\forall u^* \in P_n \cap P \subset V$$

$$\text{if } \overline{u}_n m u_n = |u_n|^2 = 1 , \quad \overline{u_{n+1}} \, m \, u_{n+1} = |u_{n+1}|^2 = 1,$$

and :

$$u^* = u_n \, \overline{u_n} \, m \, u^* + u_{n+1} \, \overline{u_{n+1}} \, m u^*$$

$$u^* = u_n a_1 + u_{n+1} a_2, \quad \text{with } a_1^2 + u_1^2 = 1 \text{ if } |u^*| = 1 .$$

Now :

$$\lambda_n = \underset{u \in V_n}{\text{Min.}} \; \overline{u} \, k \, u \quad \text{with } V_n = \left\{ u : |u| = 1, \; \overline{u_i} m u = 0, \; i \leqslant n \right\} \subset V$$

$$\lambda_n^* = \underset{u \in P, \; |u| = 1}{\text{Min.}} \; \overline{u} \, k \, u = \overline{u}^* k \, u_n^*$$

$$= \overline{u}_n k u_n \, a_1^2 + \overline{u_{n+1}} k u_{n+1} a_1^2 + 2 \overline{u}_n k u_{n+1} a_1 a_2 .$$

But :

$$k u_n - \lambda_n m u_n = 0, \quad k u_{n+1} - \lambda_{n+1} m u_{n+1} = 0 \implies \overline{u}_n k u_{n+1} = \lambda_{n+1} \overline{u}_n m u_{n+1} = 0$$

Therefore :

$$\lambda_n^* \leqslant \overline{u}^* k \, u^* = \lambda_n a_1^2 + \lambda_{n+1} a_2^2 \leqslant \lambda_{n+1} \left[a_1^2 + a_2^2 \right] = \lambda_{n+1} \quad . \quad \text{Q.E.D.}$$

<u>Remark</u> : The theorem is generalized by recurrence, in the case of a finite number of constraints of the preceding type.

5. – <u>METHODS OF DYNAMIC REDUCTION</u>

The study of the static or dynamic behaviour of a discretized structure, may lead to voluminous calculations exceeding the capacity of the computation programmes or computers available. The methods of analysis described below are methods of reduction, namely methods of reducing the number of degrees of freedom. One may use the decomposition of a structure into substructures, each of which is represented in modal form, and forming the subject of separate analyses which are less voluminous than in the case of the complete structure. Assembly of the substructure then enables us to represent the complete structure, which can be calculated in an economic manner. Decomposition into sub-structures can be natural to a greater or lesser degree.

1) The mass coupling method[9,10]

To establish our ideas, let us take a structure Ω, decomposed into two substructures, Ω_1 and Ω_2, connected by a common boundary Σ_c. We represent each substructure, for example Ω_1, using the Ritz method, by a linear combination of two types of kinematically admissible modes : firstly the "boundary modes" of a static type, obtained by unit displacement of each degree of freedom of the boundary Σ_c, namely :

$$S_1 = \left[S_{11} \; S_{12} \; S_{13} \; \cdots \; S_{1n_1} \right] ,$$

and secondly the eigenmodes with the boundary supposed to be clamped,

$$S_2 = \left[S_{21} \; S_{22} \; S_{23} \; \cdots \; S_{2n_2} \right]$$

for which we will take a limited number n_2.

The boundary modes are static. They concern substructure Ω_1 in its entirety, but they are due solely to the displacements of the boundary Σ_c, in sufficient number to represent correctly any shape of Σ_c. They can be calculated conveniently using the finite element method, giving a unit value to each nodal degree of freedom in succession, with the others having zero value. But we can also select a truncated functional basis, discretized or not, to represent the displacements of Σ_c. The representation of these displacements of Σ_c must be the same for Ω_1 and Ω_2, so as to provide for a convenient assembly of Ω_1 and Ω_2, simply equalizing the coefficients of the two respective linear combinations.

Thus any shape of Ω_1 (or Ω_2) will be correctly represented, in any shape of the complete structure Ω, by a linear combination of static shape functions, due to deformations of the common boundary Σ_c, and the eigenmodes with a clamped boundary, namely :

$$(4.23) \qquad U = \underset{\sim}{U}(M_o, t) = S \, q = S_1 \, {}^1q + S_2 \, {}^2q \; ; \; {}^1q \in R^{n_1}, \; {}^2q \in R^{n_2}$$

The stiffness and mass matrices, K and M, correspond to generalized co-ordinates q, such that, using the notations of this chapter :

$$K = \overline{S} \, k \, S = \overline{K}$$

$$M = \overline{S} \, m \, S = \overline{M} \quad \left(= \int_{\Omega_1} \rho \, \overline{S(M_o)} \, S(M_o) d\Omega\right).$$

We then have the following result :

Theorem :

The boundary modes are elastically decoupled from the eigenmodes with a clamped boundary, namely :

$$(4.24) \qquad {}^1K_2 = \overline{{}^2K_1} = 0, \text{ where } {}^1K_2 = \overline{S_1} \, k \, S_2 .$$

This result is maintained after assembly.

In fact let us suppose that, S_1 being already known, we have to calculate a "boundary mode",

$$(4.25) \qquad \underset{\sim}{U}(M_o) = S_1 {}^1q \, , \, M_o \in \Omega_1 \, ,$$

namely a shape function of substructure Ω_1, due to a given shape of boundary $\underset{\sim}{U}(M_o) = s_1 {}^1q$ where $M_o \in \Sigma_c$. This is based on the assumption that we take ${}^1q$ (in this formula, s_1 is a shape basis of Σ_c ; ${}^1q$ is given, but can have any value).

The result is (4.25), namely : ${}^2q = 0$.

In this calculation, where only kinematic boundary conditions are given, the solution satisfies the principle of virtual work, hence, in succession:

$$\overline{Sq} \, k \, S\delta q = \overline{Q}\delta q = 0 \quad , \quad \forall \delta q \quad \left(\delta {}^1q = 0, \, {}^1q \text{ given}\right)$$

$$\overline{q} \, \overline{S} \, k \, S \, \delta q = 0 \, , \quad \forall \delta q \text{ such that } \delta {}^1q = 0$$

$$\overline{q}K_2 \delta {}^2q = 0, \quad \forall \delta {}^2q$$

$${}^2K_1 {}^1q + {}^2K_2 {}^2q = 0, \quad \text{with } {}^2q = 0 \quad (\text{see } 4.25).$$

Therefore :

$$^2K_1 \, ^1q = 0, \quad \forall \, ^1q,$$

therefore :

$$^2K_1 = 0 = \overline{^2K_1}, \quad \text{whence} \quad (4.24).$$

This result is maintained after assembly of Ω_1 and Ω_2. We therefore have for Ω_2 :

$$U = S_3 \, ^3q + S_4 \, ^4q \; ; \; ^3q \in \mathbb{R}^{n_1}, \; ^4q \in \mathbb{R}^{n_4},$$

where S_3 is the line of the boundary modes of Ω_2. The deformation energy of Ω_2, taking account of (4.24) for Ω_2, will be :

$$w_2 = \frac{1}{2} \overline{^3q} \; ^3K_3 \, ^3q + \frac{1}{2} \overline{^4q} \; ^4K_4 \, ^4q$$

Assembly is achieved by writing :

$$^1q = \, ^3q.$$

Thus the total deformation energy for the assembled structure is written:

$$w_1 + w_2 = \frac{1}{2} \overline{^1q} \left[^1K_1 + \, ^3K_3 \right] \, ^1q + \frac{1}{2} \overline{^2q} \; ^2K_2 \, ^2q + \frac{1}{2} \overline{^4q} \; ^4K_4 \, ^4q,$$

which clearly demonstrates that elastic couplings between boundary modes and eigenmodes with fixed boundary, are zero for Ω. Q.E.D.

It should also be noted that matrices $^2K_2$ and $^4K_4$ are diagonal, by virtue of what has been said in the preceding paragraphs.

Numerous variants can be introduced to accelerate convergence, or obtain a higher degree of accuracy, in particular on the stresses of common boundaries, for a given number of basis modes. We can incorporate advantageously static modes calculated specially for certain types of loading, for example dynamic, to take account of the effect of certain masses, concentrated or not. We can also introduce symmetries, where appropriate, without difficulty. The order reduction made possible by this method, can reach a factor of 10, or even 100. But the number of modes to be introduced to obtain a given final degree of precision, is unknown at the outset, and methods of selection have been proposed, to assist in selection of the modes introduced[11]. Finally we note that variants have also been proposed, for replacement of fixed boundary eigenmodes by free substructure eigenmodes, with judicious correction of their stiffness and mass matrices[12]. In fact these free modes only supply mediocre convergence, as they correspond to zero loading on the common boundaries, contrary to what occurs in the deformations of the assembled substructures.

2) The stiffness coupling method[13]

This is a dynamic substructuring method, consisting in decomposing the structure into a number of main substructures, the latter being interconnected by elements, or coupling substructures, having smaller dimensions and no mass. The masses of these small intermediate substructures are added judiciously to those for the main substructures. They therefore only intervene by their stiffnesses. Furthermore, the main substructures are each represented in the truncated base of their free eigenmodes.

Let us take for example a structure Ω, decomposed into two main substructures, Ω_1 and Ω_2, connected by a coupling substructure Ω_c. Ω_c will thus have two common boundaries, Σ_{c1} and Σ_{c2}, with Ω_1 and Ω_2.

Discretization into finite elements gives, for example, the following nodal unknowns :

$$\begin{bmatrix} {}^1q \\ {}^2q \end{bmatrix} \quad \text{for } \Omega_1 \text{, where } {}^2q \text{ represents the degrees of freedom of } \Sigma_{c1}$$

$$\begin{bmatrix} {}^3q \\ {}^4q \end{bmatrix} \quad \text{for } \Omega_2 \text{, where } {}^3q \text{ represents the degrees of freedom of } \Sigma_{c2}.$$

Substructures Ω_1 is represented in the truncated basis of its free eigenmodes, such that :

$$(4.26) \qquad \begin{bmatrix} {}^1q \\ {}^2q \end{bmatrix} = \begin{bmatrix} {}^1S_1 \\ {}^2S_1 \end{bmatrix} {}^1q' \, .$$

Likewise for substructure Ω_2 :

$$(4.27) \qquad \begin{bmatrix} {}^3q \\ {}^4q \end{bmatrix} = \begin{bmatrix} {}^3S_2 \\ {}^4S_2 \end{bmatrix} {}^2q' .$$

In that bases the kinetic energy of Ω is written, using obvious notations :

$$E = \frac{1}{2} \overline{{}^1q'} \, {}^1\mu_1 \, {}^1q' + \frac{1}{2} \overline{{}^2q'} \, {}^2\mu_2 \, {}^2q'$$

where ${}^1\mu_1$ and ${}^2\mu_2$ are the diagonal matrices of the generalized modal masses, while the deformation energy will be :

$$W = \frac{1}{2} \overline{{}^1q'} \, {}^1\gamma_1 \, {}^1q' + \frac{1}{2} {}^2q' \, {}^2\gamma_2 \, {}^2q'$$

$$+ \frac{1}{2} \begin{bmatrix} \overline{{}^2q} & \overline{{}^3q} \end{bmatrix} \begin{bmatrix} {}^2K_{c2} & {}^2K_{c3} \\ {}^3K_{c2} & {}^3K_{c3} \end{bmatrix} \begin{bmatrix} {}^2q \\ {}^3q \end{bmatrix} ,$$

where K_c is the stiffness matrix of the coupling substructure. The last term of W is written, with the basis change (4.26) and (4.27) :

$$\frac{1}{2} \overline{{}^1q'} \, \overline{{}^2S_1} \, {}^2K_{c2} \, {}^2S_1 \, {}^1q' + \overline{{}^1q'} \, \overline{{}^2S_1} \, {}^2K_{c3} \, {}^3S_2 \, {}^2q' +$$

$$+ \frac{1}{2} \overline{{}^2q'} \, \overline{{}^3S_2} \, {}^3K_{c3} \, {}^3S_2 \, {}^2q' .$$

Clearly, the coupling of the two substructures Ω_1 and Ω_2, in generalized unknowns ${}^1q'$ and ${}^2q'$, is solely of the stiffness type.

This method, which is relatively easy to apply, naturally presents the disadvantages already mentioned with respect to the utilization of free eigenmodes.

A variant would consist in not introducing a coupling element Ω_c, and taking account of assembly of Ω_1 with Ω_2, by means of a Lagrangian multiplier λ, namely adding to the total deformation energy the term :

$$\left[\; \overline{^2q} - \overline{^3q}\; \right]\lambda$$

namely, with (4.26) and (4.27) :

$$\left[\; \overline{^1q'\; ^2s_1} - \overline{^2q'\; ^3s_2}\; \right]\lambda,$$

which gives a generalized stiffness matrix relative to the unknowns $^1q'$, $^2q'$ and λ, and a generalized mass matrix as follows :

$$\begin{bmatrix} ^1\gamma_1 & 0 & \overline{^2s_1} \\ 0 & ^2\gamma_2 & -\overline{^3s_2} \\ ^2s_1 & -^3s_2 & 0 \end{bmatrix}, \quad \begin{bmatrix} ^1\mu_1 & 0 & 0 \\ 0 & ^2\mu_2 & 0 \\ 0 & 0 & 0 \end{bmatrix}.$$

The last lines allow a reduction on $^1q'$ and $^2q'$, but the method only presents an advantage if an efficient reduction algorithm is available.

Concerning the preceding substructuring methods, we can make the following general comment : the greater the number of substructures for a given main structure, the greater is the number of common boundaries, and consequently the greater the number of boundary degrees of freedom. The number of eigenmodes to be introduced can drop as a result, while the boundary modes or degrees of freedom increases. In the extreme case, we come back to the finite elements method, where boundary modes only are used.

3) The Guyan method[14,15]

This method widely used for high order discretized systems, is based on a static reduction of the degrees of freedom, this reduction also being applied to the mass matrix. For preference, it is applied to the search for the first eigenmodes of a structure, frequently the most important, which can be reproduced relatively correctly, as we know, by judiciously distributed discrete static loads.

The choice of dynamically loaded degrees of freedom to be retained, depends in fact on a physical analysis of the dynamic problem concerned.

Let K and M be the matrices of stiffness and mass, calling ^1q the column of degrees of freedom retained, and ^2q the column of the degrees of freedom which we wish to eliminate. If the latter are not loaded statically, the equation for the static problem is written :

$$
\begin{bmatrix} ^1K_1 & ^1K_2 \\ ^2K_1 & ^2K_2 \end{bmatrix} \begin{bmatrix} ^1q \\ ^2q \end{bmatrix} = \begin{bmatrix} ^1Q \\ 0 \end{bmatrix} .
$$

Hence the relation already indicated in chapter II :

$$
^2q = - \, ^2K_2^{-1} \, ^2K_1 \, ^1q
$$

which gives the following change of basis :

$$
\begin{bmatrix} ^1q \\ ^2q \end{bmatrix} = \begin{bmatrix} 1 \\ - \, ^2K_2^{-1} . \, ^2K_1 \end{bmatrix} \, ^1q
$$

or : $q = S . \, ^1q$.

After this change of basis, the matrices K and M become :

$$
K' = \overline{S} \, K \, S, \quad M' = \overline{S} \, M \, S,
$$

namely :

$$
\begin{bmatrix}
K' = \, ^1K_1 - \, ^1K_2 \, ^2K_2^{-1} \, ^2K_1 \\[2ex]
M' = \, ^1M_1 - \, ^1M_2 \, ^2K_2^{-1} \, ^2K_1 - \, ^1K_2 \, ^2K_2^{-1} \left[^2M_1 - \, ^2M_2 \, ^2K_2^{-1} \, ^2K_1 \right].
\end{bmatrix}
$$

It is worth noting that the reduction of the mass matrix is a relatively costly operation. Interesting studies have been made on the approximations obtained by this method, which is thus justified in the most frequent cases subject to certain criteria concerning the choice of the degrees of freedom eliminated[16-19].

General remark

The substructuration methods have been the subject of an excellent synthesis by Morand[23].

Let us still consider a structure Ω, decomposed into two substructures Ω_1 and Ω_2, connected along their common boundary Σ_c, and any kinematically admissible displacement U of Ω. We can then write :

$$U = \begin{bmatrix} U_1 \\ U_2 \end{bmatrix} \quad , \quad \forall\, U \in \mathcal{C} \, ,$$

where U_1 and U_2 are the displacements of Ω_1 and Ω_2 respectively, in the kinematically admissible displacement U.

In this displacement, the boundary Σ_c undergoes a displacement u_c, which is the trace of U_1 or U_2 on Σ_c.

Let us assume Ω_2 separated from Ω_1. The displacement u_c induces a displacement U_{1c} statically on Ω_1, so that U can be written :

$$U = \begin{bmatrix} U_1 - U_{1c} \\ 0 \end{bmatrix} + \begin{bmatrix} U_{1c} \\ U_2 \end{bmatrix} \, ,$$

and this $\forall U$ K.A.

Thus the class $\mathcal{C}_1$ of displacements U_1 (any kinematically admissible) of Ω_1, can be decomposed into a direct sum of two classes, one comprising displacements U_{1c} induced statically by any displacements of Σ_c on Ω_1, and the other comprising displacements $U_1 - U_{1c}$.

The latter are in fact any displacements of Ω_1 with boundary Σ_c fixed, as the traces of U_1 or U_{1c} on Σ_c, are equal by construction.

The operation can be repeated on Ω_2, and we can write :

$$U = \begin{bmatrix} U_1 - U_{1c} \\ U_2 - U_{2c} \end{bmatrix} + \begin{bmatrix} U_{1c} \\ U_{2c} \end{bmatrix} \, , \quad \text{or } \mathcal{C} = \mathcal{C}_E \oplus \mathcal{C}_c \, .$$

These formulae enable us to represent $U_1 - U_{1c}$ (resp. $U_2 - U_{2c}$) in a total basis for the eigenmodes of Ω_1 (resp. Ω_2), with fixed boundary, and U_{1c} (resp. U_{2c}) in a total basis for the "static boundary modes" induced by a total basis of boundary shape functions.

The mass coupling method is thus rigourously demonstrated.

The other two methods of reduction described above can be brought back to this method, subject to certain approximations which are easy to explain.

Let us consider the formula :

$$U = \begin{bmatrix} U_1 - U_{1c} \\ 0 \end{bmatrix} + \begin{bmatrix} U_{1c} \\ U_2 \end{bmatrix},$$

The Guyan method equates for example to considering domains the union of which constitute in fact a substructure Ω_1, and of which the boundaries common with $\Omega_2 = \Omega - \Omega_1$ have Σ_c as their union, then representing U by :

$$U = \begin{bmatrix} U_{1c} \\ U_2 \end{bmatrix},$$

thus ignoring the contribution of $U_1 - U_{1c}$ in Ω_1. This corresponds to assuming that Ω_1 only undergoes displacements generated statically by Ω_2 and Σ_c. In other words, the inertial forces on Ω_1 are disregarded in this representation.

These considerations enable us to make a judicious choice of substructures and useful approximations.

They are generalized in particular to systems comprising a number of coupled media, for example solids and fluids[24-27].

6. - <u>RESPONSES TO FORCED EXTERNAL EXCITATIONS</u>

Let us consider a conservative <u>linear discretized structure</u>, subjected to a generalized external load $\underset{\sim}{Q}(t)$. Its equation is written [see (4.1)] :

$$M \, \underset{\sim}{\ddot{q}}(t) + K \, \underset{\sim}{q}(t) = \underset{\sim}{Q}(t) \qquad \underset{\sim}{q}(t) \in \mathbf{R}^N .$$

If the structure is dissipative, the power dissipated for a linear structure is a quadratic form of the velocity, namely :

$$\frac{1}{2} \, \overline{\dot{q}(t)} \, B \, \dot{q}(t) \; , \; B = \overline{B}, \quad \text{positive definite.}$$

The damping force is therefore $- B \, \dot{q}(t)$, and the preceding equation
becomes classically, with given initial conditions :

$$(4.28) \qquad M \, \ddot{q}(t) + B \, \dot{q}(t) + K \, q(t) = Q(t) \, .$$

There are numerous computational algorithms for obtaining the response
$q(t)$ to the excitation $Q(t)$. Distinction is made between <u>direct integration methods</u>, which we shall consider briefly in the next chapter, and which
are generally based on step by step integration of (4.28) with respect to
time, and <u>modal methods</u> in which integration is made after representation
of system (4.28) in the truncated basis of the eigenmodes for the associated conservative system.

We will limit ourselves at this point to giving a few indications
concerning these modal methods, by reason of the useful and classical
theoretical results which they can be used to obtain.

Let S be the basis, truncated at order n, of the eigenmodes for the
conservative system associated with (4.28), namely the eigenmodes of the
system :

$$\left[- \omega^2 M + K \right] q = 0$$

where we have put :

$$q(t) = qe^{j\omega t} ,$$

so that :

$$S = \left[S_1 \; S_2 \; \ldots \; S_n \right], \; S_i \in \mathbb{R}^N, \; i = 1, 2, \ldots, n.$$

In S, q is represented by a column $q' \in \mathbb{R}^n$, such that :

$$q = S \, q',$$

and (4.28) becomes :

$$(4.29) \quad \begin{cases} \mu \ddot{\underset{\sim}{q}}'(t) + \beta \dot{\underset{\sim}{q}}'(t) + \gamma \underset{\sim}{q}'(t) = \underset{\sim}{Q}'(t) \\[2mm] \text{with :} \\[2mm] \mu = \overline{S}\, M\, S, \quad \beta = S\, \overline{B}\, S, \quad \gamma = S\, K\, \overline{S}, \quad Q' = \overline{S}\, Q. \end{cases}$$

The matrices μ and γ are real, positive diagonals, as we saw in paragraphs 2 and 3. On the other hand, the generalized damping matrix β is
generally not diagonal. The diagonal terms are the respective damping factors
for the different modes, and the non-diagonal terms are the dissipative
inter-mode couplings.

To simplify, we will assume the system initially at rest :

$$\underset{\sim}{q}(o) = \dot{\underset{\sim}{q}}(o) = 0,$$

A Laplace transform immediately gives, for (4.28) :

$$\mathcal{L}(q) = \left[Mp^2 + Bp + K \right]^{-1} \mathcal{L}(Q),$$

where $p \in \mathbb{C}$ and where $\mathcal{L}(q)$ is the transform of $q(t)$. The formula gives the
transfer function of the system, but inversion is practically impossible for
systems even of low rank. Furthermore, under the same conditions, (4.29)
gives :

$$\mathcal{L}(q') = \left[\mu p^2 + \beta p + \gamma \right]^{-1} \mathcal{L}(Q').$$

Inversion is immediate if β is diagonal. A hypothesis commonly used for
simplification consists precisely in assuming matrix β to be diagonal
(Basile's hypothesis), which does not lead to any notable error when the
generalized damping factors are small, as we can observe under experimental
conditions. In fact in this case, the eigenmodes can easily be isolated
experimentally, by the use of an appropriate harmonic excitation to compensate the dissipated energy. The modal admittance for a mode with index i
is then :

$$(4.30) \qquad {}^{i}A_{i}(\omega) = \frac{1}{{}^{i}\mu_{i}\left[\omega_{1}^{2} - \omega^{2}\right] + j\, {}^{i}\beta_{i}\cdot\omega} = \frac{{}^{i}q'_{i}}{{}^{i}Q'_{i}} \quad \in \mathbb{C},$$

where ω_{i} is the eigenpulsation of the mode, and $j = \sqrt{-1}$.

Let us now consider a <u>linear continuous structure</u>, and any excitation of surface density $\underset{\sim}{p}(M_o,t)$ on the boundary Σ_o. Let us call $\underset{\sim}{y}(M,t)$ the response at point M, and to simplify, in a given direction $(y \in \mathbb{R})$.

We shall call $\underset{\sim}{h}(M,M_o,t)$ the impulse response, namely the response at M due to a time Dirac distribution $\underset{\sim}{p}(M_o) \, \delta_{t=o} \, d\Sigma_o$. The response $d\underset{\sim}{y}(M,t)$, in the convolution algebra $\mathcal{D}'^{+}$, is written :

$$d\underset{\sim}{y}(M,t) = \underset{\sim}{h}(M,M_o,t) \underset{t}{*} \underset{\sim}{p}(M_o,t)d\Sigma_o,$$

hence :

$$\underset{\sim}{y}(M,t) = \int_{\Sigma_o} \underset{\sim}{h}(M,M_o,t) \underset{t}{*} \underset{\sim}{p}(M_o,t)d\Sigma_o.$$

Let us now assume that the time functions are temperate distributions[20], (for example : summable, bounded, locally summable, or slowly increasing functions, namely below any power of t, square summable functions, bounded support distribution, etc.), and we obtain, by Fourier transform :

$$(4.31) \qquad \underset{\sim}{Y}(M,\omega) = \int_{\Sigma_o} \underset{\sim}{H}(M,M_o,\omega) . \underset{\sim}{P}(M_o,\omega) \, d\Sigma_o,$$

where Y, H and P are the respective Fourier transforms of y, h and p. Interpretation of function H is immediate, if we select :

$$(4.32) \qquad \left[\begin{array}{l} \underset{\sim}{p}(M_o,t) = \delta_{M_o} \, e^{j\omega t} \\[2em] \underset{\sim}{P}(M_o,\omega) = \delta_{M_o} \delta_\omega \\[2em] \underset{\sim}{Y}(M,\omega) = \underset{\sim}{H}(M,M_o,\omega) \end{array} \right. \left. \begin{array}{l} \\[2em] \\ \end{array} \right\} \Rightarrow$$

Functions $\underset{\sim}{H}(M,M_o,\omega)$ is calculated easily by the modal method. In fact in the modal basis truncated at order n :

$$\underset{\sim}{y}(M,t) = \underset{\sim}{S}(M) \, \underset{\sim}{q}(t) = \underset{\sim}{S_i}(M) . \, \underset{\sim}{{}^i q}(t) \quad (i = 1,2,\ldots,n).$$

The virtual work of the external forces can be written, for the component y :

$$\int_{\Sigma_o} \underset{\sim}{p}(M_o,t) \, \delta\underset{\sim}{y}(M_o,d)d\Sigma_o = \int_{\Sigma_o} \underset{\sim}{p}(M_o,t) . \, \underset{\sim}{S}(M_o) \, \underset{\sim}{\delta q}(t) \, d\Sigma_o = \bar{Q} \, \delta q.$$

The generalized external load Q, the conjugate of q, is written :

$$Q(t) = \int_{\Sigma_o} \overline{S(M_o)}\, p(M_o,t)\,d\Sigma_o .$$

The equation of the structure in the truncated basis is therefore :

$$(4.33) \qquad \mu\ddot{q} + \beta\dot{q} + \gamma q = Q = \int_{\Sigma_o} \overline{S(M_o)}\, p(M_o,t)\,d\Sigma_o .$$

Now if :

$$p(M_o,t) = \delta_{M_o}\, e^{j\omega t}$$

we have :

$$\mu\ddot{q} + \beta\dot{q} + \gamma q = \overline{S(M_o)}\, e^{j\omega t}$$

Under these conditions, the response to a harmonic force concentrated at M_o, assuming β to be diagonal, is written :

$$y(M,t) = S(M)\, q(t) = S_i(M)\, {}^iA_i\, S_i(M_o)\, e^{j\omega t} \quad , \; i = 1, \ldots, n)$$

with

$${}^iA_i = \cfrac{1}{{}^i\mu_i\left[\omega_i^2 - \omega^2\right] + j\, {}^i\beta_i\, \omega}, \; (i \text{ fixed}) .$$

Now, by (4.32) :

$$(4.34) \qquad H(M,M_o,\omega) = S_i(M)\, {}^iA_i\, \overline{S_i(M_o)} .$$

Substituted into (4.32), this formula enables us to calculate the response y, by an inverse Fourier transform.

In a case where the <u>excitation</u> $p(M_o,t)$ <u>is simply periodic</u>, its expansion into a Fourier series immediately gives :

$$p(M_o,t) = \sum_{m=-\infty}^{+\infty} P_m(M_o)\, e^{j\left[m\omega t + \varphi_m\right]} .$$

Equation (4.33) then gives the response :

$$y(M,t) = \sum_{m=-\infty}^{+\infty} \underset{\sim}{S_i}(M)\ {}^i A_i(\omega) \int_{\Sigma_o} \overline{\underset{\sim}{S_i}(M_o)}\ \underset{\sim}{P_m}(M_o)d\Sigma_o\ e^{j\left[m\omega t + \varphi_m\right]}.$$

In the case where $\underset{\sim}{p}(M_o,t)$ is a random stationary excitation[21] with a crossed power density spectrum between two points M_o and M_o',

$$\mathcal{S}_p(M_o,M_o',\omega),$$

calculation of the space-time intercorrelation function of the response $\underset{\sim}{y}(M,t)$, leads, by Fourier transform, to the direct power spectrum of the response. In fact we had :

$$\underset{\sim}{y}(M,t) = \int_{\Sigma_o} \underset{\sim}{h}(M,M_o,t)\ \underset{t}{*}\ \underset{\sim}{p}(M_o,t)d\Sigma_o.$$

Let us call :

$$E\left(\underset{\sim}{y_1}(t)\cdot\underset{\sim}{y_2}(t+\tau)\right) = R_{y_1 y_2}(\tau),$$

the crosscorrelation function between two functions, y_1 and y_2. We have :

$$E\left(\underset{\sim}{y}(M,t)\cdot\underset{\sim}{y}(M',t+\tau)\right) =$$

$$E\left(\int_{\Sigma_o x\Sigma_o} \underset{\sim}{h}(M,M_o,t\ \underset{t}{*}\ \underset{\sim}{p}(M_o,t)\cdot\underset{\sim}{h}(M',M_o',t+\tau)\ \underset{t}{*}\ \underset{\sim}{p}(M_o',t+\tau)d\Sigma_o\ d\Sigma_o'\right).$$

Developing by integrals on $\mathbb{R}$, we easily find the convolution product of the distribution functions :

$$R_y(M,M',\tau) = E\left(\int_{\Sigma_o x\Sigma_o} \underset{\sim}{h}(M,M_o,-t)\ \underset{t}{*}\ \underset{\sim}{h}(M',M_o',t)*R_p(M_o,M_o',\tau)d\Sigma_o\ d\Sigma_o'\right).$$

Taking the Fourier transform of the two members, and taking account of the fact that the Fourier transform of $h(-t)$, is the conjugate of the transform of $h(t)$, we obtain :

$$\mathcal{S}_y(M,M',f) = \int_{\Sigma_o x\Sigma_o} \overline{\underset{\sim}{H}(M,M_o,f)}\cdot\underset{\sim}{H}(M',M_o',f)\cdot\mathcal{S}_p(M_o,M_o',f)d\Sigma_o\ d\Sigma_o',$$

with $\omega = 2\pi f$. We need then only merge M and M', and replace H by its value (4.34). Noting that modes $S_i(M)$ are real scalars in the direction initially selected, the direct power spectrum density is then written :

$$\mathfrak{I}_y(M,M,f) = \sum_{ij} \underset{\sim i}{S}_i(M)\underset{\sim j}{S}_j(M)\ \overline{{}^iA_i(f)}\ {}^jA_j(f)\ \times$$

$$\int_{\Sigma_o \times \Sigma_o} \underset{\sim i}{S}_i(M_o)\underset{\sim j}{S}_j(M_o')\ \mathfrak{I}_p(M_o,M_o',f)\,d\Sigma_o\,d\Sigma_o' \ .$$

This formula can also be written, with i, j = 1, 2, ..., n :

$$\mathfrak{I}_y(M,f) = \sum_i |{}^iA_i(f)|^2 \left[\underset{\sim i}{S}_i(M)\right]^2 Q_{ii}(f)$$

$$+ \sum_{i \neq j} \overline{{}^iA_i(f)}\ {}^jA_j(f)\ \underset{\sim i}{S}_i(M)\underset{\sim j}{S}_j(M)Q_{ij}(f) \ .$$

The second sum of the right hand side member gives the interaction of the different modes with each other. If these modes are sufficiently far apart in eigenfrequency, terms ${}^iA_i$ and ${}^jA_j$ are small and the interaction is low. The corresponding terms can be neglected. The situation is not the same if the frequencies are close, and their presence then severely complicates calculations. At all events, these methods still require knowledge or estimation of damping factors. Furthermore, for wide spectral band ramdom excitation processes, the modal frequency density, namely the number of modes per hertz, frequently becomes high and determination of the modes becomes inaccurate, or even impossible. Other non-modal, semi-empirical methods have recently been developed, making it possible to overcome these difficulties[22].

CHAPTER V

NON-LINEAR DEFORMATIONS - BUCKLING

1. FORMULATION FROM THE NATURAL STATE FOR A HYPERELASTIC MEDIUM
2. FORMULATION FROM A PRESTRESSED STATE FOR AN ARBITRARY MEDIUM
3. DISCRETIZATION
4. METHODS OF SOLUTION
5. STATIC BUCKLING

1. - FORMULATION FROM THE NATURAL STATE FOR A HYPERELASTIC MEDIUM

We will consider a continuous material medium, satisfying the general hypotheses of paragraph 1 of Chapter I. As a reminder, the definition of a natural state, as adopted in (1.37), is as follows :

A natural state is one with no external loads, from which we measure stresses and deformations, and from which we express or measure the constitutive law of the medium, namely the relation between stress and deformation.

Let Ω_o be the domain of E_3 occupied by the medium in the natural state, with boundary Σ_o. Let f_o be the volume density of given external loads, and F_o be the surface density of given external loads on a part Σ_{oF} of Σ_o. We are also given displacements U_d on the complementary part Σ_{oU} of Σ_{oF}.

In this paragraph, we shall assume the medium to be hyperelastic. Therefore by hypothesis, there exists a deformation volume density α. We shall also adopt the same definition for kinematically admissible displacements U, as in Chapter I.

In any deformed state, the medium occupies a domain Ω, of which point M, the transform of point M_o of Ω_o, is such that :

$$M = \underset{\sim}{M}(M_o, t)$$

The deformation D at M is given by (1.3) :

$$(5.1) \qquad D = \frac{1}{2}\left[\frac{\overline{\partial M}}{\partial M_o} \frac{\partial M}{\partial M_o} - 1_{E_3} \right] = \overline{D}.$$

We also assume that the deformation volume density α is approximated by a quadratic form of the deformation D (see Chapter I, paragraph 5.3), namely :

$$(5.2) \qquad \alpha = \frac{1}{2}\underset{\sim}{\alpha}(D)(D).$$

Putting then :

$$M = M_o + U$$

$$\frac{\partial M}{\partial M_o} = 1_{E_3} + \frac{\partial U}{\partial M_o} ,$$

we then obtain classically :

$$(5.3) \qquad D = \frac{1}{2}\left[\frac{\partial U}{\partial M_o} + \overline{\frac{\partial U}{\partial M_o}} + \overline{\frac{\partial U}{\partial M_o}}\,\frac{\partial U}{\partial M_o}\right] = \underset{\sim}{D}(U).$$

Taking a kinematically admissible virtual displacement field, with value δU at point M of the deformed state, and referred to Ω_o by reciprocal images, the total virtual deformation energy will be :

$$\delta w = \delta \int_{\Omega_o} \frac{1}{2}\,\underset{\sim}{\alpha}(D)\,(D)\,d\Omega_o.$$

$$(5.4) \qquad \begin{cases} \delta w = \displaystyle\int_{\Omega_o} \underset{\sim}{\alpha}(D)\,(\delta D)\,d\Omega_o \\[2em] \text{with :} \\[1em] \delta D = \dfrac{1}{2}\left[\dfrac{\partial \delta U}{\partial M_o} + \overline{\dfrac{\partial \delta U}{\partial M_o}} + \overline{\dfrac{\partial \delta U}{\partial M_o}}\,\dfrac{\partial U}{\partial M_o} + \overline{\dfrac{\partial U}{\partial M_o}}\,\dfrac{\partial \delta U}{\partial M_o}\right], \quad [\text{see } (A.35)]. \end{cases}$$

Retaining the non-linear terms, to take account of possible large displacements, we then find :

$$(5.5.) \qquad \begin{aligned} &\underset{\sim}{\alpha}(D)(\delta D) \\[1em] &= \frac{1}{4}\,\underset{\sim}{\alpha}\left(\frac{\partial U}{\partial M_o} + \overline{\frac{\partial U}{\partial M_o}} + \overline{\frac{\partial U}{\partial M_o}}\,\frac{\partial U}{\partial M_o}\right)\left\{\frac{\partial \delta U}{\partial M_o} + \overline{\frac{\partial \delta U}{\partial M_o}} + \overline{\frac{\partial \delta U}{\partial M_o}}\,\frac{\partial U}{\partial M_o} + \overline{\frac{\partial U}{\partial M_o}}\,\frac{\partial \delta U}{\partial M_o}\right) \\[1em] &= \frac{1}{4}\,\underset{\sim}{\alpha}\left(\frac{\partial U}{\partial M_o} + \overline{\frac{\partial U}{\partial M_o}}\right)\left(\frac{\partial \delta U}{\partial M_o} + \overline{\frac{\partial \delta U}{\partial M_o}}\right) + \frac{1}{4}\,\underset{\sim}{\alpha}\left(\frac{\partial U}{\partial M_o} + \overline{\frac{\partial U}{\partial M_o}}\right)\left(\overline{\frac{\partial \delta U}{\partial M_o}}\,\frac{\partial U}{\partial M_o} + \overline{\frac{\partial U}{\partial M_o}}\,\frac{\partial \delta U}{\partial M_o}\right) \\[1em] &\quad + \frac{1}{4}\,\underset{\sim}{\alpha}\left(\overline{\frac{\partial U}{\partial M_o}}\,\frac{\partial U}{\partial M_o}\right)\left(\frac{\partial \delta U}{\partial M_o} + \overline{\frac{\partial \delta U}{\partial M_o}}\right) + \frac{1}{4}\,\underset{\sim}{\alpha}\left(\overline{\frac{\partial U}{\partial M_o}}\,\frac{\partial U}{\partial M_o}\right)\left(\overline{\frac{\partial \delta U}{\partial M_o}}\,\frac{\partial U}{\partial M_o} + \overline{\frac{\partial U}{\partial M_o}}\,\frac{\partial \delta U}{\partial M_o}\right). \end{aligned}$$

In the right hand side member of this formula, the first term shows a linear term with respect to U. The second and third terms are quadratic, and the fourth is cubic with respect to U. This formula thus enables us to define a stress C', by duality, as already stated in Chapter I, the conjugate of the virtual deformation δD, and in the deformed state, such that :

$$(5.6) \qquad \underset{\sim}{\tilde{C}}' = \underset{\sim}{\alpha}(D) = \frac{1}{2}\underset{\sim}{\alpha}\left(\frac{\partial U}{\partial M_o} + \frac{\overline{\partial U}}{\partial M_o} + \frac{\overline{\partial U}}{\partial M_o}\frac{\partial U}{\partial M_o}\right)$$

where $\tilde{C}'$ is the transpose of C' in the Euclidian deformation space $\vec{E}_6$, with by definition [see (1.6)] :

$$\tilde{C}'D = T_r(C'D) \ , \quad C' = \overline{C'}$$

Thus :

$$(5.7) \qquad \delta w = \int_{\Omega_o} \underset{\sim}{\alpha}(D)\delta D d\Omega_o = \int_{\Omega_o} T_r(C'\delta D)d\Omega_o = \int_{\Omega_o} \tilde{C}'\delta D d\Omega_o$$

It should be remembered that C' is the Piola-Kirchhoff stress.

Since :

$$\delta D = \frac{1}{2}\delta\left[\frac{\overline{\partial M}}{\partial M_o}\frac{\partial M}{\partial M_o} - {}^1 E_3\right]$$

$$= \frac{1}{2}\left[\frac{\overline{\partial\delta M}}{\partial M_o}\frac{\partial M}{\partial M_o} + \frac{\overline{\partial M}}{\partial M_o}\frac{\partial\delta M}{\partial M_o}\right]$$

$$= \frac{1}{2}\left[\frac{\overline{\partial\delta U}}{\partial M_o}\frac{\partial M}{\partial M_o} + \frac{\overline{\partial M}}{\partial M_o}\frac{\partial\delta U}{\partial M_o}\right],$$

we obtain :

$$\delta w = \int_{\Omega_o} T_r\left(C'\frac{\overline{\partial M}}{\partial M_o}\frac{\partial\delta U}{\partial M_o}\right)d\Omega_o.$$

The principle of virtual work gives :

$$\int_{\Omega_o} T_r\left(C'\frac{\overline{\partial M}}{\partial M_o}\frac{\partial\delta U}{\partial M_o}\right)d\Omega_o - \int_{\Omega_o} \overline{f}_o\delta U d\Omega_o - \int_{\Sigma_{oF}} \overline{F}_o\delta U d\Sigma_o = 0.$$

$$\forall\delta U \ \text{K.A.}$$

This is the total Lagrangian formulation.

The local equilibrium equations can be written immediately :

$$(5.8) \quad \left[\begin{array}{l} \operatorname*{div}_{M_o} \left[C' \dfrac{\overline{\partial M}}{\partial M_o} \right] + \overline{f_o} = 0 \qquad \text{in } \Omega_o \\[2em] \dfrac{\overline{\partial M}}{\partial M_o} \overline{C'} \, n_o = F_o \qquad \text{on } \Sigma_{oF} \end{array} \right.$$

with $C' = \overline{C'}$, in the case of a non-polarized medium considered.

The stress $C' \dfrac{\overline{\partial M}}{\partial M_o}$ is not Hermitian.

We can easily find the functional spaces to which the various fields introduced must belong, for the quantities introduced to have a meaning, naturally without prejudice to the existence or uniqueness of the solution in these spaces.

2. – FORMULATION FROM A PRESTRESSED STATE FOR AN ARBITRARY MEDIUM

The natural state considered in paragraph 1 is none other than a conventional reference state, from which we make all subsequent measurements. It goes without saying that practical considerations limit the arbitrariness of a state of this type, to a substantial extent. Thus in a fabricated body, it is not normally possible to extract test-samples for the measurement of the constitutive law, without modifying the local state of the medium in the sample. We therefore make do with approximations, usually assuming that the reference state considered is a quasi-natural state, having the definition adopted above.

In the following, we shall approach the same question, but taking as our reference state, not a "natural" state but any other state, prestressed with respect to the former. In other words, the natural state still remains the general reference state for all measurements, but the formulation is made from another state where a prestress field exists.

We shall now call $\mathcal{E}_{oo}$ the natural state, and $\mathcal{E}_o$ the prestressed state. The prestressed field can be a self-stressed state with respect to $\mathcal{E}_{oo}$, with no external loads, or a state prestressed under the action of fields f_o and F_o, defined on Ω_o and Σ_{oF} respectively, corresponding to $\mathcal{E}_o$.

Let therefore $\overline{\Omega_{oo}}$ be the domain occupied by the medium in state $\mathcal{E}_{oo}$, and $\overline{\Omega_o}$ the domain which it occupies in state $\mathcal{E}_o$, with $\overline{\Omega}$ the domain which it occupies in a subsequent state $\mathcal{E}$ under the action of a force volume density f defined in Ω_o, and a force surface density F defined on Σ_{oF}.

Under these conditions, point $M_{oo} \in \Omega_{oo}$ goes to $M_o \in \Omega_o$ and then to $M \in \Omega$.

For example, let us consider the Cauchy stress C in state $\mathcal{E}$, (see Chapter I, paragraph 4), and the field δM of kinematically admissible displacements from this state (C and δM are defined on $\overline{\Omega}$).

The virtual deformation work is the opposite of the quantity :

$$\delta w = \int_{\Omega} T_r (C \frac{\partial \delta M}{\partial M}) \, d\Omega \, , \quad \text{with} \quad C = \overline{C}$$

Related to the reference state $\mathcal{E}_o$ (prestressed), it is written :

$$\delta w = \int_{\Omega_o} T_r (C \frac{\partial \delta M}{\partial M_o} \frac{\partial M_o}{\partial M}) \, \det (\frac{\partial M}{\partial M_o}) \, d\Omega_o .$$

In this formula, all quantities are defined on Ω_o, by direct images.

It is also written   [see (A.6)] :

$$(5.9) \qquad \delta w = \int_{\Omega_o} T_r (\frac{\partial M_o}{\partial M} C \frac{\partial \delta M}{\partial M_o}) \, \det (\frac{\partial M}{\partial M_o}) \, d\Omega_o .$$

In this formula, stress $\frac{\partial M_o}{\partial M} C \det (\frac{\partial M}{\partial M_o})$ called the Boussinesq stress or Piola-Lagrange stress, is not Hermitian (as an endomorphism of $\vec{E}_3$). Incidentally (5.9) is not symmetrical for M and δM.

It should also be noted that by application of the Stokes formula, (5.9) gives :

$$(5.10) \qquad \delta w = \int_{\Omega_o} T_r (\frac{\partial M_o}{\partial M} C \frac{\partial \delta M}{\partial M_o}) \, \det (\frac{\partial M}{\partial M_o}) \, d\Omega_o$$

$$= \int_{\Omega_o} - \operatorname*{div}_{M_o} \left[\frac{\partial M_o}{\partial M} C \det (\frac{\partial M}{\partial M_o}) \right] \delta M \, d\Omega_o + \int_{\Sigma_o} \overline{\delta M} \frac{\partial M_o}{\partial M} C \det (\frac{\partial M}{\partial M_o}) \, n_o d\Sigma_o ,$$

where n_o is the external unitary normal to Σ_o.

We can incidentally find again this Boussinesq stress by the following method :

Let n be the external unitary normal to Σ, in state $\mathcal{E}$, and n_o the external unit normal to Σ_o in state $\mathcal{E}_o$. $\forall$ The differential fields dM defined on Ω, the volume element $d\Omega$, such that :

$$d\Omega = \overline{dM}\, nd\Sigma = vol(d_1 M)(d_2 M)(dM)$$

is also written [see (A.28) and (A.2)] :

$$d\Omega = \overline{dM}\, nd\Sigma = \det\left(\frac{\partial M}{\partial M_o}\right) vol(d_1 M_o)(d_2 M_o)(dM_o)$$

$$= \det\left(\frac{\partial M}{\partial M_o}\right) d\Omega_o$$

$$= \det\left(\frac{\partial M}{\partial M_o}\right) \overline{dM_o}\, n_o\, d\Sigma_o$$

$$= \det\left(\frac{\partial M}{\partial M_o}\right) \cdot \overline{\frac{\partial M_o}{\partial M}}\, dM \;\; n_o d\Sigma_o, \forall dM.$$

Whence :

$$nd\Sigma = \det\left(\frac{\partial M}{\partial M_o}\right) \overline{\frac{\partial M_o}{\partial M}}\, n_o d\Sigma_o,$$

and :

$$Cnd\Sigma = \left[C\, \overline{\frac{\partial M_o}{\partial M}}\, \det\left(\frac{\partial M}{\partial M_o}\right) \right] n_o\, d\Sigma_o$$

$$= \left[\overline{\frac{\partial M_o}{\partial M}}\, C\, \det\left(\frac{\partial M}{\partial M_o}\right) \right] n_o\, d\Sigma_o.$$

Returning to formula (5.9), as :

$$\frac{\partial M}{\partial M_o}\, \frac{\partial M_o}{\partial M} = {}^1 E_3$$

we obtain :

$$\delta w = \int_{\Omega_o} T_r \left(\frac{\partial M_o}{\partial M} \, C \, \frac{\overline{\partial M_o}}{\partial M} \, \frac{\overline{\partial M}}{\partial M_o} \, \frac{\partial \delta M}{\partial M_o} \right) \det \left(\frac{\partial M}{\partial M_o} \right) \, d\Omega_o .$$

Whence :

$$(5.11) \qquad \begin{cases} \delta w = \displaystyle\int_{\Omega_o} T_r \left(C'' \, \frac{\overline{\partial M}}{\partial M_o} \, \frac{\partial \delta M}{\partial M_o} \right) d\Omega_o \\[2ex] \text{with} \\[2ex] C'' = \dfrac{\partial M_o}{\partial M} \, C \, \dfrac{\overline{\partial M_o}}{\partial M} \, \det \left(\dfrac{\partial M}{\partial M_o} \right) = \overline{C''} . \end{cases}$$

This is the Piola-Kirchhoff stress, which is Hermitian and the dual of the virtual deformation :

$$\delta D = \frac{1}{2} \left[\frac{\overline{\partial M}}{\partial M_o} \, \frac{\partial \delta M}{\partial M_o} + \frac{\overline{\partial \delta M}}{\partial M_o} \, \frac{\partial M}{\partial M_o} \right]$$

defined from state $\mathcal{E}$, and achieving equilibrium with the external loads in this state.

Furthermore, we easily find the relation between the stress C'' and the Piola-Kirchhoff stress C' previously defined. In fact :

$$\delta w = \int_{\Omega_o} T_r \left(C'' \, \frac{\overline{\partial M}}{\partial M_o} \, \frac{\partial \delta M}{\partial M_o} \right) d\Omega_o = \int_{\Omega_{oo}} T_r \left(C'' \frac{\partial M}{\partial M_{oo}} \, \frac{\overline{\partial M_{oo}}}{\partial M_o} \, \frac{\partial \delta M}{\partial M_{oo}} \, \frac{\overline{\partial M_{oo}}}{\partial M_o} \right.$$

$$\left. \times \det \left(\frac{\partial M_o}{\partial M_{oo}} \right) \, d\Omega_{oo} \right.$$

$$= \int_{\Omega_{oo}} T_r \left(\frac{\partial M_{oo}}{\partial M_o} \, C'' \, \frac{\overline{\partial M_{oo}}}{\partial M_o} \, \frac{\partial M}{\partial M_{oo}} \, \frac{\partial \delta M}{\partial M_{oo}} \right) \det \left(\frac{\partial M_o}{\partial M_{oo}} \right) \, d\Omega_{oo} .$$

Hence :

$$(5.12) \qquad \begin{cases} \delta w = \displaystyle\int_{\Omega_{oo}} T_r \left(C' \, \delta D \right) d\Omega_{oo} \\[2ex] \text{with} \\[2ex] C' = \dfrac{\partial M_{oo}}{\partial M_o} \, C'' \, \dfrac{\overline{\partial M_{oo}}}{\partial M_o} \, \det \left(\dfrac{\partial M_o}{\partial M_{oo}} \right) = \overline{C'} \end{cases}$$

Finally we can deduce from (5.11) the local equilibrium equations in the state $\mathcal{E}$, referred to the prestress state $\mathcal{E}_o$. We prefer to write the weak formulation for equilibrium, from the principle of virtual work :

$$\delta w - \delta \mathcal{T}_e = 0, \qquad \forall \delta M \text{ K.A.,}$$

where $\delta \mathcal{T}_e$ is the virtual work of the given external loads,

namely :

$$(5.13) \qquad \int_{\Omega_o} T_r (C'' \, \overline{\frac{\partial M}{\partial M_o}} \, \frac{\partial \delta M}{\partial M_o}) \, d\Omega_o \; - \int_{\Omega_o} \overline{f} \delta M \, d\Omega_o \; -$$

$$-\int_{\Sigma_{oF}} \overline{F} \delta M \, d\Sigma_o \; = \; 0, \; \forall \delta M \text{ K.A.,}$$

Let us put then :

$$M = M_o + U,$$

The unknown real displacement field U is defined on $\overline{\Omega_o}$, and classically we obtain :

$$\frac{\partial M}{\partial M_o} = {}^1 E_3 + \frac{\partial U}{\partial M_o}$$

and (5.13) is written :

$$(5.14) \qquad \int_{\Omega_o} T_r (C'' \left[{}^1 E_3 + \overline{\frac{\partial U}{\partial M_o}} \right] \frac{\partial \delta U}{\partial M_o}) \, d\Omega_o \; - \int_{\Omega_o} \overline{f} \, \delta U \, d\Omega_o \; - \int_{\Sigma_{oF}} \overline{F} \delta U \, d\Sigma_o \; = 0$$

$$\forall \delta U \quad \text{K.A.}$$

Let us put then :

$$\begin{cases} f = f_o + f_I \\[2ex] F = F_o + F_I. \end{cases}$$

This defines f and F, assuming the densities f and F defined in the state $\mathcal{E}_o$. For simplification's sake, we have assumed that F_o and F_1 are defined on the same set. Let us also put :

(5.15) $C'' = C_o'' + C_I''$, with $C_o'' = C_o$

where C_o is the Cauchy stress in the equilibrium state $\mathcal{E}_o$.

It will be remembered that C_o is the stress which balances the external actions f_o and F_o. It is clear that C_o'' is stress C_o to which C'' is reduced, when $f_1 = F_1 = U = 0$.

For equilibrium of state $\mathcal{E}_o$, we obviously have :

$$(5.16) \qquad \int_{\Omega_o} T_r(C_o \frac{\partial \delta U}{\partial M_o}) \, d\Omega_o - \int_{\Omega_o} \overline{f}_o \, \delta U d\Omega_o - \int_{\Sigma_{oF}} \overline{F}_o \, \delta U d\Sigma_o \; = \; 0 \; \forall \, \delta U \; K.A.$$

Taking account of (5.15) and (5.16), (5.14) gives :

$$(5.17) \qquad \left[\begin{array}{l} \int_{\Omega_o} T_r \left(\left[C_I'' + C_o \frac{\overline{\partial U}}{\partial M_o} + C_I'' \frac{\overline{\partial U}}{\partial M_o} \right] \frac{\partial \delta U}{\partial M_o} \right) \, d\Omega_o - \int_{\Omega_o} \overline{f_I} \delta U \, d\Omega_o \\[2em] \hspace{6em} - \int_{\Sigma_{oF}} \overline{F_I} \, \delta U \, d\Sigma_o = 0 \\[1.5em] \forall \, \delta U \; K.A. \end{array} \right.$$

This equation (5.17) has been established with no hypothesis relating to the constitutive law of the medium, nor to the amplitude of the displacements U. This is the updated Lagrangian formulation.

If the medium is hyperelastic, and has an elastic energy volume density α, we generally have :

$$\tilde{C}'' = \frac{\partial \alpha}{\partial D} \; .$$

If the deformations from state $\mathcal{E}_o$ remain small, with displacements U having any value, noting that these two hypotheses remain compatible, we can make a limited expansion of α with respect to D, from the deformation value D_o existing at each point M_o of the prestressed state $\mathcal{E}_o$, namely [see (A.38)] :

$$\alpha = \overset{\sim}{\alpha}(D_o) + \overset{\sim}{\alpha}'(D_o)(D_I) + \frac{1}{2}\overset{\sim}{\alpha}''(D_o)(D_I)(D_I) + 0(|D_I|^2),$$

$$\text{with } \frac{|0(|D_I|^2)|}{|D_I|^2} \to 0 \; \text{ when } \; |D_I| \to 0.$$

Limiting ourselves to the quadratic term, we find :

$$\tilde{C}'' = \underset{\sim}{\alpha}'(D_o) + \underset{\sim}{\alpha}''(D_o)(D_I) = \tilde{C}_o + \tilde{C}''_I$$

or :

$$\left[\begin{array}{l} \tilde{C}_o = \underset{\sim}{\alpha}'(D_o) \\[2ex] \tilde{C}''_I = \underset{\sim}{\alpha}''(D_o)(D_I) = \tilde{C}_I . \end{array}\right.$$

$\tilde{C}_o = \alpha'(D_o)$ is non-zero as $\mathcal{E}_o$ is no longer the natural state, and C''_I, which we now shall call C_I, is linear in D_I in this approximation.

Strictly in the case of a hyperelastic body, we have :

$$w = \int_{\Omega_{oo}} \alpha(D)d\Omega_{oo} = \int_{\Omega_o} \alpha(D) \left[\det\left(\frac{\partial M_o}{\partial M_{oo}}\right)\right]^{-1} d\Omega_o$$

where :

$$D = \frac{1}{2}\left[\frac{\partial M}{\partial M_{oo}}\frac{\partial M}{\partial M_{oo}} - {}^1E_3\right]$$

or also :

$$D = \frac{1}{2}\left[\frac{\partial M_o}{\partial M_{oo}}\frac{\partial M}{\partial M_o}\frac{\partial M}{\partial M_o}\frac{\partial M_o}{\partial M_{oo}} - {}^1E_3\right]$$

Let us put :

$$D_o = \frac{1}{2}\left[\frac{\partial M_o}{\partial M_{oo}}\frac{\partial M_o}{\partial M_{oo}} - {}^1E_3\right]$$

and

$$D_I = \frac{1}{2}\left[\frac{\partial M}{\partial M_o}\frac{\partial M}{\partial M_o} - {}^1E_3\right]$$

We obtain :

$$\frac{\partial M}{\partial M_o}\frac{\partial M}{\partial M_o} = 2D_I + {}^1E_3$$

NON-LINEAR DEFORMATIONS - BUCKLING

Whence :

$$D = \frac{1}{2}\left[\frac{\overline{\partial M}_o}{\partial M_{oo}}\left[2D_I + 1_{E_3}\right]\frac{\partial M_o}{\partial M_{oo}} - 1_{E_3}\right] = D_o + \frac{\overline{\partial M}_o}{\partial M_{oo}} D_I \frac{\partial M_o}{\partial M_{oo}} \cdot$$

Thus :

$$\alpha = \underset{\sim}{\alpha}'(D_o)\left(\frac{\overline{\partial M}_o}{\partial M_{oo}} D_I \frac{\partial M_o}{\partial M_{oo}}\right) + \frac{1}{2}\underset{\sim}{\alpha}''(D_o)\left(\frac{\overline{\partial M}_o}{\partial M_{oo}} D_I \frac{\partial M_o}{\partial M_{oo}}\right)\left(\frac{\overline{\partial M}_o}{\partial M_{oo}} D_I \frac{\partial M_o}{\partial M_{oo}}\right) +$$

$$+ \ldots \ldots$$

$$\frac{1}{2}\tilde{D}_I\, A\, D_I = \frac{1}{2}\underset{\sim}{\alpha}''(D_o)\left(\frac{\overline{\partial M}_o}{\partial M_{oo}} D_I \frac{\partial M_o}{\partial M_{oo}}\right)\left(\frac{\overline{\partial M}_o}{\partial M_{oo}} D_I \frac{\partial M_o}{\partial M_{oo}}\right)$$

If the medium is not hyperelastic, but simply elastic, from state $\mathcal{E}_o$ we have :

$$C_I = \underset{\sim}{C}_I(D_I) \qquad (\underset{\sim}{C}_I \text{ for example linear})$$

with :

$$D_I = \frac{1}{2}\left[\frac{\partial U}{\partial M_o} + \frac{\overline{\partial U}}{\partial M_o} + \frac{\overline{\partial U}}{\partial M_o}\frac{\partial U}{\partial M_o}\right] \cdot$$

And if the medium is linear and elastic from $\mathcal{E}_o$, the function $\underset{\sim}{C}_I$ is linear in D_I. Under these conditions, we can make conspicuous the following stress terms in (5.17) :

$$(5.18) \quad \begin{cases} \dfrac{1}{2}\underset{\sim}{C}_I\left(\dfrac{\partial U}{\partial M_o} + \dfrac{\overline{\partial U}}{\partial M_o}\right), & \text{linear in U} \\[2ex] \dfrac{1}{2}\underset{\sim}{C}_I\left(\dfrac{\overline{\partial U}}{\partial M_o}\dfrac{\partial U}{\partial M_o}\right), & \text{quadratic in U} \\[2ex] \dfrac{1}{2}\underset{\sim}{C}_I\left(\dfrac{\partial U}{\partial M_o} + \dfrac{\overline{\partial U}}{\partial M_o}\right)\dfrac{\overline{\partial U}}{\partial M_o}, & \text{quadratic in U} \\[2ex] \dfrac{1}{2}\underset{\sim}{C}_I\left(\dfrac{\overline{\partial U}}{\partial M_o}\dfrac{\partial U}{\partial M_o}\right)\dfrac{\overline{\partial U}}{\partial M_o}, & \text{cubic in U.} \end{cases}$$

The term $C_o\dfrac{\overline{\partial U}}{\partial M_o}$ still remains linear in U and C_o, irrespective of the behaviour of the medium fron the state $\mathcal{E}_o$. In addition, we find terms of the same order as in paragraph 1, but naturally different.

It is easy to proceed to various orders of approximation from

(5.17) and (5.18)

An interesting case is that of a pressurized tank, where :

$$F = p\, n \qquad \text{on } \Sigma ,$$

p is a pressure in the final state $\mathcal{E}$. This term is also linearized [see (2.20) and (A.91) extended on $U \in \vec{E}_3$] :

$$\left[\begin{array}{l} p\, n\, d\Sigma = p\left[n_o - \dfrac{\overline{\partial U}}{\partial M_o}\, n_o\right]\left[1 + \widehat{\underset{\Sigma_o}{\operatorname{div}}\, U}\right] d\Sigma_o \\[2em] p\, n\, d\Sigma = P_o\, n_o d\Sigma_o + P_I\, n_o d\Sigma_o \\[2em] \text{with } P_I n_o = - P_o \dfrac{\overline{\partial U}}{\partial M_o}\, n_o + P_o\, \widehat{\underset{\Sigma_o}{\operatorname{div}}\, U}\, n_o . \end{array} \right.$$

The pressure p_I is therefore linear in U, and gives a first order stiffness variation, due to the rotation of the unit normal n_o, in passing from state $\mathcal{E}_o$ to state $\mathcal{E}$, and to the extension of the boundary surface Σ_o [*)]

Finally we note that (5.17) takes us back to the formulation from the natural state, making :

$$C_o = f_o = F_o = 0$$

and replacing mutatis mutandis Ω_o by Ω_{oo}.

The formulation (5.17) is important, in particular in applications relating to static and dynamic buckling. In the latter case, it supplies the weak equation for the perturbed system directly, provided the inertial terms are added.

3. – DISCRETIZATION[9-14]

Discretization of a non-linear problem is achieved by the usual discretization methods of any weak formulation, in particular by the finite element

---

*) Expressing p as a function of its value in state $\mathcal{E}_o$, it can be demonstrated that $p_I\, \overline{n}_o\, \delta U$ derives from a quadratic form[48].

method. We shall consider successively the two approaches forming the
subject of the two preceding paragraphs, limiting ourselves to hyperelastic
media, but we know that extension to other media is relatively easy.

In the formulation from the natural state, the discretization of the
virtual deformation energy :

$$\delta w = \int_{\Omega_o} \underset{\sim}{\alpha}(D)\,(\delta D)\,d\Omega_o,$$

where α is replaced by (5.5), gives, after assembly of the elements :

$$(5.19) \qquad \delta w = \overline{\delta q}\; H(q) \;, \qquad q \in \mathbb{R}^N$$

where q is the column of the N degrees of freedom of the discretized structure, representing the unknown field U, and where H(q) is a column of R^N
of the third degree in q, as shown by (5.5). Thus H(q) is decomposed into
a sum of 3 columns, such that :

$$(5.20) \qquad H(q) = K_1 q + H_2(q) + H_3(q) = \frac{\overline{\partial w}}{\partial q}.$$

K_1 is a constant matrix. This is the usual stiffness matrix for the
linearized problem (see chapter II), obtained by neglecting non-linear
deformations.

$H_2(q)$ and $H_3(q)$ are homogeneous columns of degree 2 and degree 3 respectively, in q.

The discretization of the virtual work of the external forces gives a
quantity :

$$\overline{\delta q}\; Q, \qquad Q \in \mathbb{R}^N,$$

where Q is a column of $\mathbb{R}^N$, representing the generalized external load
corresponding to the generalized displacement q.

The discretized equilibrium equation for a non-linear static problem is
then written :

$$(5.21) \qquad \begin{bmatrix} H(q) \;=\; Q\,. \\ \text{with } (5.20). \end{bmatrix}$$

In the second approach, where the formulation is made from a prestressed state, the discretization of equation (5.17), taking account of (5.18) gives for the deformation energy :

$$(5.22) \qquad \delta w = \delta q L(q) \quad , \quad q \in \mathbb{R}^N ,$$

where q is again the column of the N degrees of freedom of the discretized problem, but here representing the unknown displacement field U from the prestressed state $\mathcal{E}_o$, and where L(q) is a column of $\mathbb{R}^N$ which can be decomposed, in accordance with (5.17) and (5.18), into a sum of four terms :

$$(5.23) \qquad L(q) = K_1' q + K_2' q + L_2(q) + L_3(q) = \overline{\frac{\partial w}{\partial q}} .$$

K_1' and K_2' are the matrices of $\mathbb{R}^N$, such that :

$$K_1' q \text{ represents the term obtained from } \frac{1}{2} \underset{\sim}{C}_I \left(\overline{\frac{\partial U}{\partial M}}_o + \overline{\frac{\partial U}{\partial M}}_o \right)$$

$$K_2' q \text{ represents the term obtained from } C_o \overline{\frac{\partial U}{\partial M}}_o .$$

K_1' is again, and very generally (linear elasticity from the natural state), the matrix of elastic stiffness, merging with preceding matrix K_1.

K_2' is a stiffness matrix which depends linearly on the prestressed field C_o in state $\mathcal{E}_o$. Once this prestress is determined and known, the matrix K_2' no longer depends on the constitutive law of the medium, and for this reason is referred to as the <u>geometrical stiffness matrix</u>. The prestress C_o must be determined by prior calculation, non-linear where appropriate if the calculation problem of the state $\mathcal{E}_o$ is itself non-linear[17]. It should be noted that the stiffness K_2' can be substantially larger than the natural stiffness K_1'. This is the case in particular of the bending stiffness of inflated structures or tensioned cables.

Columns $L_2(q)$ and $L_3(q)$ are respectively second order and third order homogeneous columns in q, obtained from the quadratic and cubic terms of (5.18).

We then come to the case of the discretized equation for a static problem, in the form :

$$(5.24) \qquad \begin{bmatrix} L(q) = Q' \\ \text{with (5.23).} \end{bmatrix}$$

In the light of the preceding remarks, equations (5.21) and (5.24) can be written, respectively :

$$(5.25) \qquad K_1 q + K_2(q)(q) + K_3(q)(q)(q) = Q$$

$$(5.26) \qquad K_1' q + K_2' q + L_2(q)(q) + L_3(q)(q)(q) = Q'$$

where K_1, K_1' and K_2' are constant matrices of $\mathbb{R}^N$, and where the forms K_2 and L_2 are quadratic in q, with K_3 and L_3 cubic in q.

Resolution of equation (5.25), corresponding to the first approach, is preferred in particular in the case of linear elasticity problems, as it implies only one discretization of the medium in the natural state. More-over, the components of terms K_2 and K_3 of $\mathbb{R}^N$, appearing in (5.25), can be stored once and for all.

Equation (5.26) corresponds to the second approach, and can be used in a step by step analysis, discretizing the medium at each step, in each successive state. This procedure is relatively cumbersome, and can only be considered for a reasonably limited number of steps. It is applicable in particular for non-elastic problems, using highly automated programmes. To avoid this constraint, we know that we can retranslate equation (5.26) from natural state $\mathcal{E}_{oo}$, or at each step from state $\mathcal{E}_o$, without subsequent different discretization, in a total Lagrangian formulation.

Case of an elastoplastic body[10-16] or non-linear elastic body

In paragraph 2.7 we already gave certain details concerning the analysis of structures subjected to elastoplastic behaviour. At this point, we will add supplementary clarification concerning the discretization of a structu-re of this type (or element of a structure, if only part of the latter is liable to work in the plastic domain).

We have seen that a step by step method, with iterations in each step, is well suited for an incremental hypoelastic behaviour of the type [see (2.24)] :

$$(5.27) \qquad d\sigma = \left[\mathcal{A} - \alpha \mathcal{A}^*(\sigma) \right] d\varepsilon, \quad d\sigma, \ d\varepsilon \in \mathbb{R}^6 .$$

We then proceed as follows[4,9] : we assume that the increments of the displacement field are sufficiently small to linearize the stress and the deformation, in each finite element e.

Let S be a local reference system, particular to element e, with C and D as the stress and deformation in Ω_e, represented respectively by matrices σ and ε for the step considered.

To simplify, we will assume reference system S to be unitary. To establish our ideas, if the elements are three-dimensional we have :

$$\overline{S}S = 1_{R^N}$$

$$C = S\sigma\overline{S}, \quad D = S\varepsilon\overline{S}.$$

For a displacement dU, or a virtual displacement δU, [see (A.35)] :

$$\begin{bmatrix} dD = dS\varepsilon\overline{S} + Sd\varepsilon\overline{S} + S\varepsilon\overline{dS} = D_I \quad [cf. \ (5.17)] \\[2mm] \delta D = \quad S\varepsilon\overline{S} + S\delta\varepsilon\overline{S} + S\varepsilon\overline{\delta S} \\[2mm] dC = dS\sigma S + Sd\sigma\overline{S} + S\sigma\overline{\delta S} = C_I \quad [cf. \ (5.17)]. \end{bmatrix}$$

The increment of the virtual volume density of deformation energy is then written, in state $\mathcal{E}_o$:

$$(5.28) \qquad T_r(dC\delta D) = T_r(C_I \delta D)$$

$$= T_r\left(\left[dS\sigma\overline{S} + Sd\sigma\overline{S} + S\sigma\overline{dS} \right] \left[\delta S\varepsilon\overline{S} + S\delta\varepsilon\overline{S} + S\varepsilon\overline{\delta S} \right] \right).$$

Now matrices $\overline{S}dS$ and $\overline{S}\delta S$ are skew-symmetrical, and there exist two columns $d\theta$ and $\delta\theta$ of $\mathbb{R}^3$, representing the real and virtual rotations of S in the reference system S, such that [see (A.20)] :

$$(5.29) \qquad \begin{bmatrix} \overline{S}dS = i(d\theta) \\[2mm] \overline{S}\delta S = i(\delta\theta). \end{bmatrix}$$

(5.28) is now written, with (5.29) :

$$(5.30) \qquad T_r(dC\delta D) = T_r\left(d\sigma\delta\varepsilon - 2\ i(\delta\theta)d\sigma\varepsilon + 2\ i(d\theta)\sigma\delta\varepsilon + 2\ i(d\theta)\ \sigma i(\delta\theta)\varepsilon \right.$$

$$\left. - 2\ \varepsilon i(\delta\theta)i(d\theta)\sigma \right).$$

If we call $^e q$ the column of the nodal coordinate of element e in S, and

$d^e q$ its increment corresponding to dU, with $^e q$ its virtual increment, we obtain a matrix ℓ such that :

$$d\theta = \ell \, d^e q$$
$$\delta\theta = \ell \, \delta^e q.$$

(5.30) is then written in the form of a sum of terms, which can be written :

$$\overline{\delta^e q} \quad {}^e k_e \, d^e q$$

using law (5.27), and with σ and ε updated.

We then proceed to the representation of $^e q$ in a reference system common to all the elements, by a matrix $^e S'$, updated at each step :

$$d^e q = {}^e S' dq' \ , \quad \delta^e q = {}^e S' \delta q' ,$$

then the assembly is executed in the usual way (Chapter II, paragraph 1.1).

The generalized external loadings are calculated without difficulty, and each incremental solution is found by an iteration (see paragraph 2.7). This method can also be used profitably in the case of an elastic structure with linear or non-linear elasticity, provided account is taken of the prestress terms at each step, for large displacements.

4. – <u>METHODS OF SOLUTION</u>

Various methods can be used to solve equation (5.21) (or (5.24)), in other words to find the static equilibrium solution for the discretized non-linear elastic problem, in which the generalized external load Q (or Q') is given. For example[9], we can quote the minimization, fixed point, Newton-Raphson, simplex, step by step, and evolution methods, etc.

The accuracy of these methods naturally depends on the degree of fineness of the discretization and on the algorithms used, but also on the degree of non-linearity of the problem, and on the amplitude of the solution. When this amplitude is proved to be large, it is always preferable to use an incremental process, during which the external load is introduced step by step, executing the iterations particular to the method selected, at each step.

It should be remembered that the minimization methods consist in finding the solution q which makes the total energy $w = \overline{Q}q$ minimum, or at least stationary where there is a possibility of instability, either by the gradient or conjugate gradient method.

More generally, as the non-linear equation to be solved is written in the form :

$$(5.31) \qquad \underset{\sim}{f}(q) = 0, \quad f \in \mathbb{R}^N$$

we can also find q which satisfies

$$\inf \left[\overline{\underset{\sim}{f}(q)} \; \underset{\sim}{f}(q) \right],$$

without recourse to the notion of energy. The minimum is then zero. The fixed point method applied to (5.31), which is then written :

$$(5.32) \qquad \left[1_{\mathbb{R}^N} + \underset{\sim}{f} \right] (q) = q,$$

requires application $1_{\mathbb{R}^N} + \underset{\sim}{f}$ to be contractive (with $\mathbb{R}^N$ obviously complete). Applied to (5.25), (5.32) can be written :

$$q = K_1^{-1} \left[Q - K_2(q)(q) - K_3(q)(q)(q) \right].$$

In this form, and associated with an incremental process, this method is recommended by certain writers[18] as being more efficient than the Newton-Raphson method, and its convergence has been studied[19]. The Newton-Raphson method is in widespread use, and is an iterative method consisting in solving equation (5.31) from an approximate solution q_o, corrected by an unknown value Δq.

The equation :

$$\underset{\sim}{f}(q_o + \Delta q) = 0$$

is replaced by the approximate equation :

$$\underset{\sim}{f}(q_o) + \underset{\sim}{f}'(q_o) \, \Delta q = 0, \text{ which gives :}$$

$$\Delta q = - \left[\underset{\sim}{f}'(q_o) \right]^{-1} \underset{\sim}{f}(q_o), \text{ and a corrected value :}$$

$$q_1 = q_o + \Delta q.$$

We then look for a new value Δq, such that :

$$\underset{\sim}{f}(q_1 + \Delta q) = 0$$

and

$$\Delta q = - \left[\underset{\sim}{f}{}'(q_1)\right]^{-1} \underset{\sim}{f}(q_1)$$

and so on.

This method requires the calculation of the "gradient" $\underset{\sim}{f}{}'(q_i)$ on each iteration, which is a costly operation. A classical variant consists in using the same value for "gradient" $\underset{\sim}{f}{}'(q_i)$, during a certain number of iteration steps, which avoids this drawback but increases the number of steps (modified Newton-Raphson method). The method may not achieve convergence if the initial value q_o is badly chosen. Anyhow, it is represented as a particular case of the fixed point method. In fact (5.31) is also written:

$$a(q)\ \underset{\sim}{f}(q) = 0,$$

where $a(q)$ is a regular application, as we are looking for :

$$q = q + \Delta q, \quad \text{quand} \quad q \to 0$$

$$= q + a(q)\ \underset{\sim}{f}(q)$$

with, here :

$$\Delta q = a(q)\ f(q) = - \left[\underset{\sim}{f}{}'(q)\right]^{-1} \underset{\sim}{f}(q),$$

whence the theorem of convergence.

The Newton-Raphson method is generally associated with a step by step method, and applied at each step. It can also be elegantly coupled with a Ritz method[50].

Methods obtained from the first approach, with reference to the natural state, are pure Lagrangian methods. Those from the second approach can pass as "Eulerian" methods. Strictly, a formulation of this type would consist in calculating a structure, the matter of which would "flow" through the interfaces of the elements, the latter remaining fixed. However, in formulation from a prestress state, we assume implicitly that the modelling is in a way updated[12] (updated Lagrangian method).

It should also be remembered that "evolution" methods in static problems, consist in replacing equation (5.31) by an evolution equation, for example :

$$\dot{q} + c\, \underset{\sim}{f}(q) = 0,$$

where c is a judiciously selected diagonal "damping" matrix.

The method consists in finding a stationary solution for this equation. Certain writers have applied methods of this type with success, as they apparently give calculations which are not expensive.

This leads us to say a few words on the subject of dynamic problems.

<u>Dynamic problems</u> : Let us take the case of a dynamic problem in linear elasticity and non-linear deformations. The problem is presented in the following discretized form :

$$(5.33) \qquad M\, \underset{\sim}{\ddot{q}}(t) + B(\underset{\sim}{\dot{q}}(t)) + H(\underset{\sim}{q}(t)) = \underset{\sim}{Q}(t),$$

where H and B are generally non-linear.

Interpolation methods are of the step by step type, the time interval being decomposed into a certain number of intervals $\Delta t = h$.

Numerous <u>calculation schemes</u> have been proposed. Certain are considered as optimum, and are incorporated in available computation programmes[20,21].

These schemes are compared, from the point of view of <u>accuracy</u> and <u>stability</u>, using the example of a harmonic oscillator with one degree of freedom. The distinction is made between <u>implicit</u> and <u>explicit</u> methods. Generally, if we write an integration scheme for a differential equation of the variable $\underset{\sim}{y}(t)$, in the form :

$$y_{n+1} = y_n + h\underset{\sim}{\phi}\,(y_{n-p}, \ldots, y_n, y_{n+1}, h)$$

where y_n is the value of y at step n, the scheme is said to be explicit if y_{n+1} does not appear in ϕ, or implicit in the contrary case. The first non-zero term of the Taylor expansion of $y_{n+1} - y_n - h\phi$ gives the order of the scheme.

Explicit schemes are in widespread use for reasons of simplicity[22], and even accuracy, an implicit scheme only giving the value of y_{n+1} after solution of an equation.

The _stability_ of a scheme, linked to the general notion of the stability of a system, is defined by the fact that the solution at step n, is continuous with respect to a perturbation of initial data, and numerical perturbations due to calculation of the preceding steps.

Stability can be _conditional_ or _unconditional_, according to whether it depends on the dimension of the step or not. Conditional stability is generally obtained for a sufficiently small step value, which can increase the computation costs. Conditionally stable schemes are preferred for this reason, for calculation of wave propagation where small steps are generally necessary.

Calculation schemes are most often based on a hypothesis of interpolation of the acceleration inside a step, or on a time finite difference method. These methods are frequently initialized by different methods – For example, we present three methods among those in most widespread use.

Wilson and Clough method[23]

In equation (5.33), $\ddot{\underset{\sim}{q}}(t)$ is interpolated linearly in time interval h, giving :

$$\ddot{\underset{\sim}{q}}(t) = \ddot{q}_{n-1} + \left[\ddot{q}_n - \ddot{q}_{n-1}\right]\frac{t}{h}, \qquad 0 \leqslant t \leqslant h$$

$$(5.34) \qquad \dot{q}_n = \int_{[n-1]h}^{nh} \ddot{q}\,dt = \dot{q}_{n-1} + \left[\ddot{q}_{n-1} + \ddot{q}_n\right]\frac{h}{2}$$

$$(5.35) \qquad q_n = \int_{[n-1]h}^{nh} \dot{q}\,dt = q_{n-1} + h\dot{q}_{n-1} + \frac{h^2}{3}\ddot{q}_{n-1} + \frac{h^2}{6}\ddot{q}_n .$$

(5.33) gives, with (5.34) and (5.35)

$$(5.36) \qquad \ddot{q}_n = R(q_{n-1}) + Q_n,$$

whence q_n by application of (5.34) and (5.35).

This method is conditionally stable, and its _Wilson-Farhoomand variant_[24] (θ-method) can be used to make it unconditionally stable. This is based on a linear extrapolation of the acceleration, in the form :

(5.37) $\ddot{q}_{n+1} = \left[1 - \dfrac{1}{\theta}\right] \ddot{q}_n + \dfrac{1}{\theta} \ddot{\underset{\sim}{q}}(t)$, $t_n \leqslant t \leqslant t_n + \theta h$.

We find :

$$(5.38)\quad \left[\begin{array}{l} \dot{\underset{\sim}{q}}(t) = \dot{q}_n + \dfrac{\theta h}{2}\left[\ddot{q}_n + \ddot{\underset{\sim}{q}}(t)\right] \\[2em] \underset{\sim}{q}(t) = q_n + \theta h\,\dot{q}_n + \dfrac{\theta^2 h^2}{3}\ddot{q}_n + \dfrac{\theta^2 h^2}{6}\ddot{\underset{\sim}{q}}(t) . \end{array} \right.$$

We write equilibrium at time $t = t_n + \theta h$.   (5.38) gives $\dot{\underset{\sim}{q}}(t)$ and $\ddot{\underset{\sim}{q}}(t)$ as a function of $\underset{\sim}{q}(t)$, and values for the preceding steps, which are placed in (5.33) to give an equation for $\underset{\sim}{q}(t)$, which is then solved. $\ddot{\underset{\sim}{q}}(t)$ is deduced by (5.38), and finally we calculate $\ddot{q}_{n+1}$ by (5.37). The values of q_{n+1} and $\dot{q}_{n+1}$, are then calculated from (5.38), for $\theta = 1$.   It can be demonstrated that the method is unconditionally stable if :

$$\theta \leqslant 1,37.$$

Newmark method[25] (or β-method)

From the Taylor series expansion, it is assumed that :

$$(5.39)\quad \left[\begin{array}{l} \dot{q}_{n+1} = \dot{q}_n + h\left[\left[1-\gamma\right]\ddot{q}_n + \gamma\,\ddot{q}_{n+1}\right] \\[2em] q_{n+1} = q_n + h\dot{q}_n + h^2\left[\left[\dfrac{1}{2} - \beta\right]\ddot{q}_n + \beta\,\ddot{q}_{n+1}\right] . \end{array} \right.$$

Optimum values are selected for the parameters β and γ, to obtain a scheme which is both stable and accurate. One demonstrates that the scheme is stable if the values adopted are :

$$\gamma = \dfrac{1}{2} , \quad \beta = \dfrac{1}{4} , \quad \beta \geqslant \dfrac{1}{4}\left[\dfrac{1}{2} + \gamma\right]^2 .$$

A generalization of the Wilson and Newmark schemes has been proposed by Argyris et al.[26], where the interpolation of the acceleration is cubic. This scheme is extremely accurate although uneconomic, and can serve as a reference in certain calculations.

Houbolt method[27]

This method uses a finite differences scheme (backward differences[28]) :

$$(5.40) \quad \begin{cases} \ddot{q}_{n+1} = \dfrac{1}{h^2} \left[2q_{n+1} - 5q_n + 4q_{n-1} - q_{n-2} \right] \\[2ex] \dot{q}_{n+1} = \dfrac{1}{6h} \left[11q_{n+1} - 18q_n + 9q_{n-1} - 2q_{n-2} \right]. \end{cases}$$

Putting (5.40) in (5.33), we obtain a non-linear equation for q_{n+1}, on which the Newton-Raphson method can be used. This is an unconditionally stable method, but one which presents a high degree of numerical damping, which can be of interest in damping high frequency vibrations where the latter are considered non important. A variant[29] uses a linear extrapolation of the non-linear part of (5.33), considered as the second member. The advantage lies in a high degree of simplification of the method, and compensation of the numerical damping already mentioned. Furthermore, the initialization must be executed by another routine.

All these methods are relatively costly, and it is preferable in problems of free or forced vibration, where the spectrum is relatively narrow, to use dynamic reduction methods of the modal type (see Chapter IV), and in certain cases, using substructuring procedures, to operate on the degrees of freedom of the non-linear substructures, reducing the other. However, it should be pointed out that modal methods are extremely inaccurate in cases where the excitations is highly localized, and poorly represented in a truncated modal base, and when its frequency spectrum is wide.

5. - STATIC BUCKLING

5.1 - General [30-32]

We will start with a qualitative description of static buckling. Buckling, or the unstable state of a structure, occurs when at least one dimension is small with respect to the others, and when external loads give rise to stresses which work in the direction of the large dimensions. The structure is deformed in the direction of the large dimensions, which we will call u, and not in the others, up to the point of buckling. At the point of buckling, displacements appear in the direction of the small dimensions, which we will call w.

The stress field before buckling can also be due to reaction forces, resulting from imposed displacement fields, or thermal fields associated with certain boundary conditions.

Buckling instabilities can be global, namely concerning the structure in its entirety, or local, namely confined to sometimes small regions of the structure. In this case, local and global buckling conditions can be uncoupled, but intermediate cases can give more or less pronounced couplings[49].

According to the particular case, the critical point of buckling, in the space of the states, is either a bifurcation point, or a limit point (collapse, snap).

Bifurcation points are characterized by the possibility of the existence, after the critical point, of several distinct solutions. The stable post-buckling configuration, if it exists, generally corresponds to a state of minimum energy.

In the classic cases of beams, plates or shells, bifurcation points are illustrated by the following figures :

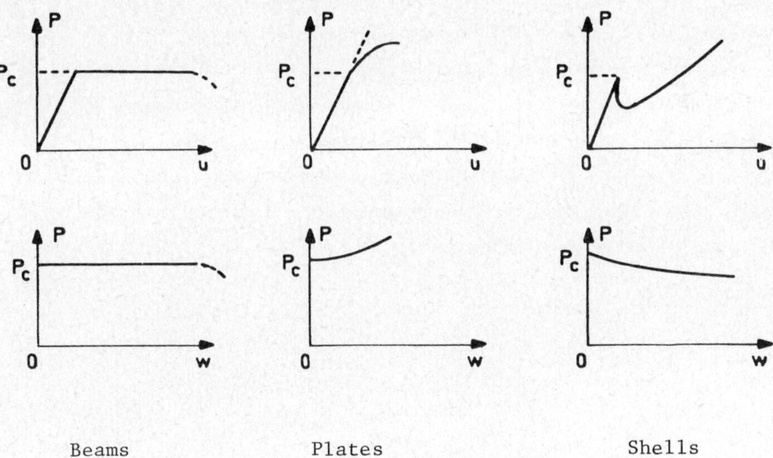

Beams   Plates   Shells

In numerous cases, as soon as the buckling load P_c is reached, displacements w occur and the deformation energy, due up to the point only the displacement u (traction and compression) decreases, while the energy due to displacements w increases (the compression stiffness decreases). In certain cases, the postbuckling equilibrium load P goes below the critical load P_c.

If load P is maintained constant and equal to P_c, the difference between the
equilibrium load and the applied critical load increases in absolute value
terms, and the phenomenon becomes explosive. This is the case for example
with beams and shells.

In the case of plates, the slope of trajectory (u,P) can remain positive
after the critical point, as a result of edge conditions which can produce an
increase in stresses in direction u, so that total deformation energy again
increases. This is also the case with curved panels. The post-buckling
equilibrium load is then greater than the critical load, again giving
stable configurations (case of membranes under shear stress, for example).

Analysis of the post-buckling conditions is therefore of prime importance,
from the fact that post-buckling configurations can provide stable equilibrium, enabling a structure to bear loads over critical load values, in
particular with reference to local buckling, without excessive deformation.

Nevertheless, this analysis is extremely delicate, and has been the subject of extensive work since the famous study by Koiter (1945)[33].

The aerospace industry, due to its increasing requirements for improved
stiffness/weight ratios, has been a permanent source of stimulation for
researchers in the study and forecasting of this phenomenon, and for design
departments in research aiming the rearrangement of elements, and the combination of stiffeners, to take advantage thereof[34], despite the fact that
"plates supported at their edges were able to sustain loads far in excess
of buckling load was considered with suspicion and did not upset the belief
that the loads should be kept below the critical load". (Van der Neut).

Furthermore, the phenomenon is greatly complicated, up to the point of
changing its nature, by the inevitable presence of imperfections. There
are so-called differences which exist between nominal data and actual data.
These data are first of a geometric nature : in this case, actual execution
may vary slightly with respect to the drawing, mean surfaces not being
strictly plane, cylindrical or other, thicknesses strictly constant, symmetries rigorously observed, etc. These data also concern loadings which are
frequently poorly known, and can present differences with respect to expected directions or points of application, etc. Finally the data are of a
physical nature : thus unforeseen local heterogeneities (weaknesses, cracks,

etc.) can exist, as also in the mechanical characteristics. Differences of a mixed type, existing between nominal boundary conditions and actual conditions (imperfect clamping, supports with friction, etc.), are also added to the preceding list.

In his fundamental study, Koiter applied his method of perturbation to the case of imperfections. This method was used, with variants, by other writers, and was confirmed by remarkable experimental works[35].

To summarize, the existence of bifurcations is subject to very precise conditions, as we shall see, according to which the generalized loading must be orthogonal, in the meaning of the selected metric tensor, to a certain critical configuration. These conditions are favoured by strict symmetries. Any difference, even slight, with respect to these conditions excludes the existence of "bifurcation points", transforming them to another buckling configuration, referred to as a "limit point". This phenomenon is illustrated in classic cases by the following figures :

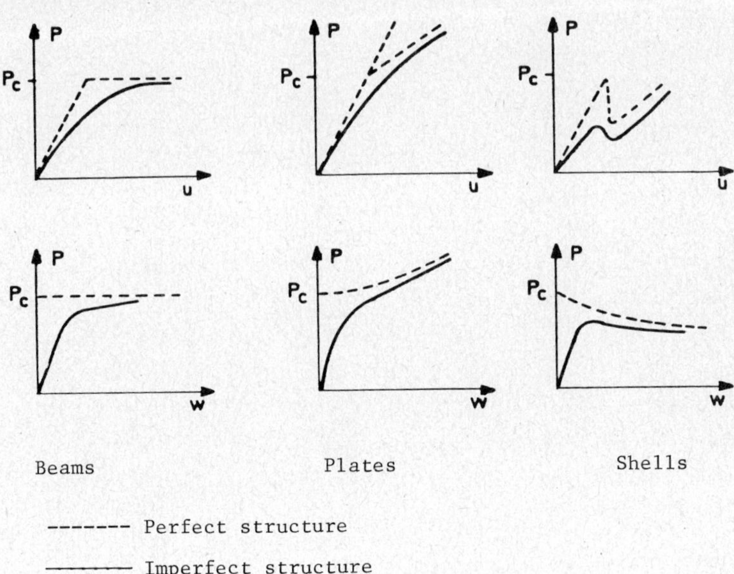

------- Perfect structure

——— Imperfect structure

These figures demonstrate that the critical load can be reduced or even eliminated, by the presence of imperfections, with considerable practical consequences.

5.2 - The criterion of static stability[36]

The analysis of buckling problems is essentially governed by the study
and selection of stability criteria. We will limit ourselves in the following pages to giving certain details concerning the most frequently used
static buckling criterion, the energy criterion.

In 1788, Lagrange put forward a criterion for conservative discrete systems , later designated the "energy criterion", according to which the
necessary and sufficient condition for a system of this type to be in a
stable state of equilibrium, is that its total potential energy presents a
local minimum for this state. If this energy is maximum, the equilibrium is
unstable. Based on linearization, and therefore relating to infinitely small
movements, this criterion was criticized by Dirichlet (1846) (as the asymptotic expansions anticipate stability), who extended it to bounded perturbations.

The most important generalization of this criterion, as far as we are
concerned, was the following extension to elastic bodies : the necessary
and sufficient conditions for an elastic body to be in a stable state of
equilibrium, under conservation loading and isothermal conditions, is that
its total potential energy has a weak relative minimum for the kinematically
admissible virtual displacements.

Even limited to these express conditions, the Lagrange-Dirichlet criterion is still the subject of numerous criticism, as in fact it is not based
on any precise definition of stability, and is deficient for continuous
systems (unless we admit the definition of Hadamard, which is adapted to
this criterion). A static test, limited to conservative loadings and kinematically admissible virtual displacements, it does not apply to dynamic
cases, non-conservative forces, real displacements or finite displacements
(in addition, it assumes a non-neutral state of equilibrium, namely a non
isolated state).

In 1892, Liapounov[37] gave a precise definition of stability, also presenting fundamental methods and results to which all subsequent works
referred. Under certain conditions, and according to Liapounov's definition
of stability, the energy criterion constitutes a necessary, but not sufficient condition for static continuous systems. Furthermore, even today,

there does not exist a set of stability conditions which are both necessary and sufficient.

Proposed by Koiter[33], and in the majority of practical applications, as a necessary and sufficient condition for continuous systems, the necessary and sufficient conditions for the energy V to attain a weak local minimum at equilibrium, can easily be translated for discrete systems by the condition :

$$\delta^2 \ V \geqslant 0,$$

where $\delta^2 \ V$ is the second variation of V for kinematically admissible virtual displacements δU. But this condition, written as Fréchet differentials, is not sufficient for continuous systems, as demonstrated by certain litigious cases[38].

In the following paragraphs, we shall adopt the energy criterion.

5.3 - Application of the energy criterion[33,38]

As in the preceding paragraphs, calling w the deformation energy in the state of equilibrium $\mathcal{E}$ studied, and U the actual displacement field of the system, assumed to be conservative, from reference state $\mathcal{E}_{00}$, if the system is in equilibrium under the action of a loading field f on Ω and on part Σ_F of boundary Σ of domain Ω occupied by the body, the minimum energy criterion gives[*], $\blacktriangledown$ U K.A. :

$$(5.41) \quad \begin{cases} \underset{\sim}{w}(U+\delta U) - \underset{\sim}{w}(U) - \overline{f}\delta U > 0 \Longleftrightarrow \text{stability of equilibrium in } \mathcal{E} \\[2mm] \underset{\sim}{w}(U+\delta U) - \underset{\sim}{w}(U) - f\delta U < 0 \Longleftrightarrow \text{instability of equilibrium in } \mathcal{E} \ . \end{cases}$$

The problem consists in replacing this local criterion by practical local criteria, which can be used in current applications, where we assume w to be continuous, and even of class C^m up to an order which is as high as necessary.

---

[*] The bar is used here to indicate transposition in the functional vector space.

A limited Taylor-McLaurin expansion of $(5.41)_1$ gives [see (A.38)]

$$(5.42) \quad \begin{cases} \underset{\sim}{w}(U + \delta U) - \underset{\sim}{w}(U) - \overline{f}\delta U = \underset{\sim}{w}'(U)(\delta U) - \overline{f}\delta U + \frac{1}{2}\underset{\sim}{w}''(U)(\delta U)(\delta U) + 0(|\delta U|^p) \\[2mm] \text{with} \\[2mm] \dfrac{|0(|\delta U|^p)|}{|\delta U|^2} \to 0 \text{ when } |\delta U| \to 0, \quad \forall \delta U \text{ K.A.} \end{cases}$$

Now by the principle of virtual work, the static equilibrium verifies
the equation :

$$\underset{\sim}{w}'(U) - \overline{f} = 0$$

(5.42) is then written :

$$\underset{\sim}{w}(U + \delta U) - \underset{\sim}{w}(U) - \overline{f}\delta U = \frac{1}{2}\underset{\sim}{w}''(U)(\delta U)(\delta U) + 0(|\delta U|^p) \quad \forall \delta U \text{ K.A.}$$

Then replacing δU by $\dfrac{\delta U}{|\delta U|}|\delta U|$, dividing by $|\delta U|^2$, and making $|\delta U|$ tend
towards zero, the stability criterion now becomes :

$$(5.43) \quad \begin{bmatrix} \dfrac{1}{2}\underset{\sim}{w}''(U) \left(\dfrac{\delta U}{|\delta U|}\right)\left(\dfrac{\delta U}{|\delta U|}\right) > 0 \\[3mm] \forall \delta U \text{ K.A.} \end{bmatrix} \Longleftrightarrow \begin{bmatrix} \text{stability of equilibrium in } \mathcal{E} \end{bmatrix}$$

If there exists a solution $U = U_o$, such that the system is stable in the
reference state $\mathcal{E}_{oo}$, remaining so up to solution U_o (in state $\mathcal{E}_o$), and
if :

$$(5.44) \quad \exists \dfrac{\delta U}{|\delta U|} \neq 0, \text{ K.A.} \quad \underset{\sim}{w}''(U_o)\left(\dfrac{\delta U}{|\delta U|}\right)\left(\dfrac{\delta U}{|\delta U|}\right) = 0 \Longleftrightarrow \mathcal{E}_o \text{ critical state,}$$

solution U_o corresponds to a <u>critical state</u> $\mathcal{E}_o$ or a neutral state of equilibrium. Under these conditions, (5.41) gives that stability of $\mathcal{E}_o \Longrightarrow$

$$(5.45) \quad \begin{bmatrix} \left[\underset{\sim}{w}(U+\delta U) - \underset{\sim}{w}(U) - \overline{f}\delta U = \dfrac{1}{3!}\underset{\sim}{w}'''(U)(\delta U)(\delta U)(\delta U)\right. \\[3mm] \left. + \dfrac{1}{4!}\underset{\sim}{w}^{(IV)}(U)(\delta U)(\delta U)(\delta U)(\delta U) + 0(|\delta U|^p) \geqslant 0, \right. \\[3mm] \text{with :} \\[3mm] \dfrac{|0(|\delta U|^p)|}{|\delta U|^4} \to 0 \text{ when } |\delta U| \to 0 \end{bmatrix}$$

If now in state $\mathcal{E}_o$:

$$(5.46) \quad \left[\begin{array}{l} \exists \delta U \neq 0 \text{ K.A.} \mid \underset{\sim}{w}''(U_o)(\delta U)(\delta U) = 0 \\[2ex] \underset{\sim}{w}'''(U_o)(\delta U)(\delta U)(\delta U) \neq 0, \end{array} \right.$$

putting $\delta U = \dfrac{\delta U}{|\delta U|} |\delta U|$, dividing by $|\delta U|^3$, and making $|\delta U|$ tend towards zero, we see that the system is unstable in $\mathcal{E}_o$, as condition (5.45) cannot be satisfied.

This is the most frequent case, in which the system passes from a stable state to an unstable state, via the critical state, and reciprocally.

On the other hand, if for solution U_o :

$$(5.47) \quad \left[\begin{array}{l} \exists \delta U \neq 0 \text{ K.A.} \mid \underset{\sim}{w}''(U_o)(\delta_1 U)(\delta_1 U) = 0, \\[2ex] \mid \underset{\sim}{w}'''(U_o)(\delta_1 U)(\delta_1 U)(\delta_1 U) = 0, \end{array} \right.$$

it is clear that :

$$\underset{\sim}{w}^{IV}(U_o)(\delta_1 U)(\delta_1 U)(\delta_1 U)(\delta_1 U) \geq 0$$

is a necessary condition for stability. It can be formed a finer necessary condition, which is also sufficient[33]. In fact we seek to achieve condition (5.41), $\forall \delta U = \delta_1 U + \delta_2 U$, where $\delta_2 U \in$ the class of the $\delta_2 U$ orthogonal to $\delta_1 U$, ($|\delta U|$ being sufficiently small).

Taking account of (5.47), the problem comes down to finding the local minimum for :

$$\frac{1}{2} \left[\underset{\sim}{w}''(U_o)(\delta_2 U)(\delta_2 U) + \underset{\sim}{w}'''(U_o)(\delta_1 U)(\delta_1 U)(\delta_2 U) + \right.$$
$$\left. + \underset{\sim}{w}'''(U_o)(\delta_1 U)(\delta_2 U)(\delta_2 U) \right] + \ldots .$$

A necessary condition is given for $\delta_2 U$, satisfying :

$$2\underset{\sim}{w}''(U_o)(\delta_2 U) + \underset{\sim}{w}'''(U_o)(\delta_1 U)(\delta_1 U) + 2\underset{\sim}{w}'''(U_o)(\delta_1 U)(\delta_2 U) = 0.$$

If we keep only the first two terms, we find a quadratic solution $\delta_2 U_m$ in $\delta_1 U$ which makes the last term negligible. This necessary condition then becomes, only keeping the terms of lowest degree in $\delta_1 U$:

$$\left[-\frac{1}{2} \underset{\sim}{w}''(U_o)(\delta_2 U_m)(\delta_2 U_m) + \frac{1}{4!} \underset{\sim}{w}^{IV}(U_o)(\delta_1 U)(\delta_1 U)(\delta_1 U)(\delta_1 U) \geqslant 0 \right.$$

$$\left. \text{where } \delta_2 U_m = -\frac{1}{2}\left[\underset{\sim}{w}''(U_o)\right]^{-1} \cdot \underset{\sim}{w}'''(U_o)(\delta_1 U)(\delta_1 U) \, , \right.$$

($\underset{\sim}{w}''(U)$ being invertible in the subspace orthogonal to $\delta_1 U$).

The sufficient condition for stability in U_o is obviously that the left hand side member be still positive definite.

We shall apply the preceding conditions to a discretized system ; it will result in practical methods for finding the critical states of a system, and in particular the critical buckling loads.

The first formulation written from the natural state $\mathcal{E}_{oo}$, led to equilibrium equation (5.21) :

(5.48) $H(q) = Q$ with   (5.20)

where $q \in \mathbb{R}^N$ represents the unknown displacement field U.

Under the same conditions, the critical state $\mathcal{E}_o$ corresponds to the equilibrium solution U_o, represented by column q_o, and the first member of (5.44), a quadratic form in δU, is written :

$$(5.49) \quad \left[\begin{array}{l} \delta^2 w = \left[\dfrac{\partial^2 w}{\partial U \partial U}\right]_o (\delta U)(\delta U) = \left[\dfrac{\partial^2 w}{\partial q \partial q}\right]_o (\delta q)(\delta q) \\[12pt] \qquad = \overline{\delta q} \; K \; \delta q \end{array} \right.$$

with a matrix K defined on $\delta q \in \mathbb{R}^N$.

Now we had the formulae (5.20) and (5.25) :

$$(5.50) \quad \left[\begin{array}{l} \delta w = \dfrac{\partial w}{\partial q}\, \delta q = \overline{\delta q}\, H(q) = \overline{\delta q}\left[K_1\, q + H_2(q) + H_3(q) \right] \\[14pt] \qquad\qquad\qquad = \overline{\delta q}\left[K_1\, q + K_2(q)(q) + K_3(q)(q)(q) \right] \end{array} \right.$$

where K_2 and K_3 are tensors of $\mathbb{R}^N$, respectively bilinear symmetrical and trilinear symmetrical. Thus with (5.49) :

$$\delta^2 w = \overline{\delta q} \; K \; \delta q = \overline{\delta q} \; [K_1 \; \delta q + 2K_2(q)(\delta q) + 3K_3(q)(q)(\delta q)]$$

Whence :

$$K = K_1 + 2K_2(q) + 3K_3(q)(q).$$

Condition (5.44) is therefore written, after discretization :

the state $\mathcal{E}_o$, corresponding to solution q_o, is critical if :

$$(5.51) \quad \left[\begin{array}{l} \exists \, \delta q \neq 0 \quad \Big|\Big[K_1 + 2K_2(q_o) + 3K_3(q_o)(q_o) \Big] \delta q = 0 \\[2ex] \text{or otherwise} \\[2ex] \det (K_1 + 2K_2(q_o) + 3K_3(q_o)(q_o)) = 0. \end{array} \right.$$

The resolution of this equation can be conducted in the following manner:
one solves equation (5.48) in the form :

$$(5.52) \qquad H(q) = \lambda Q^* = Q, \quad \lambda \text{ scalar parameter,}$$

looking for a critical solution q_o, corresponding to the value λ_o of λ,
proceeding by incrementation of λ and successive iterations, until (5.51)
is checked. For this purpose, we calculate the value of λ for each value
of $\det(K)$, and we look for the value of λ corresponding to cancellation of
$\det(K)$. It can be demonstrated that this value can be found by interpolation of $\det(K)$, using the positive or negative values found when the critical state corresponds to a bifurcation. In the case of a limit point, we
are obliged to proceed by extrapolation, the curve representing $\det(K)$
plotted against λ having an infinite tangent on passing through zero[48].

When the non-linearities are moderate, another method consists in treat-
ing the non-linear terms of (5.52) as imperfections. Now we shall see that
the linear, or linearized, case corresponds practically always to a bifurcation. These imperfections will therefore transform the critical bifurcation point into a limit point, the critical load of which will be all the
more close to the actual critical load that the solution approaches the exact
solution.

When the displacement q remains moderate, and this is precisely the case
with the step by step method where we proceed by a series of small loading
increments, the linearization of the problem can provide major simplification, without notable error.

In fact let us linearize equation (5.48), from a solution q_o, Q_o assumed
to be exact. Calling ε the linearization symbol, and q_I, Q_I the solution
linearized to the first order we obtain :

$$q = q_o + \varepsilon q_I, \quad Q = Q_o + \varepsilon Q_I, \quad \varepsilon^2 = 0^{51}.$$

(5.25) gives :

$$(5.53) \quad \begin{bmatrix} K_1 \, q_o + K_2(q_o)(q_o) + K_3(q_o)(q_o)(q_o) = Q_o \\[2mm] \left[K_1 + 2K_2(q_o) + 3K_3(q_o)(q_o)\right] q_1 = Q_1 \end{bmatrix}$$

The resolution of equation $(5.53)_2$ is obtained by inversion of the
matrix :

$$H'(q_o) = K_1 + 2K_2(q_o) + 3K_3(q_o)(q_o),$$

if the latter is regular. In certain cases (see § 5.4), q_I corresponds to
the direction taken by the solution from q_o, namely giving the tangent to
the solution. It is clear that instability occurs at point q_o which makes
this linearized matrix singular, thus coinciding with criterion (5.51).

A number of approximations of $(5.53)_2$ are possible, according to the
case.

a) One linearizes from $q_o = Q_o = 0$. In this case $w'(o) = 0$, and the
corresponding state is the natural state. $(5.53)_2$ becomes :

$$(5.54) \quad K_1 \, q_I = Q_I.$$

This is the conventional linearization of elasticity with small deformations and small displacements. The natural state is stable by hypothesis,
and K is positive definite. Furthermore, a second order linearization, with
loading Q_I unchanged, would give :

$$K_1 \, q_{II} + 2K_2(q_I)(q_I) = 0.$$

b) One linearizes from q_o, Q_o which are non-zero, although q_o is taken
to be small. In the left hand side member of equation $(5.53)_2$, the quadratic term $3K_3(q_o)(q_o)$ in q_o can be neglected in a first approximation, with
respect to the other two. But in equation $(5.53)_1$, the non-linear terms can
also be neglected with respect to the linear term in q_o. (5.53) is then

written :

$$(5.55) \qquad \begin{cases} K_1 \, q_o = Q_o \\[2mm] \left[K_1 + 2K_2(q_o) \right] \, q_I = Q_I. \end{cases}$$

Under these conditions, the condition of instability is written :

$$K_1 + 2K_2(q_o) \text{ singular.}$$

This is the <u>linear or classic buckling approximation</u>. K_1 is still the mechanical stiffness matrix for the linearized problem, and $2K_2(q_o)$, a linear matrix in q_o, is a "<u>geometrical" stiffness matrix</u> and owes its existence to a certain prestress field due to loading Q_o[39,10].

It is appropriate to point out that the solution of $(5.55)_2$ does not correspond, in contrast to $(5.53)_2$, to a strict linearization, which must be made from an exact q_o, Q_o solution of the non-linear problem, and which is not the case here.

Nevertheless, this approximation enables us to find, classically, an approximate critical load Q_c corresponding to a certain direction of loading Q_o^*, in the Q load space . In fact let us put :

$$Q_o = \lambda Q_o^* \, , \quad \lambda \, \epsilon \, \left[0, \, + \infty \right[$$

$(5.55)_1$ gives :

$$K_1 q_o^* = Q_o^*,$$

This equation provides q_o^* ; and $(5.55)_2$ gives :

$$\left[K_1 + 2 \, \lambda \, K_2(q_o^*) \right] \, q_I = Q_I.$$

The smallest eigenvalue λ_c of $K_1 + 2 \lambda K_2(q_o^*)$ gives the smallest critical load :

$$Q_c = \lambda_c \, Q_o^*.$$

This approximation always leads to a case of instability by bifurcation, as there always exist two solutions at point $\lambda_c \, q_o^* = q_o$: solution λq_o corresponding to the fundamental path, and solution λq_I to the bifurcated path in the case of a symmetrical bifurcation [see § 5.4), formula (5.63)].

c) The displacement q_o is not sufficiently small to apply the preceding
method of approximation. The approximation of linear buckling can lead to
serious errors, with overestimation or underestimation of buckling loads.
As the critical state is a priori unknown, generally it cannot be postulat-
ed that it corresponds to a small value q_o. It does happen frequently that
the preceding approximate solution translates local buckling, in other
words a local weakness in the structure, compensated by a redistribution of
stresses and the still unaltered resistive strength of the adjacent region.

The introduction of non-linear terms can indicate the start of buckling,
and in fact lead to critical loads which are substantially larger than
those of the preceding approximation[38]. The art of the engineer has in fact
made it possible to take account of this phenomenon, by introducing stiffeners in the structure, which permit this redistribution[34]. It also happens,
as mentioned in paragraph 5.1, that the structure achieves a stable post-
buckling state after buckling, which must be calculated.

Equation (5.53) or (5.55), even when limited to appropriately loaded
parts of the structure, provides the designer with an indication of these
local buckling points, provided the latter are adequately uncoupled from
global instability, the latter being obtained from a more complex
calculation.

The problem then consists in finding the unknown critical load $Q = Q_c$,
such that :

$$(5.56) \quad \left[\begin{array}{l} H(q) = Q, \quad \text{or} \quad K_1 q + K_2(q)(q) + K_3(q)(q)(q) = Q \\[2ex] \det(H'(q)) = 0, \quad \text{or} \quad \det(K_1 + 2\,K_2(q) + 3\,K_3(q)(q)) = 0. \end{array} \right.$$

We can put :

$$Q = \lambda Q_o^*$$

to find :

$$Q_c = \lambda_c\, Q_o^*,$$

and use one of the methods described in paragraph 4.

The most efficient method consists in executing a step by step calculat-
ation[40-44], giving λ incremental values, which can be done incidentally by
using the preceding approximations, provided the increments are sufficiently small, employing the modified Newton-Raphson method for example. For
each value of λ, we test the sign of $\det(H'(q))$, λ_c corresponding to the
cancellation of the determinant.

To determine <u>post-buckling states</u>, certain writers[44] recommend a modification of loading Q_o, to avoid calculation difficulties in the neighbourhood of a critical bifurcation point, replacing Q_o by $\lambda Q_o^* + \eta Q_1^*$ instead of
λQ_o^*, η being small. The fictitious loading ηQ_1^* introduces an artificial
imperfection, which supresses the bifurcation found previously.

We then calculate the solution for η constant and λ bracketing λ_c, and
recalculate the different states for $\eta = 0$ and $\lambda > \lambda_c$, giving trajectories
(q,λ) approximately parallel. It should be noted that in this case, Q_1^*
must be orthogonal to Q_o^*, as we shall see in the following paragraph, and
stressing that these post-buckling states are only of interest if stable,
this can be determined by local considerations (namely in the vicinity of
the critical state).

Let us now consider the second approach, from a prestressed state. This
approach leads to the discretized equation (5.26), namely :

(5.57) $K_1' q + K_2' q + L_2(q)(q) + L_3(q)(q)(q) = Q'$

or $L(q) = Q'$,

where K_1' and K_2' are constant matrices (in linear elasticity), and L_2 and L_3
are symmetrical tensors of $\mathbb{R}^N$, of second and third order respectively. It
will be remembered that K_2' is a matrix of geometrical stiffness due to
the prestressed field C_o in the reference state $\mathcal{E}_o$, and that the discretized quantities are obtained by integration on the prestress domain Ω_o, and
not on the domain Ω_{oo} corresponding to the natural state $\mathcal{E}_{oo}$ (furthermore,
it frequently happens that no distinction has to be made between these two
domains for these integrations).

Linearization leads to formulae similar to those for the preceding case.

But in fact, two problems arise :

- determination of the load Q'_o, giving the prestress C_o corresponding to a buckling instability;

- or if the load Q'_o corresponds to a stable state $\mathcal{E}_o$, determination of the load Q'_1 which added to Q'_o, gives a state $\mathcal{E}$, unstable by buckling.

These two problems are different. In fact <u>in the first case</u>, the direction of Q_o is assumed to be given, and buckling is calculated as in the first approach, replacing Q'_o by λQ^*_o, and executing a calculation, linearized or not. As before, linearized calculation gives linear buckling. The stiffness matrix K'_2 is still linear in C_o, namely Q_o, and can be replaced by matrix $\lambda K'^*_2$ where K'^*_2 corresponds to the generalized load Q'^*_o. The problem thus linearized comes down to one of eigenvalues.

Strictly, C_o will be obtained by a non-linear calculation, for example step by step with iteration. Examples demonstrate that if C_o is obtained by a linear calculation, this can lead to major errors[17].

For this problem, the test for finding the critical point, is still calculation of the sign of :

$$\det(K'_1 + K'_2 + 2\,L_2(q_o) + 3\,L_3(q_o)(q_o))$$

for the value $q_o = 0$ corresponding to the state $\mathcal{E}_o$ in question, namely :

$$\text{Sign}(\det(K'_1 + K'_2)).$$

In other words, the calculation of L_2 and L_3 is only useful for the calculation of C_o, but if C_o is given (possibly by another process), this calculation is useless.

In a non-linear calculation, we can again proceed by incrementation of load Q'_o weighted by a scalar multiplier λ, and completed by the calculation of the eigenvalue λ of the matrix :

$$K'_1 + \lambda K'_2$$

until the eigenvalue $\lambda = 1$ is obtained[45].

The <u>second problem</u> corresponds to that of a structure prestressed by a load Q'_o, to which we add a load Q'_1, the direction of which may be different. In this case also, the prestress field and the geometric stiffness matrix

must be calculated first. The critical buckling load Q'_{1c} is then calculated by the methods indicated for the first approach, testing the sign of :

$$\det(K'_1 + K'_2 + 2\,L_2(q_1) + 3\,L_3(q_1)(q_1))$$

calculated for a value $q_1 \neq 0$, corresponding to the state $\mathcal{E}$ under study.

It should be noted that this case is very frequent, corresponding for
example to that of inflated structures, or structures stiffened by tensioned cables (prestressed concrete, etc.), for which the prestress field C_o is
a self-stressed field where $Q'_o = 0$.

The geometric stiffness matrix K'_2 is often obtained in the case of
shell or beam elements, by using customary classical simplifications for
this type of structure, where flexural displacements are frequently pre-
dominant. Now these simplifications are often abusive, and can lead to
serious errors. Thus K'_2 is obtained from the term :

$$(5.58) \qquad \int_{\Omega_o} T_r(C_o\, \frac{\overline{\partial U}}{\partial M_o}\, \frac{\partial \delta U}{\partial M_o})d\Omega_o = \overline{\delta q}\, K'_2\, q$$

of formula (5.17).

If we take as displacements U only flexural displacements, we cannot
take account, for example, of rigid body rotations of the structure or
the element concerned. This simplification therefore introduces stiffnesses
which are not only superfluous but harmful, which can thus supply, in a
problem of natural vibrations for a prestressed structure, rigid body modes
with eigenfrequencies other than zero. The repercussions on the calculation
of a critical state are no less obvious.

Furthermore, it should be noted that gradient $\dfrac{\overline{\partial U}}{\partial M_o}$ can be decomposed
classically [see (A.17)], into a deformation term D_L and an anti-
Hermitian rotation term $- i(\Omega)$:

$$\frac{\overline{\partial U}}{\partial M_o} = D_L - i\,(\Omega).$$

This rotation term is generally important, in the case of structures
comprising thin elements subject to small deformations but large displacements.

If we compare (5.58) with the natural stiffness term :

$$\int_{\Omega_o} T_r(\underset{\sim}{C}_I(D_L)\frac{\partial \delta U}{\partial M_o})d\Omega_o = \overline{\delta q}\ K'_{\mid}\ q$$

[see (5.18)], we observe that the prestress C_o is comparable with the linear operator $\underset{\sim}{C}_I$, namely playing the part of anisotropic characteristics of the material. The components of C_o are generally much smaller in absolute value than those of $\underset{\sim}{C}_I$, and :

$$(5.59) \qquad \int_{\Omega_o} T_r(\underset{\sim}{C}_I(D_L)\,\delta D_L)d\Omega_o \ \gg \ \int_{\Omega_o} T_r(C_o\,D_L\,\frac{\partial \delta U}{\partial M_o})d\Omega_o,$$

In fact in cartesian coordinates, D_L is represented by $^j\varepsilon_\ell$, and $C_I = \underset{\sim}{C}_I(D_L)$ by :

$$^i\sigma_k = A^{\ell\,i}_{\,jk}\cdot{}^j\varepsilon_\ell,$$

and C_o by $^i\sigma_{oj}$. We then have :

$$T_r(C_I\delta D_L) = A^{\ell\,i\ j}_{\,jk}\varepsilon_\ell\cdot\delta^\kappa\varepsilon_i \gg T_r(C_oD_L\delta D_L) = {}^i\sigma_{oj}\cdot{}^j\varepsilon_\ell\cdot\delta^\ell\varepsilon_i$$

$$= {}^i\sigma_{oj}\cdot{}^j\varepsilon_\ell{}^\ell\big|_k\delta^k\varepsilon_i$$

as :

$$A^{\ell i}_{\,jk} \gg {}^i\sigma_{oj}{}^\ell\big|_k \qquad \text{in general.}$$

This explains the absence of buckling in three-dimensional bodies, where the rotation Ω is generally negligible (we only keep the component D_L in the term $\frac{\overline{\partial U}}{\partial M_o}$). This is not so in the case of thin bodies, as we have said, where, even when (5.59) applies, the term :

$$-\int_{\Omega_o} T_r(C_o\ i(\Omega)\ \frac{\partial \delta U}{\partial M_o}\ )d\Omega_o$$

may not only be important but preponderant, thus being responsible for buckling instability[8].

5.4 - <u>Limit point and bifurcation point</u> [39,46-48]

We indicated in paragraph 5.1 that the critical point of instability can be a limit point or a bifurcation point, according to the case.

It should be remembered that a critical bifurcation point is characterized by the existence of two possible solutions at this point, or again two possible paths in the graph $\{q, \lambda\}$, the structure assumed to be discretized for simplification's sake, and λ being, as before, a loading scalar such that :

$$Q = \lambda Q^* \; ;$$

while only one path is possible at a limit point.

In the present paragraph, we will give certain indications on this subject, recapitulating certain fundamental results.

A study of the different possible cases can be made in a local way, namely by examination of solutions in the vicinity of a critical point (q_o, λ_o).

For example, if we linearize the discretized equilibrium equation (5.21), we find system $(5.53)_2$:

$$(5.60) \quad \left[\begin{array}{l} Kq_I = Q_I \\[1em] \text{with :} \\[1em] K = K_1 + 2\, K_2(q_o) + 3\, \underset{\sim}{K}_3(q_o)(q_o) = \underset{\sim}{K}(q_o) \\[1em] q = q_o + q_I, \quad \lambda = \lambda_o + \lambda_I, \quad (\text{or } Q = Q_o + Q_I). \end{array} \right.$$

If q_o is not a critical point, the linearized solution from q_o is given by :

$$q_I = K^{-1}\, Q_I = \frac{\text{Adj}(K)}{\det(K)}\, Q_I \, .$$

If matrix K is singular for $q = q_o$:

$$\det\,(\underset{\sim}{K}(q_o)) = 0,$$

q_o is a critical point, and $\exists$ a mode $q_o^M \neq o$ such that :

$$\left[\underset{\sim}{K}(q_o) \right] q_o^M = 0 .$$

To simplify, let us assume that the kernel of K is unidimensional[*], and let us normalize mode q_o^M, such that :

$$\left| q_o^M \right| = 1$$

In this case, $\mathrm{val}\,(\mathrm{Adj}\,\underset{\sim}{K}(q_o))$ is unidimensional and subtended by q_o^M, and :

$$(5.61) \quad \begin{cases} \mathrm{Adj}\,(K(q_o)) = \alpha q_o^M \overline{q_o^M} \\[2mm] \text{with :} \\[2mm] \alpha = T_r(\mathrm{Adj}(\underset{\sim}{K}(q_o))) \quad , \text{ if } \left| q_o^M \right| = 1, \text{ hence } \alpha > 0. \end{cases}$$

In fact :

$$\mathrm{Adj}(K).K = \underline{\det(K)} = 0 \iff \exists q_o^M \neq 0 \text{ such that } Kq_o^M = 0, \; (q_o^M \in \mathrm{Ker}(K))$$

Let us put :

$$\pi = q_o^M \overline{q_o^M}, \quad \pi' = \underline{1} - \pi \quad , \text{ with } \left| q_o^M \right| = 1.$$

If $K = \overline{K}$, we have :

$$K\pi = \pi K = 0 \quad \text{and} \quad \forall \pi'V \neq 0 \Longrightarrow K\pi'V \neq 0 \quad \text{aş } \pi'V \notin \mathrm{Ker}(K).$$

In addition :

$$K\pi'V = [\pi + \pi'] K\pi'V = \pi'K\pi'V, \; \forall V$$

and :

$$K\pi'V = 0 \Longrightarrow \pi'V = 0 .$$

Now :

$$\mathrm{Adj}(K).K\pi'V = 0 \quad \forall \pi'V \neq 0 \Longrightarrow \mathrm{Adj}(K)\pi'K\pi'V = 0 \quad , \quad \forall \pi'V \neq$$

$$\Longrightarrow \mathrm{Adj}(K)\pi'W = 0, \forall \pi'W \neq 0$$

$$\Longrightarrow \pi'W \in \mathrm{Ker}(\mathrm{Adj}(K)), \forall \pi'W \neq 0 \ .$$

Or otherwise :

$$\mathrm{val}(\pi') = \mathrm{Ker}(\mathrm{Adj}(K)) \ .$$

Therefore :

$$\mathrm{Adj}(K) = \alpha q_o^M \, \overline{q_o^M} \ \text{and} \ \ \alpha = T_r(\mathrm{Adj}(K)) \ .$$

But $(5.60)_1$ gives :

$$\mathrm{Adj}(K).Kq_I = \mathrm{Adj}(K) \ Q_I$$

$$\det(K).q_I = \mathrm{Adj}(K) \ Q_I = 0 \ ; \ \text{furthermore} \ \left\{ \begin{array}{l} K = \overline{K} \\[4pt] d \, \det(K) = T_r(\mathrm{Adj}(K)dK) \\[6pt] \dfrac{\partial \, \det(K)}{\partial K} = \mathrm{Adj}(K) \ ; \end{array} \right.$$

and taking account of (5.61) :

$$\alpha \ q_o^M \, \overline{q_o^M} \ Q_I = 0,$$

or again :

$$(5.62) \qquad \beta \ q_o^M = 0 \ \ \text{with} \ \beta = \alpha \ \overline{q_o^M} \ Q_I \ .$$

The different cases of instability are then deduced from condition (5.62).

a) $\beta = 0$, namely $Q_I = \lambda_I \ Q_o^*$ is such that :

$$\overline{q_o^M} \ Q_I = 0 \ .$$

Q_I is orthogonal to mode q_o^M. In this case, <u>q_o is a bifurcation point</u> : a detailed study of the linearized problem demonstrates that there exist two distinct paths at q_o, these being q_{I1} and q_{I2}, corresponding to two loading values, λ_{I1} and λ_{I2}.

For a given value λ_I in the neighborhood of λ_o, plane $\lambda_I = c^{te}$ cuts the graph on these two paths at two points, such that the direction of the segment which joins them in $\mathbb{R}^N$ is precisely q_o^M. The two values λ_{I1} and λ_{I2}

are given by a second degree equation, the constant term of which is equal to :

$$(5.63) \qquad \gamma = \overline{q_o^M} \left[\frac{\partial K}{\partial q}\right]_o (q_o^M)(q_o^M).$$

If coefficient γ is non-zero, the bifurcation point is asymmetrical.

If coefficient γ is zero, one of the solutions, namely λ_{I1} is zero, <u>the bifurcation point is then symmetrical</u> with respect to the fundamental path, and the direction of the bifurcated path is q_o^M.

<u>The stability of the different paths</u> is deduced by a more detailed study, by linearization at the higher orders. Let us put :

$$\det(K) = D$$

$$q_I = q_I^* \, x$$

where q_I^* is the direction of the bifurcated path. We call q_F the fundamental path described from a zero loading.

In the case of an asymmetrical bifurcation, and assuming the fundamental path to be stable for $\lambda \in \left[0, \lambda_o\right[$, we find :

$$\frac{\lambda_I}{x} > 0 \Longrightarrow \left[\frac{dD}{dx}\right]_o > 0.$$ D increases with x, and vanishes for

x = 0. An unstable path becomes stable, and vice versa.

If $\gamma = 0$, the bifurcation is symmetrical. One finds that $\left[\frac{dD}{dx}\right]_o = 0$. Therefore D keeps the same sign on either side of the critical point : a stable path remains stable and an unstable path remains unstable, when passing through the critical point. In this case :

$$\lambda_{II} = \lambda^* \, x^2 .$$

More precisely :

$$\lambda^* > 0 \Longrightarrow \left[\frac{d^2 D}{dx^2}\right]_o > 0,$$ the bifurcated path is stable and remains so.

$$\lambda^* < 0 \Longrightarrow \left[\frac{d^2 D}{dx^2}\right]_o < 0,$$ the bifurcated path is unstable and remains so.

The preceding results are summarized in the following figures :

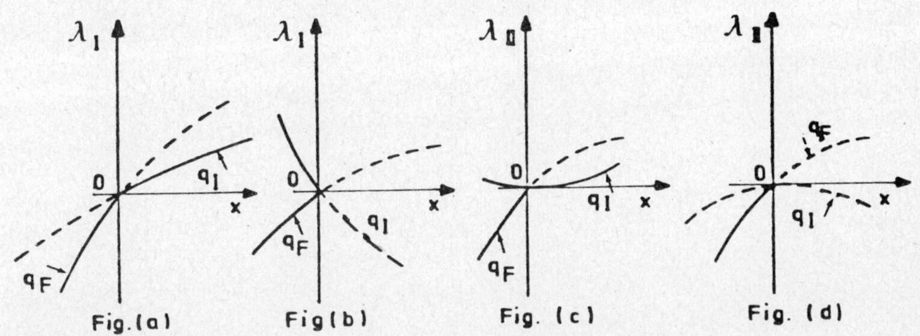

------------ Stable path

----------- Unstable path

(a), (b) $\Longrightarrow$ asymmetrical bifurcations

(c), (d) $\Longrightarrow$ symmetrical bifurcations

Let us remember also that the study of imperfect systems is deduced from that of perfect systems, by linearization in the neighbourhood of the perfect case.

The various cases indicated by the preceding figures are illustrated in the figures below, in which the evolution of a perfect system is shown by continuous lines, and that of an associated imperfect system by dashed lines. The scalar parameter η characterizes the degree and direction of the imperfection.

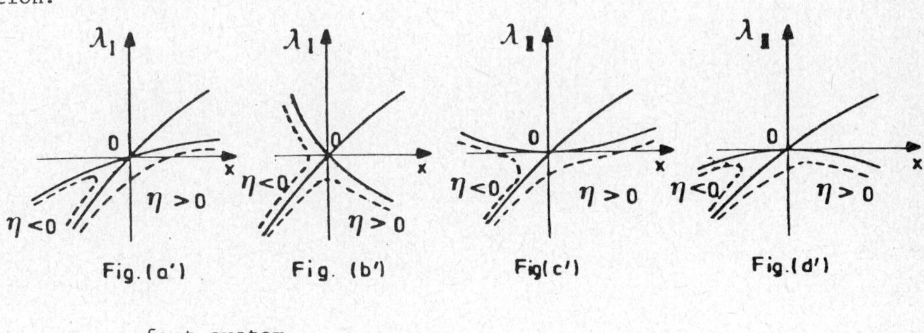

------------ perfect system

----------- imperfect system

(a', b') $\Longrightarrow$ asymmetrical bifurcation of perfect system

(c' d') $\Longrightarrow$ symmetrical bifurcation of perfect system

b) $\beta \neq 0$, namely $\overline{q_o^M Q_I} \neq 0$. The right hand side member of equation $(5.60)_1$ is not orthogonal to q_o^M.

If the scalar γ defined above in (5.63), is such that :

$$\gamma \neq 0,$$

the system passes from a stable to an unstable configuration, via the critical point, and it can be demonstrated that the slope of $\frac{dD}{d\lambda}$ becomes infinite at this point. This case is illustrated by the following figure :

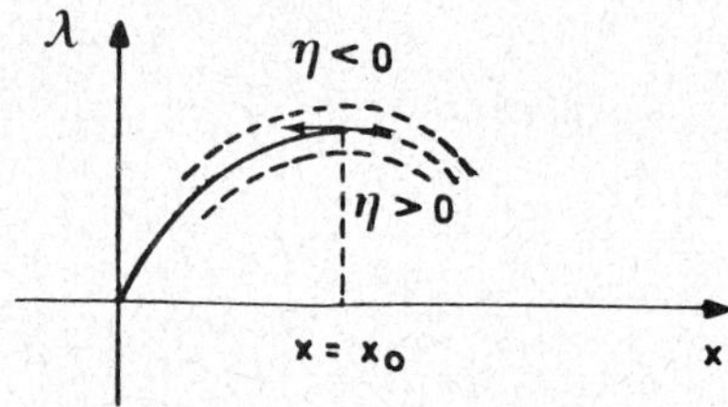

CHAPTER VI

SHELL THEORY

1. GENERAL
2. EQUILIBRIUM EQUATIONS
3. DEFORMATIONS
4. STRESSES
5. VARIATIONAL PRINCIPLES
6. LINEAR CONSTITUTIVE LAWS
7. SHELLS OF REVOLUTION
8. DISCRETIZATION

1. – GENERAL[1-7]

1.1 – Introduction

A shell is a continuous, three-dimensional material structure, one dimension of which, the thickness is small with respect to the two others. A theory of shells is one intended to translate the three-dimensional behaviour of a structure of this type, by means of surface fields, and using approximations which are expressed in the form of kinematic, dynamic or energy hypotheses.

The theory was born practically from the historical works Love[1] in 1888. It has been marked out by the works of American, Russian, German and Dutch schools, and many others.

This theory has been and still is marked by numerous difficulties : apart from the geometrical difficulties inherent in a surface theory, where the fields are defined on a curved space, namely the utilization of an appropriate differential geometry, difficulties are related to the expression of constitutive laws between surface quantities, these being invariant and coherent mechanical laws. Controversies have generally arisen from the different nature of the hypotheses introduced. Another category of difficulties arose from the search for an evaluation of the approximations made in each theory, with respect to the three-dimensional theory considered as exact. In certain cases, writers have been able to give approximate values for the errors committed on certain quantities, according to characteristic nondimensional parameters such as $\frac{h}{L}$ or $\frac{h}{R}$, ratios of thickness to a main dimension, or to the minimum curvature radius, or also such as a norm relating to deformation. Other writers have even been able to find solution bounds.

Other specific difficulties arose in the calculation of solutions by approximate methods, in particular with the finite element method, relating to the establishment of satisfactory (curved) shell elements.

The problems studied, both theoretical and practical, are those classically encountered in the calculation of structures, namely static or dynamic problems, problems of instability, etc., the solution of which naturally suffered from all the difficulties mentioned.

It goes without saying that experimental problems have not been spared their share of difficulties and particular controversies.

Another category of problems still little explored, particularly in the non-linear field, concerned those problems which specialists in theoretical mechanics posed to the specialists in applied mathematics, namely problems of the existence and possibly the uniqueness of solutions[9], as also those of the convergence of approximate solutions.

This long search was marked by certain events remarkable in our eyes, since the time of Love : the static-geometrical analogy of Gol'denweizer[5] and Lur'e[8], and its application by means of the complex variable[4,10] ; the introduction of the symmetrical stresses by Sanders[11] and Léonard[12]; the theory of plane   stress of Koiter[13], and its comparison with that of Love; the method of asymptotic expansions of Gol'denweizer[5,14,15]; the theorems of approximation and error estimates of Koiter[16] or others[17] ; the theory of Cosserat surfaces by Green and Naghdi[18], or da Silva and Cohen[19] ; the development of non-linear theories[11,20,8,6]; and finally, the enormous extension of practical applications, by means of the finite element of finite difference discretization methods.

Naturally we shall not go into all these problems. We shall content ourselves with a description of various aspects of the theory, according to the hypotheses adopted, giving certain indications relating to problems of discretization in view of practical applications.

1.2 - Definitions

A shell is a three-dimensional medium, assumed to be non-polarized and embedded in E_3, which can be considered to be generated by a small segment of variable length, the middle of which (for example) describes a surface Σ_m, while remaining normal to this surface. Σ_m is called the "middle surface".

We assume that Σ_m, with generic point m, is a differential manifold with dimension 2, embedded in E_3, of class C^2, compact, canonically oriented and Riemannian[21].

Σ_m is parameterized by a complete atlas of a finite number of maps, with the topology of an embedded manifold.

Σ_m can have an edge or not. Its edge, called C_m, can also be a pseudo-edge, comprising a number of closed, unidimensional pseudo-manifolds with empty intersections, namely a number of disjointed curves with singular points.

If Σ_m has a non-empty edge, it is part of an oriented manifold $\tilde{\Sigma}_m$.

As Σ_m is Riemannian by hypothesis, an Euclidian tangent linear manifold $\vec{E}_{2m}$, exists at each point, namely a field of differentiable fundamental metric tensors g, defined on $\vec{E}_2$, and induced by the metric tensor of $\vec{E}_3$, and a field of differentiable Hermitian projectors Π (orthogonal projection), applying $\vec{E}_3$ onto $\vec{E}_2$, and defining the Riemannian connection[22,23].

Let then ϕ be a mapping of an open set $\mathcal{O}$ of $\mathbb{R}^2$ onto Σ_m, we obtain :

$$m = \phi(X), \quad X \in \mathcal{O} \subset \mathbb{R}^2.$$

$\forall$ the field f of tangent vectors V :

$$V = f(m).$$

We shall put, $\forall$ the field of differentials $dm \in E_{2,m}$,

$$\hat{d}V = \Pi dV,$$

$\hat{d}V$ is the covariant differential corresponding to dm.

In addition, the natural basis of $\vec{E}_{2,m}$ at m is defined by :

$$S = \frac{\partial m}{\partial X},$$

Thus :

$$dm = \frac{\partial m}{\partial X} dX = SdX, \quad dX = \begin{bmatrix} d^1X \\ d^2X \end{bmatrix} \in \mathbb{R}^2$$

$$\partial_\alpha m = \frac{\partial m}{\partial X} \partial_\alpha X = \frac{\partial m}{\partial^\alpha X} = S_\alpha \ , \ \text{with} \ \partial_\alpha X = |_\alpha \ ; \ \alpha = 1,2.$$

Therefore :

$$S = \begin{bmatrix} S_1 & S_2 \end{bmatrix} = \begin{bmatrix} \partial_1 m & \partial_2 m \end{bmatrix}$$

$$g_{\alpha\beta} = g(S_\alpha)(S_\beta) = \overline{S_\alpha}\, S_\beta = {}^\alpha\overline{S}\, S_\beta = {}^\alpha G_\beta$$

with :

$$G = \overline{S}\, S.$$

and $\forall V_1, V_2 \in \vec{E}_2$:

$$V_1 = SX_1, \quad V_2 = SX_2, \quad \overline{V_1}\, V_2 = g(V_1)(V_2) = \overline{X_1}\, G\, X_2 .$$

G is the "first fundamental form" of Σ_m. This is the de Gram matrix for basis S.

Now let vol be the (constant) gauge for $\vec{E}_3$. This induces on $\vec{E}_{2,m}$, a gauge vol_2, thus defined due to the Euclidian structure of E_3 :

Taking N as the unit normal at m to Σ_m, we have :

$$N = \frac{i(S_1)(S_2)}{|i(S_1)(S_2)|} .$$

Now $[\text{see } (A.21)_1]$:

$$|i(S_1)(S_2)|^2 = \overline{i(S_1)(S_2)}\, i(S_1)(S_2) = \det(\overline{S}\, S)$$

and as E_2 and E_3 are real and positive :

$$\det(S) = \pm \sqrt{\det(\overline{S}\, S)} .$$

If we adopt the positive determination for the direct bases, we have :

$$(6.1) \qquad i(S_1(S_2) = N \det(S) = N\, \mathrm{vol}_2(S_1)(S_2) .$$

Under these conditions, the basis $[S\ N]$ of $\vec{E}_3$ at m is direct, and :

$$(6.2) \qquad \mathrm{vol}(S_1)(S_2)(N) = \overline{N}\, i(S_1)(S_2) = \mathrm{vol}_2(S_1)(S_2) > 0.$$

The orientation of the edges is achieved, without difficulty, by conti-nuity. We also define on edge C_m, the normal ν to C_m at m in $\vec{E}_2$, ν being unit and external to Σ_m in $\underset{\sim}{\Sigma}_m$.

Under the preceding conditions, we can state that the shell occupies a volume $\overline{\Omega}$, generated by the segment with length h, the middle of which describes Σ_m of the preceding type.

The manifold $\overline{\Omega} = \left[-\dfrac{h}{2}, \dfrac{h}{2}\right] \times \Sigma_m$; its edge $\partial\Omega$ is made of two surfaces $\Sigma^+ = \{\dfrac{h}{2}\} \times \Sigma_m$ and $\overline{\Sigma} = \{-\dfrac{h}{2}\} \times \Sigma_m$, and lateral edges generated by the normal segments, the middles of which describe C_m. We shall call the lateral edge Σ_L.

Natural basis of E_3

Let M be a point on the shell, situated on the normal N at m at a distance z :

$$M = m + mM = m + Nz.$$

$\forall dX \in \mathbb{R}^2$ and $dz \in \mathbb{R}$, we have :

(6.3) $dM = dm + zdN + Ndz.$

Now :

$$dm = \frac{\partial m}{\partial X}\, dX = SdX,$$

therefore :

$$dM = dm + z\,\frac{\partial N}{\partial m}\, dm + Ndz,$$

or :

$$dM = \left[1_{E_2} + z\,\frac{\partial N}{\partial m}\right] dm + Ndz$$

(6.4)
$$\begin{array}{l} dM = \mu dm + Ndz \\[2mm] \text{with :} \\[2mm] \mu = 1_{E_2} + z\,\dfrac{\partial N}{\partial m}\,. \end{array}$$

$\dfrac{\partial N}{\partial m}$ is the curvature operator of Σ_m at m. It will be remembered that :

$$\overline{N}dm = 0 \;,\; \forall dm \in \vec{E}_2$$

$$\overline{dN}dm + \overline{N}d^2m = 0$$

$$\overline{dm}\;\frac{\partial N}{\partial m}\, dm = -\,\overline{N}d^2m$$

The quantity $\overline{dm}\,\frac{\partial N}{\partial m}\,dm$ is called the __second fundamental form__ of Σ_m at m. We know that the first fundamental form and second fundamental form, namely G and $\frac{\partial N}{\partial m}$, are sufficient to define Σ_m to within an arbitrary rigid body displacement (Bonnet's theorem).

Furthermore, $\frac{\partial N}{\partial m}$ is a Hermitian endomorphism of $\vec{E}_2$, namely[23] :

$$(6.5) \qquad \frac{\partial N}{\partial m} = \overline{\frac{\partial N}{\partial m}} \ \epsilon \ \mathcal{L} \ (\vec{E}_2, \ \vec{E}_2).$$

In fact :

$$\overline{N}\,N = 1,$$

and :

$$\forall dm \ \epsilon \,\vec{E}_2, \ \overline{dN}\,N = 0 \Longrightarrow dN \ \epsilon \,\vec{E}_2,$$

but :

$$\overline{N}d_1 m = 0 \quad \forall d_1 m,$$

gives :

$$d_2\,[\overline{N}d_1 m] \ = 0 \ \ \forall d_1 m, \ d_2 m \ \epsilon \ \vec{E}_2$$

$$\overline{d_2 N}\,d_1 m + \overline{N}\,d_2\,d_1 m = 0$$

$$\overline{d_2 m}\,\frac{\partial N}{\partial m}\,d_1 m + \overline{N}\,d_2\,d_1 m = 0$$

$$\overline{d_1 m}\,\frac{\partial N}{\partial m}\,d_2 m + \overline{N}\,d_1\,d_2 m = 0 .$$

Hence :

$$\overline{d_1 m}\,\left[\frac{\partial N}{\partial m} - \overline{\frac{\partial N}{\partial m}}\right]\,d_2 m = -\,\overline{N}\,\left[d_2\,d_1 m - d_1\,d_2 m\right].$$

Now $d_2 d_1 m - d_1 d_2 m = \phi'(X)(d_2 d_1 X - d_1 d_2 X) \ \epsilon \,\vec{E}_2$, therefore :

$$d_1 m\,\left[\frac{\partial N}{\partial m} - \overline{\frac{\partial N}{\partial m}}\right]\,d_2 m = 0 \ \ \forall d_1 m, \ d_2 m,$$

whence (6.5).

Besides, it will be remembered that $\forall$ curves C on Σ_m, parameterized by the curvilinear abscissa s, oriented such that :

$$\overline{\frac{dm}{ds}}\,\frac{dm}{ds} = 1 = \overline{t}\,t \quad \text{with} \quad t = \frac{dm}{ds}$$

$$2\,\overline{\frac{d^2m}{ds^2}}\,t = 0 \,.$$

By definition, one puts :

$$\frac{d^2m}{ds^2} = n\rho$$

where n is the principle normal to C, and $\rho = \frac{1}{R}$ its curvature at m. Now :

$$\overline{N}\,\frac{d^2m}{ds^2} = \overline{N}\,n\rho = \rho\,\cos\alpha = \rho_n,$$

namely :

$$\frac{1}{R}\,\cos\alpha = \frac{1}{R_n} \quad;$$

(when C varies, while remaining tangent to t, the centre of curvature describes a circle with diameter R_n. This is Meusnier's theorem).

Furthermore :

$$\overline{N}dm = 0, \quad \forall dm \in \vec{E}_2$$

$$d[\overline{N}dm] = 0$$

$$\overline{N}\,\frac{d^2m}{ds^2} = -\,\overline{\frac{dN}{ds}}\,\frac{dm}{ds} = -\,\overline{\frac{dm}{ds}}\,\frac{\partial N}{\partial m}\,dm = \rho_n \,.$$

ρ_n, the normal curvature of Σ_m, is therefore independent of C and only depends on its tangent at m. Thus :

$$\rho_n = -\,\overline{\frac{dm}{ds}}\,\frac{\partial N}{\partial m}\,\frac{dm}{ds}$$

$$\overline{\frac{dm}{ds}}\left[\frac{\partial N}{\partial m} + \rho_n\,1_{E_2}\right]\frac{dm}{ds} = 0 \,,$$

or :

$$\overline{t} \left[\frac{\partial N}{\partial m} + \rho_n \, 1_{E_2} \right] t = 0 \, .$$

If we look for t, such that ρ_n is stationary at m, with m fixed :

$$2 \, \overline{t} \left[\frac{\partial N}{\partial m} + \rho_n \, 1_{E_2} \right] \delta t + \overline{t} \, \delta\rho_n \, t = 0 \quad \forall \delta t, \quad \text{and} \quad \delta\rho_n = 0$$

$$\left[\frac{\partial N}{\partial m} + \rho_n \, 1_{E_2} \right] t = 0 \, .$$

The eigenvalues of $\frac{\partial N}{\partial m}$ are the real principal curvatures $\rho_{n\alpha}$, and correspond to the principal orthogonal directions t_α, $\alpha = 1, 2$.

By virtue of (6.4), at point M we put :

$$\partial_1 M = \mu \partial_1 m = \mu S_1 = T_1$$

$$\partial_2 M = \partial_2 m = \mu S_2 = T_2$$

$$\partial_3 M = N = T_3$$

and the natural basis T at M will be :

$$(6.6) \qquad T = [T_1 \ T_2 \ T_3] = [\mu S_1 \ \mu S_2 \ N] = [\mu S \ N]$$

$$dM = TdY = [\mu S \ N] \begin{bmatrix} dX \\ dz \end{bmatrix}, \quad \forall dY = \begin{bmatrix} dX \\ dz \end{bmatrix} \in \mathbb{R}^3 \, .$$

Furthermore :

$$T = \frac{\partial M}{\partial Y} \, , \quad Y = \begin{bmatrix} X \\ z \end{bmatrix}, \quad X \in \mathcal{O}, \quad z \quad \left] -\frac{h}{2}, \frac{h}{2} \right[$$

<u>Remarks</u> :

The components of G are frequently called :

$$^\alpha G_\beta = \overline{S_\alpha} \, S_\beta = a_{\alpha\beta} \, , \quad \alpha,\beta = 1,2$$

The covariant components of $\frac{\partial N}{\partial m}$ are frequently called :

$$^\alpha \overline{S} \, \frac{\partial N}{\partial m} \, S_\beta = \overline{S_\alpha} \, \frac{\partial N}{\partial m} \, S_\beta = - \, b_{\alpha\beta}, \quad \alpha,\beta = 1,2 \, .$$

Volume element at M

As a result of the above, we can adopt as volume element at M :

$$(6.7) \quad \begin{cases} d\Omega = \text{vol}(d_1M)(d_2M)(d_3M) = dz \, \det(\mu) \, d\Sigma_m, \\[4pt] \text{with :} \\[4pt] d_1M = \mu d_1m, \quad d_2M = \mu d_2m, \quad d_3M = Ndz, \quad \forall d_1m, \, d_2m \in \vec{E}_2, \quad dz \in \mathbb{R} \end{cases}$$

as :

$$d\Omega = \text{vol}(\mu d_1m)(\mu d_2m)(Ndz)$$

$$= \bar{N} \, dz \, i(\mu d_1m)(\mu d_2m)$$

$$= dz \, \bar{N} \, N \, \text{vol}_2(\mu d_1m)(\mu d_2m)$$

$$d\Omega = dz \, \det(\mu) \, \text{vol}_2(d_1m)(d_2m)$$

$$d\Omega = dz \, \det(\mu) d\Sigma_m, \quad \text{with} \, \text{vol}_2(d_1m)(d_2m) = d\Sigma_m.$$

Christoffel operator

Finally, it should be remembered that $\forall$ tangent vector fields $V \in \vec{E}_{2,m}$, and $\forall$ differential field $dm \in \vec{E}_{2,m}$:

$$dm = SdX$$

$$V = SY$$

$$dV = dSY + SdY$$

$$\Pi dV = \hat{d}V = \Pi dSY + SdY.$$

$\Pi d \, S_\alpha \in \vec{E}_2$, therefore $\exists$ a matrix linearly dependent on dX, namely $\Gamma(dX)$, such that :

$$\Pi dS = S \, \Gamma(dX),$$

whence :

$$\hat{d}V = S\left[\Gamma(dX)(Y) + dY\right].$$

Thus :

$$\hat{d}V = S_\alpha \left[{}^\alpha\Gamma(|_\beta\, d^\beta X \cdot)(|_\gamma\, {}^\gamma Y) + d^\alpha Y \right] \quad ;\alpha,\ \beta,\ \gamma = 1,2$$

(6.8)
$$\hat{d}V = S_\alpha \left[{}^\alpha\Gamma_{\beta\gamma}\, d^\beta X\ {}^\gamma Y + d^\alpha Y \right].$$

The operator Γ, a tensor of $\mathbb{R}^2$, is the <u>Christoffel operator</u>, and the ${}^\alpha\Gamma_{\beta\gamma}$ are the <u>Christoffel symbols of the second kind</u>.

Now :

$$S = \frac{\partial m}{\partial X} = \phi'(X)$$

$$\Pi dS = \Pi\phi''(X)(dX) = S\,\Gamma(dX)$$

$$\hat{d}SY = \Pi\phi''(X)(dX)(Y) = S\,\Gamma(dX)(Y).$$

This formula demonstrates that ${}^\alpha\Gamma_{\beta\gamma}$ is symmetrical in β, γ. (The "torsion of connection" is zero). The Christoffel symbols of the first kind will be [see (A.87)] :

$$\gamma_{\alpha\beta\gamma} = a_{\alpha,\alpha'}\, {}^{\alpha'}\Gamma_{\beta\gamma} = {}^\alpha G_{\alpha'}\, {}^{\alpha'}\Gamma_{\beta\gamma} = \frac{1}{2}\left[a_{\alpha\beta,\gamma} + a_{\gamma\alpha,\beta} - a_{\beta\gamma,\alpha} \right].$$

1.3 – <u>General hypotheses</u>[24]

We can take successively two kinematic hypotheses respectively called "(w)" and "(wo)", namely with transverse shear or without transverse shear.

<u>Hypothesis (w) : with transverse shear</u>

Let us consider an arbitrary point M_o on the shell, in some state $\mathcal{E}_o$, defined by :

$$M_o = m_o + N_o z_o.$$

In a state $\mathcal{E}$:

$$M = m + Nz$$

the displacement W is equal to :

$$W = M - M_o = m - m_o + Nz - N_o z_o \in \vec{E}_3.$$

Let us put :

$$m - m_o = V \in \vec{E}_3, \quad \Delta N = N - N_o \in E_3, \quad \Delta z = z - z_o$$

$$W = V + \Delta N\, z_o + N_o\, \Delta z + \Delta N\, \Delta z.$$

To simplify, one hypothesis will consist in putting :

(6.9) $W = V + \Delta N\, z_o + N_o\, \Delta z,$

thus neglecting the second order variations.

The virtual displacement of M in $\mathcal{E}$ will then be :

$$\delta M = \delta m + \delta N z + N \delta z.$$

Furthermore :

(6.10) $\delta M = \delta m + \delta N z + N \delta z, \quad \text{with } \delta N \in \vec{E}_2,$

as :

$$\overline{N}\, N = 1$$

$$\overline{\delta N}\, N = 0, \quad \delta N \in \vec{E}_2$$

and likewise for the actual displacement (6.9), linearized in state $\mathcal{E}$.

Formula (6.10), also valid for a linearized actual displacement, is called
<u>hypothesis (w)</u>, with transverse shear, which will be justified in due course.

Formula (6.10) allows us to make a surface formulation for the virtual
(or actual) displacement, assuming :

(6.11) $\dfrac{\delta z}{h} = \delta a_1\, (m)\, \dfrac{z}{h} + \delta a_2\, (m)\, \dfrac{z^2}{h^2} + \dots ,$

by asymptotic expansion of δz with respect to $\dfrac{z}{h}$, up to a certain order.

In (6.10) and (6.11), the surface variables δm, δN, δa_1, δa_2, $\dots$, are
independent variables, each leading to equilibrium equations. It is clear
that the equilibrium equations generated by the variables δm and δN, will
have a formulation independent of that generated by δz.

<u>Hypothesis (wo) : without transverse shear</u>

A supplementary kinematic restriction will consist in assuming that the material normal, after deformation, remains normal to the middle surface in its deformed state. If we argue by virtual variations, the condition :

$$\overline{N}dm = 0 \qquad \forall\, dm \in \vec{E}_2$$

gives :

$$\overline{\delta N}dm + \overline{N}\delta dm = 0, \quad \forall \delta m \text{ with } m \text{ fixed.}$$

Whence :

$$\overline{\delta N}dm + \overline{N}d\delta m = 0 \quad \forall\, dm \in \vec{E}_2$$

$$(6.12) \qquad \delta N = - \overline{\frac{\partial \delta m}{\partial m}}\, N$$

and likewise for the linearized actual displacement :

$$(6.13) \qquad \Delta N = - \overline{\frac{\partial V}{\partial m_o}}\, N_o .$$

Hypothesis (6.10), accompanied by (6.11) and (6.12), will be called <u>hypothesis (wo)</u>, without transverse shear, this designation again being justified in due course. They are then translated into virtual displacements by :

$$(6.14) \qquad \left[\begin{array}{l} \delta M = \delta m - z\, \overline{\dfrac{\partial \delta m}{\partial m}}\, N + N\delta z \\[2ex] \text{with } (6.11). \end{array} \right.$$

This hypothesis $(6.14)_1$, completed by $\delta z = 0$, constitutes the famous <u>kinematic hypotheses of Kirchhoff-Love</u>.

The last hypothesis of Kirchhoff-Love consists in assuming, furthermore, that the normal stress $\sigma_{33} = 0$, which in fact represents a contradictory hypothesis with $\delta z = 0$.

<u>Remarks</u> :

We have seen that in (6.10), the two variables δm and δN were independent in hypothesis (w). They are no longer independent, as shown by (6.12), in hypothesis (wo).

Furthermore, we shall see that this hypothesis leads to a certain number
of consequences, in particular on linearized deformations which are free
from transverse shear, but also on non-linear deformations, as from a state
$\mathcal{E}_o$ to a state $\mathcal{E}_1$:

$$\Delta N = \int_o^1 \delta N .$$

To conclude, it should be noted that if we call $\delta\theta$ the linearized virtual
rotation vector of the normal, we have :

$$\delta N = i(\delta\theta)(N) = - i(N)(\delta\theta)$$

and $\delta\theta$, like δN, belongs to $\vec{E}_2$.

Now we see easily that :

(6.15) $i(N) = i_2\Pi = \Pi i_2, \quad i_2^2 = - 1_{E_2}$

where i_2 is the $+ 90°$ rotation endomorphism in $\vec{E}_2$, [see (A.24)]. Thus :

$$(6.16) \quad \begin{cases} \delta N = - i_2 \, \delta\theta \\[1ex] \text{or} \\[1ex] \delta\theta = i_2 \, \delta N . \end{cases}$$

Finally, (6.12) is a hypothesis for material normals, but always verified
for the <u>geometrical normals</u>, always orthogonal to Σ_m.

2. – <u>EQUILIBRIUM EQUATIONS</u>[24]

2.1 – <u>Equilibrium equations in the case of hypothesis (w)</u>

Let $\delta\mathcal{C}_e$ be the virtual work of the external forces applied to the shell
including inertial forces, and $- \delta w$ the virtual work of the internal forces,
for a kinematically admissible displacement field δM. The principle of
virtual work in state $\mathcal{E}$ is written :

$$\delta w = \delta \mathcal{C}_e \quad \forall \delta M. \text{ K.A.}$$

In the sequel, it will be assumed that all quantities introduced are differentiable and summable without restriction. Thus calling C the Cauchy stress at M :

$$\delta w = \int 3 \int_\Omega T_r(C \, \frac{\partial \delta M}{\partial M}) d\Omega \;, \quad C = \vec{C}.$$

Firstly, by virtue of a preceding remark, we shall assume $\delta z = 0$.

In the case of hypothesis (w), with $\delta z = 0$, we have [see (6.10)]

$$\delta M = \delta m + z \delta N.$$

$$\frac{\partial \delta M}{\partial M} = \frac{\partial \delta m}{\partial M} + \frac{\partial}{\partial M} \left[z \delta N \right]$$

and :

$$(6.17) \qquad \delta w = \int 3 \int_\Omega T_r(C \, \frac{\partial \delta m}{\partial M} + C \, \frac{\partial}{\partial M} \left[z \delta N \right]) \; vol(d_1 M)(d_2 M)(d_3 M).$$

The first term of the second member of (6.17), taking account of (6.7) and (A.4), is written :

$$\delta w_1 = \int 3 \int_\Omega T_r(C \, \frac{\partial \delta m}{\partial M}) \; vol(d_1 M)(d_2 M)(d_3 M).$$

$$\delta w_1 = \int 3 \int_\Omega \left[vol(Cd_1 \delta m)(d_2 M)(Ndz) + vol(Cd_1 M)(d_2 \delta m)(Ndz) \right]$$

as $d_3 \delta m = 0$.

Now let Π be the Hermitian projector of $\vec{E}_3$ onto $\vec{E}_2$, and $N\overline{N}$ the supplementary projector :

$$(6.18) \qquad \Pi + N\overline{N} = 1_{E_3}$$

In δw_1, the presence of vector Ndz makes it possible to retain only the tangential component πC of C, as :

$$C = \Pi C + N\overline{N} \, C.$$

Thus [see (A.18)] :

$$\delta w_1 = \int 3 \int_\Omega \mathrm{vol}(\Pi C d_1 \delta m)(d_2 M)(N dz) + \mathrm{vol}(d_1 M)(\Pi C d_2 \delta m)(N dz)$$

$$= \int 3 \int_\Omega \left[\overline{d_1 \delta m} \ C\Pi \ i(d_2 M)(N) - \overline{d_2 \delta m} \ C\Pi \ i(d_1 M)(N) \right] dz$$

$$\delta w_1 = -\int 3 \int_\Omega \left[\overline{d_1 \delta m} \ C\Pi \ i_2(d_2 M) - \overline{d_2 \delta m} \ C\Pi \ i_2(d_1 M) \right] dz$$

by reason of (6.15).

Applying Fubini's theorem, we can first integrate with respect to z, and
we then are led to put, by definition :

(6.19)

$$\overline{\delta m} \ \int_{-\frac{h}{2}}^{+\frac{h}{2}} dz \ C \ i_2 \ \mu dm = \overline{\delta m} \ \mathbf{N} i_2 \ dm$$

$$\forall \delta m \in \vec{E}_3, \ dm \in \vec{E}_2$$

$$\mathbf{N} \text{ maps } \quad \vec{E}_2 \quad \text{in} \ \vec{E}_3$$

Then :

$$\delta w_1 = -\iint_{\Sigma_m} \left[\overline{d_1 \delta m} \ \mathbf{N} i_2 \ d_2 m - \overline{d_2 \delta m} \ \mathbf{N} i_2 \ d_1 m \right]$$

Now $\forall V_1$, $v_2 \in \vec{E}_2$, [see (A.23)] :

$$\mathrm{vol}_2(V_1)(V_2) = \overline{V_2} \ i_2 \ V_1 .$$

Whence [see (A.4)] :

$$\delta w_1 = \iint_{\Sigma_m} \left[\mathrm{vol}_2(\overline{\mathbf{N}} \ d_1 \delta m)(d_2 m) + \mathrm{vol}_2(d_1 m)(\overline{\mathbf{N}} \ d_2 \delta m) \right] .$$

(6.20) $\qquad \delta w_1 = \iint_{\Sigma_m} T_r(\overline{\mathbf{N}} \frac{\partial \delta m}{\partial m}) \ \mathrm{vol}_2(d_1 m)(d_2 m)$

It should be noted that $\bar{N} \frac{\partial \delta m}{\partial m}$ is indeed an endomorphism on $\vec{E}_2$.

As for the second term of the second member of (6.17), it is written [see (A.4)] :

$$\delta w_2 = \int 3 \int_\Omega T_r \left(C \frac{\partial}{\partial M} \left[z\delta N \right] \right) \, \text{vol}(d_1 M)(d_2 M)(d_3 M)$$

$$= \int 3 \int_\Omega \left[\left[\text{vol}(Cd_1 \delta N)(d_2 M)(N) + \text{vol}(d_2 M)(Cd_2 \delta N)(N) \right] zdz \right.$$

$$\left. + \text{vol}(d_1 M)(d_2 M)(C\delta Ndz) \right].$$

Again we have :

$$C = C\Pi + C N \bar{N}$$

$$\delta w_2 = \int 3 \int_\Omega \left[\text{vol}(\Pi C\Pi d_1 \delta N)(d_2 M)(N) + \text{vol}(d_1 M)(\Pi C\Pi d_2 \delta N)(N) \right] zdz$$

$$+ \int 3 \int\int_\Omega \left[\left[\text{vol}(C \, N\bar{N}d_1 \delta N)(d_2 M)(Ndz) + \text{vol}(d_1 M)(C \, N\bar{N}d_2 \delta N)(Ndz) \right] z \right.$$

$$\left. + \text{vol}(d_1 M)(d_2 M)(C\delta Ndz) \right].$$

Now : $\bar{N}\delta N = 0 \Rightarrow \bar{N}d_1 \delta N = - \overline{d_1 N}\delta N = - \overline{d_1 N} \, i(\delta\theta)(N)$

$$\bar{N}d_2 \delta N = - \overline{d_2 N}\delta N = - \overline{d_2 N} \, i(\delta\theta)(N)$$

$$\bar{N}d_1 \delta N = \bar{N} \, i(\delta\theta)(d_1 N)$$

$$\bar{N}d_2 \delta N = \bar{N} \, i(\delta\theta)(d_2 N) \, .$$

Thus in δw_2, the last three terms of the second member are written :

$$\mathrm{vol}(C\ \overline{NN}\ i(\delta\theta)(d_1 Nz))(d_2 N)(Ndz) + \mathrm{vol}(d_1 M)(C\ \overline{NN}\ i(\delta\theta)(d_2 Nz))(Ndz)$$

$$+\ \mathrm{vol}(d_1 M)(d_2 M)(C i(\delta\theta)(Ndz))$$

$$=\ \mathrm{vol}(C\ i(\delta\theta)(d_1 \overrightarrow{mM}))(d_2 M)(d_3 M)\ +\ \mathrm{vol}(d_1 M)(C\ i(\delta\theta)(d_2 \overrightarrow{mM}))(d_3 M)$$

$$+\ \mathrm{vol}(d_1 M)(d_2 M)(C i(\delta\theta)(d_3 \overrightarrow{mM}))$$

where $\overline{NN}\ i(\delta\theta)(d_1 Nz)$ has been replaced by $i(\delta\theta)(d_1 \overrightarrow{mM})$, as $\delta\theta$ and $d_1 N \in \vec{E}_2$, and $d_1 Nz = d_1[Nz] = d_1 \overrightarrow{mM}$.

But this development is also equal to :

$$T_r(C\ i(\delta\theta)\ \frac{\partial \overrightarrow{mM}}{\partial M})\ \mathrm{vol}(d_1 M)(d_2 M)(d_3 M)$$

$$=\ T_r(C\ i(\delta\theta)(1_{E_3} - \frac{\partial m}{\partial M}))\ \mathrm{vol}(d_1 M)(d_2 M)(d_3 M)\ =\ -\ T_r(C\ i(\delta\theta)\ \frac{\partial m}{\partial M})d\Omega.$$

Therefore :

$$\delta w_2\ =\ \int 3 \int_\Omega \left[\left[(\mathrm{vol}(\Pi C \Pi d_1 \delta N)(d_2 M)(N)\ +\ \mathrm{vol}(d_1 M)(\Pi C \Pi d_2 \delta N)(N)\right] zdz\right.$$

$$\left. -\ T_r(C\ i(\delta\theta)\ \frac{\partial m}{\partial M})\mathrm{vol}(d_1 M)(d_2 M)(d_3 M)\right]$$

$$\delta w_2\ =\ \int 3 \int_\Omega \left[\left[(\mathrm{vol}(\Pi C \Pi d_1 \delta N)(d_2 M)(N)\ +\ \mathrm{vol}(d_1 M)(\Pi C \Pi d_2 \delta N))(N)\right] zdz\right.$$

$$-\ \mathrm{vol}(C\ i(\delta\theta)(d_1 m))(d_2 M)(N)dz\ -\ \mathrm{vol}(d_1 M)(C i(\delta\theta)(d_2 m)(N)dz$$

as $d_3 m = 0$.

$$\delta w_2\ =\ \int 3 \int_\Omega \left[\overline{d_1 \delta N}\ \Pi C \Pi\ i(d_2 M)(N)\ -\ \overline{d_2 \delta N}\ \Pi C \Pi\ i(d_1 M)(N)\right] zdz$$

$$-\ \int 3 \int_\Omega \left[\overline{i(\delta\theta)(d_1 m)}\ C\ i(d_2 M)(N)dz\right.$$

$$\left. -\ \overline{i(\delta\theta)(d_2 m)}\ C\ i(d_1 M)(N)dz\right]$$

$$\delta w_2 = - \int 3 \iint_{\Omega} \left[\overline{d_1 \delta N} \ \Pi C \Pi \ i_2 \ \mu d_1 m - \overline{d_2 \delta N} \ \Pi C \Pi \ i_2 \ \mu d_2 m \right] z dz$$

$$+ \int 3 \int_{\Omega} \left[\overline{i(\delta\theta)(d_1 m)} \ C \ i_2 \ \mu d_2 m - \overline{i(\delta\theta)(d_2 m)} \ C \ i_2 \ \mu d_1 m \right] dz \ .$$

Let us put then:

$$\overline{\delta\theta} \int_{-\frac{h}{2}}^{+\frac{h}{2}} i_2 \ C\Pi \ i_2 \ \mu z dz \ dm = \overline{\delta\theta} \, \mathbf{M} \, i_2 \ dm, \ \forall \delta\theta \ , \ dm \in \vec{E}_2$$

or :

$$(6.21) \quad \left[\begin{array}{l} \overline{\delta\theta} \int_{-\frac{h}{2}}^{+\frac{h}{2}} i(\vec{mM}) C \Pi \ i_2 \ \mu dz \ dm = \overline{\delta\theta} \, \mathbf{M} \, i_2 \ dm \\[2em] \forall \delta\theta \ , dm \in \vec{E}_2 \\[1em] \mathbf{M} \quad \text{maps} \quad \vec{E}_2 \quad \text{in} \quad \vec{E}_2 \, . \end{array} \right.$$

Using (6.21) and (6.19) for the second integral of δw_2, we obtain :

$$\delta w_2 = \iint_{\Sigma_m} \left[\overline{d_1 \delta N} \ i_2 \, \mathbf{M} \, i_2 \ dm - \overline{d_2 \delta N} \ i_2 \, \mathbf{M} \, i_2 \ dm \right.$$

$$\left. + \overline{i(\delta\theta)(d_1 m)} \, \mathbf{N} \, i_2 \ d_2 m - \overline{i(\delta\theta)(d_2 m)} \, \mathbf{N} \, i_2 \ d_1 m \right] \ ,$$

and [see (A.23)] :

$$\delta w_2 = \iint_{\Sigma_m} - \left[\underset{2}{\text{vol}} (\overline{i_2 \, \mathbf{M} \, \hat{d}_1 \, \delta N})(d_2 m) + \underset{2}{\text{vol}} (d_1 m)(\overline{i_2 \, \mathbf{M} \, \hat{d}_2 \, \delta N}) \right.$$

$$\left. + \underset{2}{\text{vol}} (\bar{\mathbf{N}} \, i(\delta\theta)(d_1 m))(d_2 m) + \underset{2}{\text{vol}} (d_1 m)(\bar{\mathbf{N}} \, i(\delta\theta)(d_2 m)) \right]$$

Now :

$$i(\delta\theta)(d_\alpha m) = N\ \bar{N}\ i(\delta\theta)(d_\alpha m) = -\ N\ \overline{d_\alpha m}\ i(\delta\theta)(N)$$

$$= -\ N\ \overline{d_\alpha m}\ \delta N = -\ N\ \overline{\delta N}\ d_\alpha m,\quad (\alpha = 1,2).$$

Finally :

$$(6.22)\qquad \delta w_2 = \iint_{\Sigma_m} -\ T_r(\overline{i_2 \mathbf{M}}\ \Pi\ \frac{\partial \delta N}{\partial m} - \bar{\mathbf{N}}\ N\ \overline{\delta N})\ \underset{2}{\mathrm{vol}}(d_1 m)(d_2 m).$$

Recapitulating (6.20) and (6.22), we obtain :

$$(6.23)\qquad \delta w = \iint_{\Sigma_m} T_r(\ \bar{\mathbf{N}}\ \frac{\partial \delta m}{\partial m} - \overline{i_2 \mathbf{M}}\ \Pi\ \frac{\partial \delta N}{\partial m} + \bar{\mathbf{N}}\ N\ \overline{\delta N})\,d\Sigma_m\ .$$

We then apply formula (A.95), which gives here, for example :

$$T_r(\ \bar{\mathbf{N}}\ \frac{\partial \delta m}{\partial m}) = -\ \widehat{\mathrm{div}}\ \bar{\mathbf{N}}\ \delta m + \widehat{\mathrm{div}}\left[\bar{\mathbf{N}}\ \delta m\right],$$

with :

$$\widehat{\mathrm{div}}\left[\bar{\mathbf{N}}\ \delta m\right] = T_r(\Pi\ \frac{\partial}{\partial m}\left[\bar{\mathbf{N}}\ \delta m\right]\ ),$$

thus defining $\widehat{\mathrm{div}}\ \mathbf{N}$ as a covector of $\vec{E}_3$ ($\epsilon\ \vec{E}_3^*$), or $\widehat{\mathrm{div}}\ i_2 \mathbf{M}$ as a covector of $\vec{E}_2$ ($\epsilon\ \vec{E}_2^*$). (6.23) then gives :

$$(6.24)\qquad \left[\begin{array}{l} \delta w = \iint_{\Sigma_m}\left[-\ \widehat{\mathrm{div}}\ \bar{\mathbf{N}}\ \delta m + \widehat{\mathrm{div}}\left[\bar{\mathbf{N}}\ \delta m\right]\right. \\[3em] \left. +\ \widehat{\mathrm{div}}\ \overline{i_2 \mathbf{M}}.\delta N - \widehat{\mathrm{div}}\left[\overline{i_2 \mathbf{M}}.\delta N\right] + \bar{\mathbf{N}}\ N\ \delta N\right]d\Sigma_m\ . \end{array}\right.$$

Then applying the Stokes formula, we obtain [see (A.99)] :

$$\delta w = \iint_{\Sigma_m}\left[-\ \widehat{\mathrm{div}}\ \bar{\mathbf{N}}\ \delta m + \left[\widehat{\mathrm{div}}\ i_2 \bar{\mathbf{M}} + \bar{\mathbf{N}}\ N\right]\delta N\right]d\Sigma_m$$

$$+ \int_{C_m}\left[\underset{2}{\mathrm{vol}}(\bar{\mathbf{N}}\ \delta m)(dm) - \underset{2}{\mathrm{vol}}(\overline{i_2 \mathbf{M}}\ \delta N)(dm)\right]$$

or also :

$$(6.25) \qquad \delta w = \iint\limits_{\Sigma_m} \left[- \widehat{\text{div}} \, \mathbf{N} \, \delta m + \left[\widehat{\text{div}} \, \overline{i_2 \, \mathbf{M}} + \bar{\mathbf{N}} \, \mathbf{N} \right] \delta N \right] d\Sigma_m$$

$$+ \int\limits_{C_m} \left[\overline{\delta m} \, \mathbf{N} \, i_2 \, dm - \overline{\delta N} \, i_2 \, \mathbf{M} \, i_2 \, dm \right].$$

It should be noted that on C_m :

$$i_2 dm = i_2 \frac{dm}{ds} \, ds = i_2 \, t \, ds = - \nu ds.$$

where ν is the external unit normal to C_m on Σ_m.

By integration with respect to z, the volume densities for external forces give the surface densities of resultants and couples which are conjugated with the virtual displacements δm and δN, and which we shall call $\mathbf{R}_m$ and $\mathbf{G}_m$ respectively. By integration with respect to z in Σ_L, the surface densities of the lateral forces also give curvilinear force densities on C_m, and finally (6.25) together with the principle of virtual work, give the following equilibrium equations in the case of hypothesis (w), and with $\delta z = 0$:

$$(6.26) \qquad \begin{cases} \widehat{\text{div}} \, \bar{\mathbf{N}} + \overline{\mathbf{R}_m} = 0 \\[2ex] \widehat{\text{div}} \, \overline{i_2 \, \mathbf{M}} + \bar{\mathbf{N}} \mathbf{N} + \overline{\mathbf{G}_m} = 0 \end{cases} \quad \text{on } \Sigma_m$$

$$\left[\overline{\delta m} \, \mathbf{N} - \overline{\delta N} \, i_2 \mathbf{M} \right] \nu = \left[\overline{\delta m} \, \mathbf{R}_L - \overline{\delta N} \, \mathbf{G}_L \right] t \quad \text{on } C_m \, ,$$

$$\forall \delta m \text{ and } \delta N \text{ K.A.}$$

<u>Remarks</u> :

1) The normals N in this linearization behave as rigid segments, and it was then clear that the equations would be equations of general resultants and couple resultants of the stresses distributed on each normal.

2) In $(6.26)_2$, with the natural basis S, and using (A.96) which is also valid in this case, we can write :

$$\mathrm{div}\ \overline{i_2 \mathbf{M}} = {}^\alpha s^{-1}\ \widehat{\partial}_\alpha \left[i_2 \mathbf{M} \right]$$

$$= -\ {}^\alpha s^{-1}\ \widehat{\partial}_\alpha\ \overline{\mathbf{M}}\ i_2 \quad \text{as}\ \widehat{d}i_2 = 0 \quad [\text{see (A.89)}].$$

$(6.26)_2$ is therefore written :

$$\mathrm{div}\ \overline{\mathbf{M}} + \overline{\mathbf{N}}\mathbf{N}\ i_2 + \overline{\mathbf{G}_m} i_2 = 0.$$

3) In $(6.26)_1$, $\mathbf{N}$ applies $\vec{E}_3$ in $\vec{E}_2$, and $\forall \mathrm{W}\ \mathrm{c}^t$:

$$\mathrm{div}\ \overline{\mathbf{N}}\ \mathrm{W} = \mathrm{div}\ \left[\overline{\mathbf{N}}\ \mathrm{W} \right]$$

$$= {}^\alpha s^{-1}\ \widehat{\partial}_\alpha\ \overline{\mathbf{N}}\ \mathrm{W}.$$

We still have :

$$\widehat{\mathrm{div}\ \mathbf{N}} = {}^\alpha s^{-1}\ \widehat{\partial}_\alpha\ \overline{\mathbf{N}}\ .$$

Furthermore, we refer to (A.81) to calculate the components of this convector in basis T, namely :

$$\left[\widehat{\mathrm{div}\ \mathbf{N}} \right]_i = \widehat{\mathrm{div}\ \mathbf{N}}\ T_i = {}^\alpha s^{-1}\ \widehat{\partial}_\alpha\ \overline{\mathbf{N}}\ t_i \quad ;\ \alpha = 1,\ 2\ ;\quad i = 1,2,3.$$

usually called $N^{\beta\alpha}_{\|\alpha}$ (for the contravariant tangential component).

4) As an example, if we assume $\delta z \neq 0$, with :

$$\delta z = \delta a_1(m).z,$$

an additional energy term δw_3 is added :

$$\delta w_3 = \int 3 \int_\Omega T_r (C\ \frac{\partial}{\partial M} \left[N\delta a_1\ z \right])\,d\Omega.$$

Now :

$$\frac{\partial [N\delta a_1 \cdot z]}{\partial M} = \frac{\partial N}{\partial M}\cdot\delta a_1\ z + N\ \frac{\partial \delta a_1}{\partial M}\cdot z + N\delta a_1\ \frac{\partial z}{\partial M}.$$

The first and third terms provide the contribution :

$$- \iint_{\Sigma_m} T_r (\overline{i_2 \mathbf{M}}\ \frac{\partial N}{\partial m} - \mathbf{K})\delta a_1\ d\Sigma_m$$

with :

$$K = \int_{-\frac{h}{2}}^{+\frac{h}{2}} dz \ \bar{N} \ C \ N \ \mu \ .$$

The second term supplies :

$$\iint_{\Sigma_m} T_r (\ \bar{P} \ N \ \frac{\partial \delta a_1}{\partial m}) d\Sigma_m$$

with :

$$\bar{N} \ P \ i_2 \ dm = \int_{-\frac{h}{2}}^{+\frac{h}{2}} dz \ \bar{N} \ C \ zi_2 \ \mu dm, \ \forall dm \in \vec{E}_2 \ .$$

These contributions give a third scalar equilibrium equation conjugated with δa_1, namely :

$$T_r (\overline{i_2 M} \frac{\partial N}{\partial m}) - K + \widehat{\mathrm{div}} \ [\bar{P} \ N] = - \ [\bar{N} G] \ , \ \text{on} \ \Sigma_m$$

and terms :

$$- \ \delta a_1 \ \bar{N} \ P \ \nu = \delta a_1 \ \bar{N}_L \ t, \ \text{on} \ C_m, \ \forall \delta \ a_1 \ .$$

2.2 - Equilibrium equations in the case of hypothesis (wo)

Again in the simplified case where $\delta z = 0$, we need merely take account of constraint (6.12) in expression (6.25) of δw. The term in δN in (6.25) is written [see (A.5)] :

$$-\left[\widehat{\mathrm{div}} \ \overline{i_2 M} + \bar{N} N\right] \frac{\overline{\partial \delta m}}{\partial m} \ N = - \ T_r (\ \frac{\overline{\partial \delta m}}{\partial m} \ N \left[\widehat{\mathrm{div}} \ \overline{i_2 M} + \bar{N} \ N\right])$$

$$= - \ T_r (\left[\widehat{\mathrm{div}} \ \overline{i_2 M} + \bar{N} N\right] \bar{N} \ \frac{\partial \delta m}{\partial m} \ ) \ .$$

Application of formula (A.95), and the Stokes formula, then gives :

$$- \iint_{\Sigma_m} T_r \left(\left[\overline{\widehat{\mathrm{div}\ \overline{i_2 \mathbf{M}}}} + \bar{\mathbf{N}}\, N \right] \bar{\mathbf{N}} \frac{\partial \delta m}{\partial m} \right) d\Sigma_m$$

$$= \iint_{\Sigma_m} \widehat{\mathrm{div}} \left[N\ \widehat{\mathrm{div}\ \overline{i_2 \mathbf{M}}} + \bar{\mathbf{N}}\, N\, \bar{\mathbf{N}} \right] \delta m\ d\Sigma_m$$

$$+ \int_{C_m} \overline{\delta m} \left[N\ \widehat{\mathrm{div}\ \overline{i_2 \mathbf{M}}} + N\ \bar{\mathbf{N}}\, N \right] i_2\ dm .$$

We now find, for δw :

$$\delta w = \iint_{\Sigma_m} \widehat{\mathrm{div}} \left[- \bar{\mathbf{N}} + \overline{N\ \widehat{\mathrm{div}\ \overline{i_2 \mathbf{M}}}} + \bar{\mathbf{N}}\, N\, \bar{\mathbf{N}} \right] \delta m\ d\Sigma_m$$

$$+ \int_{C_m} \left[\overline{\delta m} \left[- \mathbf{N} + N\ \widehat{\mathrm{div}\ \overline{i_2 \mathbf{M}}} + N\ \bar{\mathbf{N}} N \right] + \overline{\delta N}\ i_2 \mathbf{M} \right] i_2\ dm .$$

If we replace $\bar{\mathbf{N}}$ by $\bar{\mathbf{N}} [\Pi + N\ \bar{\mathbf{N}}]$, we obtain :

$$(6.27) \quad \left[\begin{array}{l} \delta w = - \iint_{\Sigma_m} \widehat{\mathrm{div}} \left[\Pi\, \bar{\mathbf{N}} - \overline{N\ \widehat{\mathrm{div}\ \overline{i_2 \mathbf{M}}}} \right] \delta m - \int_{C_m} \overline{\delta m} \left[\Pi\, \mathbf{N} - N\ \widehat{\mathrm{div}\ \overline{i_2 \mathbf{M}}} \right] i_2 dm \\[4ex] \qquad\qquad\qquad\qquad\qquad\qquad + \int_{C_m} \overline{\delta N}\ i_2 \mathbf{M}\ i_2\ dm . \end{array} \right.$$

We must also take account of constraint (6.12) in the third integral of (6.27), according to which the normal N must remain orthogonal to C_m. To calculate this curvilinear integral, it is convenient to take a basis of $\vec{E}_2$:

$$S' = [t\ \nu], \quad \overline{S'}\, S' = 1_{R^2} \quad (\nu \text{ provisionally internal normal})$$

on C_m. Thus :

$$\int_{C_m} \overline{\delta N}\ i_2 \mathbf{M}\ i_2\ dm = - \int_{C_m} \bar{\mathbf{N}}\ \frac{\partial \delta m}{\partial m}\ i_2 \mathbf{M}\ \nu ds .$$

Now :

$$- \bar{N} \frac{\partial \delta m}{\partial m} \, i_2 \mathbf{M} \, \nu = \left[- \frac{\partial \bar{N} \delta m}{\partial m} + \overline{\Pi \delta m} \, \frac{\partial N}{\partial m} \right] i_2 \mathbf{M} \nu$$

$$= - \frac{\partial \bar{N} \delta m}{\partial m} \, S'_\alpha \, \bar{S}'_\alpha \, i_2 \mathbf{M} \nu + \overline{\Pi \delta m} \, \frac{\partial N}{\partial m} \, i_2 \mathbf{M} \nu \, , (\alpha = 1,2)$$

$$= \partial_1 \left[\bar{N} \delta m \right] \bar{\nu} \mathbf{M} \nu - \partial_2 \left[\bar{N} \delta m \right] \bar{t} \mathbf{M} \nu + \overline{\Pi \delta m} \, \frac{\partial N}{\partial m} \, i_2 \mathbf{M} \nu$$

$$= \partial_1 \left[\bar{N} \delta m \, \bar{\nu} \mathbf{M} \nu \right] - \bar{N} \delta m \, \partial_1 \left[\bar{\nu} \mathbf{M} \nu \right] - \bar{t} \mathbf{M} \nu \, \frac{\partial \bar{N} \delta m}{\partial \nu} + \overline{\Pi \delta m} \, \frac{\partial N}{\partial m} \, i_2 \mathbf{M} \nu \, .$$

Therefore :

$$- \bar{N} \frac{\partial \delta m}{\partial m} \, i_2 \mathbf{M} \nu ds = d \left[\bar{N} \delta m \bar{\nu} \mathbf{M} \nu \right] - \bar{N} \delta m \, d \left[\bar{\nu} \mathbf{M} \nu \right] - \bar{t} \mathbf{M} \nu \, \frac{\partial \bar{N} \delta m}{\partial \nu} \, ds$$

$$+ \overline{\Pi \delta m} \, \frac{\partial N}{\partial m} \, i_2 \mathbf{M} \nu ds \, .$$

As we mentioned in paragraph 2.4, in this formula, the derivatives with respect to s alone give rise to an integration by parts on C_m, and an independent unknown $\frac{\partial \bar{N} \delta m}{\delta \nu}$ remains in addition to δm. Two cases are possible :

a) C_m has angular points, and if the term $d[\bar{N} \delta m \bar{\nu} \mathbf{M} \nu]$ shows two determinations after integration along C_m, the discontinuity must be compensated by discrete external forces.

b) C_m presents no angular points, but discrete external forces exist on C_m, which will compensate the discontinuity of this term.

Let k points of discontinuity be on C_m, we have :

$$\int_{C_m} \bar{N} \frac{\partial \delta m}{\partial m} \, i_2 \mathbf{M} \nu ds = \Sigma_k \left[\bar{\nu} \mathbf{M} \nu \bar{N} \delta m \right]_-^+ + \int_{C_m} \left[- \bar{N} \delta m \, \frac{d \left[\bar{\nu} \mathbf{M} \nu \right]}{ds} \right.$$

$$\left. - \bar{t} \mathbf{M} \nu \, \frac{\partial \bar{N} \delta m}{\partial \nu} - \overline{\Pi \delta m} \, \frac{\partial N}{\partial m} \, i_2 \mathbf{M} \nu \right] ds$$

where ν is here the external normal to C_m on Σ_m.

This formula, associated with (6.27) and the work of the external forces, easily supplies the following equilibrium equations :

$$(6.28) \quad \left[\begin{array}{l} \widehat{\mathrm{div}}\left[\overline{\Pi\,\mathbf{N}} - \mathrm{N}\,\overline{\widehat{\mathrm{div}}\,\overline{\mathbf{i}_2\,\mathbf{M}}}\right] + \overline{\mathbf{R}_{\mathrm{m}}} + \widehat{\mathrm{div}}\left[\mathbf{G}_{\mathrm{m}}\,\bar{\mathbf{N}}\right] \;,\; \text{on } \Sigma_{\mathrm{m}} \\[2ex]
\Sigma_{\mathrm{k}}\left[\bar{\nu}\mathbf{M}\nu + \bar{\mathbf{t}}\,\mathbf{G}_{\mathrm{L}}\,\mathbf{t}\right]_{-}^{+}\;\bar{\mathrm{N}}\delta\mathrm{m} = 0, \text{ at the points of discontinuity} \\[1ex]
\hspace{3cm} \text{on } \mathrm{C}_{\mathrm{m}} \\[1ex]
\overline{\Pi\delta\mathrm{m}}\left[-\Pi\,\mathbf{N} + \dfrac{\partial\mathrm{N}}{\partial\mathrm{m}}\,\mathbf{i}_2\mathbf{M}\right]\nu \;+\; \bar{\mathrm{N}}\delta\mathrm{m}\left[\widehat{\mathrm{div}}\,\overline{\mathbf{i}_2\,\mathbf{M}} - \dfrac{\partial\bar{\nu}\mathbf{M}\nu}{\partial\mathrm{m}}\,\mathbf{i}_2\right]\nu \\[2ex]
\hspace{1.5cm} -\;\bar{\mathbf{t}}\,\mathbf{M}\nu\,\dfrac{\partial\bar{\mathrm{N}}\delta\mathrm{m}}{\partial\nu} \\[3ex]
= \overline{\Pi\delta\mathrm{m}}\left[\Pi\,\mathbf{R}_{\mathrm{L}} + \dfrac{\partial\mathrm{N}}{\partial\mathrm{m}}\,\mathbf{G}_{\mathrm{L}}\right]\mathbf{t} + \bar{\mathrm{N}}\delta\mathrm{m}\left[\mathbf{G}_{\mathrm{m}}\,\mathbf{i}_2 + \bar{\mathrm{N}}\,\mathbf{R}_{\mathrm{L}} + \dfrac{\partial\bar{\mathbf{t}}\,\mathbf{G}_{\mathrm{L}}\,\mathbf{t}}{\partial\mathrm{m}}\right]\mathbf{t} \\[3ex]
\hspace{1.5cm} -\;\bar{\nu}\,\mathbf{G}_{\mathrm{L}}\,\mathbf{t}\,\dfrac{\partial\bar{\mathrm{N}}\delta\mathrm{m}}{\partial\nu}\;,\; \forall\,\Pi\delta\mathrm{m},\;\bar{\mathrm{N}}\delta\mathrm{m},\;\dfrac{\partial\bar{\mathrm{N}}\delta\mathrm{m}}{\partial\nu}\;\;\text{K.A.} \end{array} \right.$$

<u>Remarks</u> :

1) There is now only one vectorial equation of equilibrium on Σ_{m}. This is an equation which, associated with the boundary equations, directly supplies the conditions which are necessary and sufficient for equilibrium in the case of hypothesis (wo). We can also obtain this equation from $(6.26)_1$ and $(6.26)_2$, which is not surprising as hypothesis (w) also contains hypothesis (wo) as a special case. It suffices to take the divergence of $(6.26)_2$, premultiplied by N, and put :

$$\bar{\mathbf{N}} = \bar{\mathbf{N}}\,[\,\Pi + \mathrm{N}\,\bar{\mathrm{N}}\,]$$

in $(6.26)_1$. But we must keep one of these two equations in the case of hypothesis (w).

2) Stress $\bar{\mathbf{N}}\,\mathrm{N}$ is the transverse shear stress, as we shall se in the case of hypothesis (w). It no longer intervenes in the case of hypothesis (wo), as it remains a priori indeterminate (see paragraph 4.1).

3) Variation in thickness is introduced as before, when necessary.

4) Application of $\overline{\Pi\,\mathbf{N}} - \mathrm{N}\,\overline{\widehat{\mathrm{div}}\,\overline{\mathbf{i}_2\,\mathbf{M}}}$ maps $\vec{\mathrm{E}}_3$ in $\vec{\mathrm{E}}_3$, but the first term

is purely tangential, and the second converts a normal vector into a tangent vector. As for the components of the divergence of these terms, we find :

- for the tangential component :

$$\widehat{div}\left[\overline{\Pi\mathbf{N}} - \overline{N \, div \, \overline{i_2\mathbf{M}}}\right]\Pi = \widehat{div}\left[\overline{\Pi\mathbf{N}}\right]\Pi - \widehat{div}\left[\overline{N \, \widehat{div} \, \overline{i_2\mathbf{M}}}\right]\Pi$$

$$= \widehat{div}\left[\overline{\Pi\mathbf{N}}\right]\Pi + {}^{\alpha}S^{-1} \, \overline{\widehat{div} \, \overline{i_2\mathbf{M}}} \, \bar{N} \, \partial_{\alpha}\Pi \; , \; (\alpha = 1,2)$$

$$= \widehat{div}\left[\overline{\Pi\mathbf{N}}\right]\Pi - {}^{\alpha}S^{-1} \, \overline{\widehat{div} \, \overline{i_2\mathbf{M}}} \, \bar{N} \, \partial_{\alpha}\left[N \, \bar{N}\right]$$

$$= \widehat{div}\left[\overline{\Pi\mathbf{N}}\right]\Pi - {}^{\alpha}S^{-1} \, \overline{\widehat{div} \, \overline{i_2\mathbf{M}}} \, \overline{\partial_{\alpha}N}$$

$$= \widehat{div}\left[\overline{\Pi\mathbf{N}}\right]\Pi - \widehat{div} \, \overline{i_2\mathbf{M}} \, \frac{\partial N}{\partial m} \; .$$

Using conventional notations, these covariant components are written [see (A.96)] :

$$N^{\beta}_{\alpha\|\beta} + M^{\beta}_{\alpha'\|\beta} \, b^{\alpha'}_{\alpha} \; .$$

- For the normal component :

$$\widehat{div}\left[\overline{\Pi\mathbf{N}}\right]N - \widehat{div} \, N \, \widehat{div} \, \overline{i_2\mathbf{M}} \, N = - T_r\left(\overline{\Pi\mathbf{N}} \, \frac{\partial N}{\partial m}\right) - \widehat{div} \, \widehat{div} \, \overline{i_2\mathbf{M}}$$

$$= - T_r\left(\frac{\Pi\mathbf{N} + \overline{\Pi\mathbf{N}}}{2} \, \frac{\partial N}{\partial m} + \frac{\hat{\partial}}{\partial m} \, \widehat{div} \, \overline{i_2\mathbf{M}}\right)$$

$$= N^{\alpha}_{\beta} \, b^{\beta}_{\alpha} - M^{\alpha\beta}_{\|\alpha\beta}$$

using conventional notations.

5) In the non-linear case, it will be appropriate to express the equations in a preceding reference state, for example the natural state[61].

2.3 - <u>Relation of symmetry of generalized surface stresses</u>

Generalized surface stresses **N** and **M** satisfy a classical relation of symmetry, which we shall use in due course. This is derived directly from the relation of symmetry (see chapter I, paragraph 9) :

$$C = \overline{C}$$

In fact with $\delta z = 0$, let us take the expression of energy for hypothesis (w) in general :

$$\delta w = \underset{\Sigma_m}{T_r} \left(\overline{\mathbf{N}} \frac{\partial \delta m}{\partial m} - \overline{i_2 \mathbf{M}} \frac{\partial \delta N}{\partial m} + \overline{\mathbf{N}} \; N \; \overline{\delta N} \right) d\Sigma_m,$$

and let a rigid body rotation of the shell be $\delta\theta \, \boldsymbol{\epsilon} \, \vec{E}_3$, independent of M and virtual, the virtual displacement of m and N are written :

$$\delta m = i(\delta\theta)(m), \quad \delta N = i(\delta\theta)(N).$$

Now :

$$\frac{\partial \delta m}{\partial m} = i(\delta\theta) \, \Pi \;, \quad \frac{\partial \delta N}{\partial m} = i(\delta\theta)\Pi. \frac{\partial N}{\partial m} \;.$$

As the energy density is invariant in a rigid body rotation, we obtain :

$$T_r \left(\overline{\mathbf{N}} \, i(\delta\theta)\Pi - \frac{\partial N}{\partial m} \, \overline{i_2 \mathbf{M}} \, i(\delta\theta)\Pi + \overline{\mathbf{N}} \, N \, \overline{i(\delta\theta)(N)} \right) = 0, \quad \forall \delta\theta .$$

Associating the first and last terms, we have :

$$T_r \left(\left[\overline{\Pi \mathbf{N}} - \frac{\partial N}{\partial m} \, \overline{i_2 \mathbf{M}} \right] \Pi \; i(\delta\theta)\Pi \right) = 0, \quad \forall \delta\theta$$

Now :

$$\Pi \, i(\delta\theta) \, \Pi = i(N \, \overline{N} \, \delta\theta) = i_2 \, \overline{N} \, \delta\theta = i_2 \, \delta\lambda$$

putting $\overline{N}\delta\theta = \delta\lambda$.

We then have :

$$T_r \left(\left[\overline{\Pi \mathbf{N}} - \frac{\partial N}{\partial m} \, \overline{i_2 \mathbf{M}} \right] i_2 \, \delta\lambda \right) = 0, \quad \forall \delta\lambda.$$

Now $i_2\delta\lambda$ is any anti-Hermitian endomorphism of $\vec{E}_2$ [see (A.25)] . We then obtain the following condition :

$$(6.29) \qquad T_r\left(\left[\overline{\Pi \mathbf{N}} - \frac{\partial N}{\partial m} \overline{i_2 \mathbf{M}}\right] i_2\right) = 0$$

which is also written :

$$(6.30) \qquad \overline{\Pi \mathbf{N}} - \frac{\partial N}{\partial m} \overline{i_2 \mathbf{M}} = \Pi \mathbf{N} - i_2 \mathbf{M} \frac{\partial N}{\partial m} \, .$$

Using the conventional notation, this relation can be written :

$$N_{\alpha\beta} + b_{\alpha\gamma} M_\beta^\gamma = N_{\beta\alpha} + b_{\beta\gamma} M_\alpha^\gamma \, .$$

It should be noted that this relation is implied by the equilibrium equations, the rigid body rotation field being included in the general class of displacements which was considered.

3. – DEFORMATIONS[24]

3.1 – Deformation in the case of hypothesis (w), linearized case

Again with $\delta z = 0$, we shall start with the case of linearized deformation, using linearized virtual displacements for the sake of convenience.

We know that (see chapter I) :

$$\delta D_L = \frac{1}{2}\left[\frac{\partial \delta M}{\partial M} + \overline{\frac{\partial \delta M}{\partial M}}\right]$$

and :

$$\delta\left[\overline{dM}\, dM\right] = \overline{dM}\left[\frac{\partial \delta M}{\partial M} + \overline{\frac{\partial \delta M}{\partial M}}\right] dM \, .$$

$$= 2\,\overline{dM}\,\delta D_L\, dM \, .$$

Now we have formula (6.4) :

$$dM = \mu dm + N dz = dm + z dN + N dz \, .$$

Therefore :

$$\overline{dM}\,dM = [\overline{dm} + z\overline{dN} + \overline{N}dz][dm + zdN + Ndz]$$

$$(6.31)\quad
\begin{cases}
2\,\overline{dM}\,\delta D_L\,dM = \overline{dm}\left[\dfrac{\partial\delta m}{\partial m} + \overline{\dfrac{\partial\delta m}{\partial m}}\right]dm + \overline{dm}\left[\overline{\dfrac{\partial\delta m}{\partial m}}\,N + \delta N\right]dz + \\[2ex]
\qquad\qquad + \left[\overline{\delta N} + \overline{N}\,\dfrac{\partial\delta m}{\partial m}\right]dm\,dz \\[2ex]
\qquad\qquad + z\overline{dm}\left[\overline{\dfrac{\partial\delta m}{\partial m}}\dfrac{\partial N}{\partial m} + \dfrac{\partial\delta N}{\partial m} + \dfrac{\partial N}{\partial m}\dfrac{\partial\delta m}{\partial m} + \overline{\dfrac{\partial\delta N}{\partial m}}\right]dm \\[2ex]
\qquad\qquad + z^2\overline{dm}\left[\overline{\dfrac{\partial\delta N}{\partial m}}\dfrac{\partial N}{\partial m} + \dfrac{\partial N}{\partial m}\dfrac{\partial\delta N}{\partial m}\right]dm.
\end{cases}$$

Furthermore :

$$\overline{dM}\,\delta D_L\,dM = [\overline{dm}\mu + Ndz]\,\delta D_L\,[\mu dm + Ndz].$$

Decomposing D_L into its normal and tangential components :

$$\delta D_L = [\Pi + N\,\overline{N}]\delta D_L\,[\Pi + N\,\overline{N}]\;,$$

and comparing with (6.31), we easily find, $\blacktriangledown dm, \blacktriangledown Ndz$:

$$(6.32)\quad
\begin{cases}
2\Pi\delta D_L\Pi = \mu^{-1}\left[\Pi\,\dfrac{\partial\delta m}{\partial m} + \Pi\,\overline{\dfrac{\partial\delta m}{\partial m}} + z\left[\Pi\,\overline{\dfrac{\partial\delta m}{\partial m}}\dfrac{\partial N}{\partial m} + \Pi\,\dfrac{\partial\delta N}{\partial m} + \dfrac{\partial N}{\partial m}\dfrac{\partial\delta m}{\partial m}\right.\right. \\[2ex]
\qquad\qquad\qquad\qquad\qquad\qquad\qquad\qquad\qquad\qquad + \Pi\,\overline{\dfrac{\partial\delta N}{\partial m}}\Bigg] \\[2ex]
\qquad\qquad\qquad + z^2\left[\dfrac{\partial N}{\partial m}\dfrac{\partial\delta N}{\partial m} + \Pi\,\overline{\dfrac{\partial\delta N}{\partial m}}\dfrac{\partial N}{\partial m}\right]\Bigg]\mu^{-1} \\[2ex]
2\Pi\delta D_L\,N\,\overline{N} = \mu^{-1}\left[\overline{\dfrac{\partial\delta m}{\partial m}}\,N\,\overline{N} + \delta N\,\overline{N}\right] \\[2ex]
2\,N\,\overline{N}\,\delta D_L\,N\,\overline{N} = 0\;.
\end{cases}$$

Let us note that the hypothesis :

$$\delta z = \delta a_1\,z$$

would add $2\,\dfrac{\partial N}{\partial m}\,z\mu\delta a_1\mu^{-1}$ to the first term, and $\dfrac{\partial\delta a_1}{\partial m}\,Nz$ to the second, as for the third term it becomes $2\,N\,\overline{N}\,\delta a_1$ instead of zero.

It is worth noting from this moment that the second term $2\Pi\delta D_L\,N\overline{N}$, which gives the transverse shear, disappears in the case of hypothesis (wo), by virtue of (6.12).

Alternative expression of deformation

We can proceed as follows :

$$\delta M = \delta m + z\delta N, \qquad \delta z = 0$$

$$d\delta M = d\delta m + zd\delta N + \delta N dz$$

$$\frac{\partial \delta M}{\partial M}\, dM = \left[\frac{\partial \delta m}{\partial m} + z\,\frac{\partial \delta N}{\partial m}\right] dm + \delta N dz$$

$$\overline{dM}\,\frac{\partial \delta M}{\partial M}\, dM = \left[\overline{dm}\mu + dz\overline{N}\right]\left[\frac{\partial \delta m}{\partial m} + z\,\frac{\partial \delta N}{\partial m}\right] dm + \overline{dm}\mu\delta N dz$$

$$= \overline{dm}\,\mu\left[\frac{\partial \delta m}{\partial m} + z\,\frac{\partial \delta N}{\partial m}\right] dm$$

$$+ dz\,\overline{N}\left[\frac{\partial \delta m}{\partial m} + z\,\frac{\partial \delta N}{\partial m}\right] dm + \overline{dm}\mu\delta N dz$$

$$= \overline{dm}\left[\mu\Pi\,\frac{\partial \delta m}{\partial m} + z\mu\Pi\,\frac{\partial \delta N}{\partial m}\right] dm$$

$$+ dz\left[\overline{N}\,\frac{\partial \delta m}{\partial m} + z\,\overline{N}\,\frac{\partial \delta N}{\partial m}\right] dm + \overline{dm}\,\mu\delta N dz\,.$$

Hence :

$$(6.33)\quad
\begin{cases}
2\,\overline{dM}\,\delta D_L\, dM = \overline{dm}\left[\mu\left[\delta\gamma' + z\delta\kappa'\right] + \left[\overline{\delta\gamma'} + z\overline{\delta\kappa'}\right]\mu\right] dm \\[2ex]
\qquad + dz\left[\overline{N}\,\frac{\partial \delta m}{\partial m} + \overline{\delta N}\right] dm + \overline{dm}\left[\overline{\frac{\partial \delta m}{\partial m}}\,N + \delta N\right] dz \\[2ex]
\text{with}\quad \delta\gamma' = \Pi\,\frac{\partial \delta m}{\partial m}\ ,\quad \delta\kappa' = \Pi\,\frac{\partial \delta N}{\partial m}\ ,\quad \overline{\frac{\partial \delta m}{\partial m}}\,N + \delta N = \delta\gamma_3\,\boldsymbol{\epsilon}\,\vec{E}_2\,.
\end{cases}$$

Formula (6.33) enables us to calculate easily the conventional normal and tangential deformation components in the basis $T' = \overline{T}^{-1}$, supplementary to the natural basis $T = [\mu S\ N]$, or the covariant components e_{ij} in T. Thus :

$$2 \, \overline{T}_\alpha \, D_L \, T_\beta = 2e_{\alpha\beta} = \mu_\alpha^\gamma \left[\gamma'_{\gamma\beta} + z\kappa'_{\gamma\beta} \right] + \mu_\beta^\gamma \left[\gamma'_{\gamma\alpha} + z\kappa'_{\gamma\alpha} \right], \quad (\alpha,\beta,\gamma = 1,2).$$

Besides, let us put for the actual displacement :

$$\delta m = V = \Pi V + N \, \overline{N} \, V = \Pi V + Nw \,, \quad \text{where } w = \overline{N}V$$

$$W = V + z\Delta N = \Pi V + Nw + z\Delta N$$

$$\gamma' = \Pi \, \frac{\partial V}{\partial m} = \Pi \, \frac{\partial \Pi V}{\partial m} + \Pi \, \frac{\partial Nw}{\partial m}$$

$$= \frac{\hat{\partial} \Pi V}{\partial m} + w \, \frac{\partial N}{\partial m}$$

$$\gamma'_{\alpha\beta} = [\Pi V]_{\alpha \| \beta} - b_{\alpha\beta} \, w.$$

We also have, for example :

$$\overline{N} \, \frac{\partial V}{\partial m} = \overline{N} \, \frac{\partial \Pi V}{\partial m} + \overline{N} \, \frac{\partial Nw}{\partial m}$$

$$= \overline{N} \, \frac{\partial \Pi V}{\partial m} + \frac{\partial w}{\partial m} \,.$$

Now $\quad \overline{N}\Pi V = 0 \implies \overline{dN}\Pi V + \overline{N} \, d[\Pi V] = 0.$

Therefore :

$$\overline{N} \, \frac{\partial V}{\partial m} = - \, \overline{\Pi V} \, \frac{\partial N}{\partial m} + \frac{\partial w}{\partial m} \,.$$

Whence :

$$\overline{N} \, \frac{\partial V}{\partial m} \, S_\alpha = \gamma'_{3\alpha} = w,_\alpha + b_\alpha^\beta \, [\Pi V]_\beta \,, \quad (\alpha,\beta = 1,2).$$

3.2 – <u>Deformations in the case of hypothesis (wo), linearized case</u>

Introducing constraint (6.12), and still with $\delta z = 0$ to simplify matters, formula (6.31) immediately gives :

$$(6.34) \quad \left| \begin{aligned} 2 \, \overline{dM} \, \delta D_L \, dM &= \overline{dm} \left[\Pi \, \frac{\partial \delta m}{\partial m} + \overline{\Pi \, \frac{\partial \delta m}{\partial m}} \right] dm \\ &+ z\overline{dm} \left[\overline{\Pi \, \frac{\partial \delta m}{\partial m}} \, \frac{\partial N}{\partial m} + \Pi \, \frac{\partial \delta N}{\partial m} + \frac{\partial N}{\partial m} \, \Pi \, \frac{\partial \delta m}{\partial m} + \overline{\Pi \, \frac{\partial \delta N}{\partial m}} \right] dm \end{aligned} \right.$$

$$(6.34) \text{ continued} \qquad + z^2 \overline{dm} \left[\Pi \; \frac{\partial \delta N}{\partial m} \; \frac{\partial N}{\partial m} + \frac{\partial N}{\partial m} \; \frac{\partial \delta N}{\partial m} \right] dm \; .$$

We have already observed that when $\delta z = 0$, the transverse shear components are zero.

Linearized variations of curvature

These will be calculated by differentiation of the second fundamental form of Σ_m, namely :

$$\delta \left[\overline{dm} \; \frac{\partial N}{\partial m} \; dm \right] = \delta \left[\overline{dm} \; dN \right] = \overline{\delta dm} \; dN + \overline{dm} \; \delta dN$$

and as differentiation is made with m constant on Σ_m :

$$d\delta - \delta d = 0,$$

therefore :

$$\delta \left[\overline{dm} \; \frac{\partial N}{\partial m} \; dm \right] = \overline{d\delta m} \; dN + \overline{dm} \; d\delta N$$

$$= \overline{dm} \left[\frac{\partial \delta m}{\partial m} \; \frac{\partial N}{\partial m} + \frac{\partial \delta N}{\partial m} \right] dm$$

$$= \overline{dm} \left[\Pi \; \frac{\partial \delta m}{\partial m} \; \frac{\partial N}{\partial m} + \Pi \; \frac{\partial \delta N}{\partial m} \right] dm \; .$$

The variations of the curvature are therefore given (as virtual variations), in the linearized case, by the following endomorphism of $\vec{E}_2$:

$$(6.35) \qquad \delta\kappa = \Pi \; \frac{\overline{\partial \delta m}}{\partial m} \; \frac{\partial N}{\partial m} + \Pi \; \frac{\partial \delta N}{\partial m} = \overline{\delta\kappa} \; .$$

By definition, $\delta\kappa$ is necessarily Hermitian. Hence the term on z on (6.34).

Furthermore, for the tangential deformation of $\Sigma_m (z=0)$:

$$(6.36) \qquad \delta\gamma = \frac{1}{2} \left[\Pi \; \frac{\partial \delta m}{\partial m} + \Pi \; \overline{\frac{\partial \delta m}{\partial m}} \right] = \overline{\delta\gamma} \; .$$

Finally the term on z^2 in (6.34) can be calculated as follows :

$$\overline{\Pi \frac{\partial \delta N}{\partial m} \frac{\partial N}{\partial m}} + \frac{\partial N}{\partial m} \frac{\partial \delta N}{\partial m} = \delta\kappa \frac{\partial N}{\partial m} - \frac{\partial N}{\partial m} \Pi \frac{\partial \delta m}{\partial m} \frac{\partial N}{\partial m} + \frac{\partial N}{\partial m} \delta\kappa - \frac{\partial N}{\partial m} \Pi \overline{\frac{\partial \delta m}{\partial m} \frac{\partial N}{\partial m}}$$

$$= \delta\kappa \frac{\partial N}{\partial m} + \frac{\partial N}{\partial m} \delta\kappa - \frac{\partial N}{\partial m} 2\delta\gamma \frac{\partial N}{\partial m} .$$

And finally, (6.34) is written :

$$(6.37) \quad \begin{cases} 2 \overline{dM} \, \delta D_L \, dM = \overline{dm} \left[2\delta\gamma + 2z\delta\kappa + z^2 \left[\delta\kappa \frac{\partial N}{\partial m} + \frac{\partial N}{\partial m} \delta\kappa + 2 \frac{\partial N}{\partial m} \delta\gamma \frac{\partial N}{\partial m} \right] \right] dm \\[4mm] \text{with :} \\[2mm] \delta\gamma = \frac{1}{2} \left[\Pi \frac{\partial \delta m}{\delta m} + \overline{\Pi \frac{\partial \delta m}{\partial m}} \right] = \overline{\delta\gamma} \qquad \epsilon \, \mathcal{L} \, (\vec{E}_2, \vec{E}_2) \\[4mm] \delta\kappa = \overline{\Pi \frac{\partial \delta m}{\partial m} \frac{\partial N}{\partial m}} + \Pi \frac{\partial \delta N}{\partial m} = \overline{\delta\kappa} \qquad \epsilon \, \mathcal{L} \, (\vec{E}_2, \vec{E}_2) . \end{cases}$$

The preceding formula easily gives, in normal coordinates (principal lines of curvature), the three-dimensional deformations at point M of the shell of projection m on Σ_m :

$$\varepsilon = \begin{vmatrix} \dfrac{^1\gamma_1 \left[1 + z\rho_1\right] + z \cdot {}^1\kappa_1}{1 - z\rho_1} & \dfrac{^1\gamma_2 \left[1 - z^2 \rho_1 \rho_2\right] + z \cdot {}^1\kappa_2 \left[1 - z \dfrac{\rho_1 + \rho_2}{2}\right]}{\left[1 - z\rho_1\right] \left[1 - z\rho_2\right]} & 0 \\[6mm] & \dfrac{^2\gamma_2 \left[1 - z\rho_2\right] + z \cdot {}^2\kappa_2}{1 - z\rho_2} & 0 \\[6mm] 0 & 0 & 0 \end{vmatrix}$$

where ρ_1 and ρ_2 are the two principle radii of curvature of Σ_m on m.

3.3 - <u>Transverse shear deformation and normal deformation</u>

With hypothesis (w), we said that this kinematic hypothesis included transverse shear, which is obvious if we compare formulae $(6.32)_2$ and (6.34). Let us note also that $(6.32)_2$:

$$2 \ \Pi \delta D_L N \bar{N} = \mu^{-1} \left[\overline{\frac{\partial \delta m}{\partial m}} \ N + \delta N \right] \bar{N} \ ,$$

shows that this deformation is the difference between variations δN and $- \overline{\dfrac{\partial \delta m}{\partial m}} \ N$, which is precisely the variation of N when the material normal remains identical with the geometrical normal.

Furthermore, in the case of hypothesis (w), (6.23) :

$$\delta w = \iint_{\Sigma_m} T_r \left(\bar{\mathbf{N}} \ \frac{\partial \delta m}{\partial m} - \overline{i_2 \mathbf{M}} \ \frac{\partial \delta N}{\partial m} + \bar{\mathbf{N}} \ N \ \overline{\delta N} \right) \ d\Sigma_m$$

which can be written :

$$\delta w = \iint_{\Sigma_m} T_r \left(\overline{\Pi \mathbf{N} \Pi} \ \frac{\partial \delta m}{\partial m} + \bar{\mathbf{N}} \ N \ \left[\bar{N} \ \frac{\partial \delta m}{\partial m} + \overline{\delta N} \right] - \overline{i_2 \mathbf{M}} \ \frac{\partial \delta N}{\partial m} \right) \ d\Sigma_m,$$

shows that the generalized conjugated stress of this deformation is precisely $\bar{\mathbf{N}} \ N$, namely the normal component of $\bar{\mathbf{N}}$. Hypothesis (wo) thus corresponds either to taking geometrical and material normals as the same, or to neglecting the transverse shear deformation work.

As for the normal deformation, this results from the hypothesis $\delta z \neq 0$. The hypothesis $\delta z = 0$ is equivalent to neglecting the corresponding deformation work. But we can also have the same result, if we assume the three-dimensional stress of component σ_{33} to be negligible with respect to the others, which is generally well verified. Nevertheless, these two hypotheses, which provide the same result when we express energy by means of generalized deformations and stresses, have substantially different consequences on the surface constitutive law, as we shall see in paragraph 6.

In linear and isotropic elasticity, the hypothesis $\sigma_{33} = 0$ leads to :

$$(6.38) \qquad e_{33} = - \frac{\nu}{1 - \nu} \left[e_{11} + e_{22} \right]$$

where ν is here Poisson's coefficient of contraction.

<u>Remark</u>

$(6.26)_2$ shows that when $\mathbf{G}_m = 0$, the constraint reaction resulting from (6.12) is equal to $- \,\widehat{\operatorname{div}}\,\overline{i_2\,\mathbf{M}}$, as (6.26) is necessarily verified when we relax the constraint.

3.4 - <u>Linearized conditions of compatibility</u>

We will write down the conditions of compatibility relating to the generalized deformations $\delta\gamma$ and $\delta\kappa$, or γ and κ in the case of hypothesis (wo), these conditions being the same as those for variations of the first and second fundamental forms of Σ_m.

Let $\delta\theta$ be the vector of virtual rotation $\epsilon\,\vec{E}_2$ of normal N, we know that [see (6.16)] :

$$\delta\theta = i_2\,\delta N\ ,\ \epsilon\,\vec{E}_2.$$

Let us replace in the following δm and $\delta\theta$ by real variations V and $\Pi\Omega$, and let us put :

$$(6.39)\qquad
\begin{cases}
\dfrac{\partial V}{\partial m} = A \implies \gamma = \dfrac{\Pi A + \overline{\Pi A}}{2}\ ,\quad [\text{See } (6.37)]\,.\,(\,A\,\epsilon\,\mathbf{\mathcal{L}}\,(\vec{E}_2,\vec{E}_3)\,) \\[3mm]
\Pi\Omega = -\,i_2\,\overline{A}\,N,\qquad [\text{See } (6.12)].
\end{cases}$$

Now :

$$\Pi A = \frac{\Pi A + \overline{\Pi A}}{2} + \frac{\Pi A - \overline{\Pi A}}{2}\ \epsilon\,\mathbf{\mathcal{L}}\,(\vec{E}_2,\,\vec{E}_2).$$

But $\dfrac{\Pi A - \overline{\Pi A}}{2}$ is anti-Hermitian, and is written in the form [see (A.25)] :

$$\frac{\Pi A - \overline{\Pi A}}{2} = \lambda i_2\ ,\quad \lambda\,\epsilon\,\mathbb{R}.$$

Thus :

$$\Pi A = \gamma + \lambda i_2.$$

Furthermore :

$$A = \Pi A + N \overline{N} A = \gamma + \lambda i_2 - N \overline{\Pi\Omega}\, i_2$$

$$= \gamma + \lambda i(N)\Pi - N \overline{\Pi\Omega}\, i(N)\Pi, \quad (\text{as } i(N)\Pi = i_2)$$

$$= \gamma + i(N\lambda)\Pi + N \overline{N}\, i(\Pi\Omega)\Pi$$

$$A = \gamma + i(\Pi\Omega + N\lambda)\Pi .$$

Whence :

$$(6.40) \qquad \left[\begin{array}{l} A = \gamma + i(\Omega)\Pi \\[2ex] \text{with :} \\[2ex] \Omega = \Pi\Omega + N\lambda \ \in \ \vec{E}_3 \end{array} \right.$$

a formula which gives incidentally :

$$(6.41) \qquad i_2\lambda = \frac{\Pi A - \overline{\Pi A}}{2} = i_2 \overline{N} \Omega .$$

We thus come back very easily to Gol'denveizer's decomposition[5].

Furthermore, $(6.39)_1$ shows that A is a vectorial form of degree 1, integrable, or closed. We demonstrate that this closure condition corresponds to the fundamental condition :

$$(6.42)$$
$$\widehat{\mathrm{div}} \left[i_2\, \overline{A} \right] = 0$$

or :

$$(6.43) \qquad \hat{\nabla} A = 0 \qquad (\text{Poincaré's theorem})$$

where $\hat{\nabla}$ is the Riemannian manifold co-edge symbol, (see (A.82)).

In fact whatever the differentials $d_1 m$, $d_2 m \ \in \ \vec{E}_2$ at m, (6.43) is written [see (A.82)] :

$$\hat{d}_1 \left[\overline{W} A \right] d_2 m - \hat{d}_2 \left[\overline{W} A \right] d_1 m = 0, \forall \, W \, c^t \, \in \vec{E}_3$$

$$\hat{d}_1 \left[\overline{W} A \right] \overline{i_2}\, i_2\, d_2 m - \hat{d}_2 \left[\overline{W} A \right] \overline{i_2}\, i_2\, d_1 m = 0$$

or otherwise [see (A.23)] :

$$\mathrm{vol}_2(\hat{d}_1\left[i_2\ \overline{W\ A}\right])(d_2 m) + \mathrm{vol}_2(d_1 m)(\hat{d}_2\left[i_2\ \overline{W\ A}\right]) = 0$$

as $\hat{d}\ i_2 = 0$, [see (A.89)],

so that [see (A.4)] :

$$T_r\,(\frac{\hat{\partial}}{\partial m}\left[i_2\ \overline{W\ A}\right]) = 0,\ \forall\,W\ c^t,$$

and finally (6.42)   Q.E.D.

The first member of (6.42), being a co vector of $\vec{E}_3$, provides three scalar equations, which when integrated, give V.

If we now put (6.40) into (6.42), we obtain :

$$(6.44)\qquad \widehat{\mathrm{div}}\left[i_2\ \gamma\right] = \widehat{\mathrm{div}}\left[i_2\ i(\Omega)\right].$$

The tangential component of (6.44) is written :

$$\widehat{\mathrm{div}}\left[i_2\ \gamma\right]\Pi = \widehat{\mathrm{div}}\left[i_2\ i(\Omega)\right]\Pi.$$

Now with the natural basis S of $\vec{E}_2$, we easily have [see (A.96)] :

$$\widehat{\mathrm{div}}\left[i_2\ i(\Omega)\right]\Pi = {}^{\alpha}S^{-1}\ i_2\ i(\partial_\alpha\Omega)\Pi\ ,\ (\alpha = 1,2)$$

$$= {}^{\alpha}S^{-1}\ i_2\left[i(\Pi\partial_\alpha\Omega)\Pi\ + i(N\ \overline{N}\partial_\alpha\Omega)\Pi\right]$$

$$= {}^{\alpha}S^{-1}\ i_2^2\ \overline{N}\ \partial_\alpha\Omega = -\ {}^{\alpha}S^{-1}\ \overline{N}\ \frac{\partial\Omega}{\partial m}\ S_\alpha\ .$$

$$(6.45)\qquad \mathrm{div}\left[i_2 i(\Omega)\right] = -\ \overline{N}\ \frac{\partial\Omega}{\partial m}\ .$$

Thus :

$$(6.46)\qquad \left[\begin{array}{l} \mathrm{div}\left[i_2\,\gamma\right] = -\ \overline{N}\ B \\[2mm] \mathrm{with}\ : \\[2mm] B = \dfrac{\partial\Omega}{\partial m}\ . \end{array}\right.$$

Now $(6.46)_2$ shows that B is a vectorial form of degree 1 and closed, therefore according to (6.42) :

$$\widehat{\text{div}}\left[i_2\ \bar{B}\right] = 0,$$

or also :

$$\widehat{\text{div}}\left[i_2\left[\bar{B}\,\Pi + \overline{N\ \bar{N}\ B}\right]\right]$$

and taking account of $(6.46)_1$:

$$(6.47)\qquad \widehat{\text{div}}\left[i_2\left[\overline{\Pi\ B} - \overline{N\ \widehat{\text{div}}\left[i_2\ \gamma\right]\Pi}\right]\right] = 0.$$

Now $\Pi B = \Pi\dfrac{\partial\Omega}{\partial m}$ can easily be expressed as a function of γ and κ, which are reproduced below. [See (6.37)] :

$$\left[\begin{array}{l}
\gamma = \dfrac{1}{2}[\Pi A + \overline{\Pi A}]\ \ = \bar{\gamma} \\[2ex]
\kappa = \overline{\Pi A}\ \dfrac{\partial N}{\partial m} - \Pi\ i_2\ \dfrac{\partial\Pi\Omega}{\partial m} = \bar{\kappa},\quad \text{with } \Delta N = -\ i_2\ \Pi\Omega\,.
\end{array}\right.$$

In fact :

$$\Pi B = \Pi\ \frac{\partial\Omega}{\partial m} = \Pi\ \frac{\partial\Pi\Omega}{\partial m} + \Pi\ \frac{\partial N\bar{N}\Omega}{\partial m}$$

$$= \Pi\ \frac{\partial\Pi\Omega}{\partial m} + \bar{N}\Omega\ \frac{\partial N}{\partial m}\,.$$

Therefore :

$$i_2\ \kappa = i_2\ \overline{\Pi A}\ \frac{\partial N}{\partial m} + \Pi\ \frac{\partial\Pi\Omega}{\partial m}$$

$$\Pi B\ \ = i_2\kappa - i_2\ \overline{\Pi A}\ \frac{\partial N}{\partial m} + \bar{N}\ \Omega\ \frac{\partial N}{\partial m}\ ,\ \text{and with } \mathbf{(6.41)} :$$

$$= i_2\kappa - i_2\ \overline{\Pi A}\ \frac{\partial N}{\partial m} + i_2\ \frac{\Pi A - \overline{\Pi A}}{2}\ \frac{\partial N}{\partial m}\,.$$

Finally :

$$(6.48)\qquad \overline{\Pi B} = -\ \kappa i_2 + \frac{\partial N}{\partial m}\ \gamma\ i_2\,.$$

Putting (6.48) into (6.47), it is found :

$$(6.49) \qquad \widehat{\mathrm{div}}\ i_2 \left[\frac{\partial N}{\partial m}\ \gamma\ i_2 - \kappa\ i_2 - \overline{N\ \widehat{\mathrm{div}}\left[i_2\gamma\right]\Pi}\ \right] = 0 .$$

As the first member is a covector of $\vec{E}_3$, (6.49) gives three scalar conditions, which are in fact the Gauss and Mainardi-Goddazzi conditions in linearized form.

Furthermore, the normal component of (6.44), not yet used, would give :

$$(6.50) \qquad \kappa = \overline{\kappa}$$

If this relation had not been used, it would be found again.

The reciprocal is demonstrated easily by inverse calculation, using Poincaré's theorem on the existence of a vector field V twice, starting from (6.49) and (6.50), in a simply connected open set of Σ_m , and then applying (6.42).

Condition (6.49) connects the surface deformations γ and κ, but is not sufficient if the shell is multiconnected. Global conditions must then be added, which are obtained from the preceding relations, and application of de Rham's theorem[24].

3.5 – Non-linear deformations[61]

The study of instability phenomena or the problems of large displacements, or also of prestressed shells, requires the formulation of non-linear deformations as we saw in chapter V.

We shall restrict ourselves here to give the expression of these generalized surface deformations, in the case of hypothesis (wo).

Let us consider a reference state $\mathcal{E}_o$, and a deformed state $\mathcal{E}$. The tangential deformation γ and the curvature deformation κ are given by the variations of the first and second fundamental forms respectively, such that $\forall d_1 m_o,\ d_2 m_o \in \vec{E}_{20}$:

$$(6.51) \qquad \left[\begin{array}{l} \overline{d_1 m_o}\ \gamma\ d_2 m_o = \frac{1}{2}\left[\overline{d_1 m\ d_2 m} - \overline{d_1 m_o\ d_2 m_o} \right] \\[2ex] \overline{d_1 m_o}\ \kappa\ d_2 m_o = \left[\overline{d_1 m\ \frac{\partial N}{\partial m}\ d_2 m} - \overline{d_1 m_o\ \frac{\partial N_o}{\partial m_o}\ d_2 m_o} \right] \end{array} \right.$$

Formulae (6.51) then easily give :

$$\overline{d_1 m_o} \ \gamma \ d_2 m_o = \frac{1}{2} \ \overline{d_1 m_o} \left[\overline{\frac{\partial m}{\partial m_o}} \ \frac{\partial m}{\partial m_o} - 1_{E_{20}} \right] d_2 m_o .$$

Hence with :

$$m = m_o + V$$

$$(6.52) \quad \begin{cases} \gamma = \frac{1}{2} \ \pi_o \left[\frac{\partial V}{\partial m_o} + \overline{\frac{\partial V}{\partial m_o}} + \overline{\frac{\partial V}{\partial m_o}} \ \frac{\partial V}{\partial m_o} \right] \pi_o = \frac{1}{2} \left[\overline{\frac{\partial m}{\partial m_o}} \ \Pi \ \frac{\partial m}{\partial m_o} - 1_{E_{20}} \right] = \overline{\gamma} \\[3em] \delta \gamma = \frac{1}{2} \left[\Pi \ \frac{\partial \delta m}{\partial m} + \overline{\Pi \ \frac{\partial \delta m}{\partial m}} \right] = \overline{\delta \gamma} \end{cases}$$

$$\overline{d_1 m_o} \ \kappa \ d_2 m_o = \overline{d_1 m_o} \left[\overline{\frac{\partial m}{\partial m_o}} \ \frac{\partial N}{\partial m} \ \frac{\partial m}{\partial m_o} - \frac{\partial N_o}{\partial m_o} \right] d_2 m_o ,$$

$$(6.53) \quad \begin{cases} \kappa = \overline{\frac{\partial m}{\partial m_o}} \ \frac{\partial N}{\partial m} \ \frac{\partial m}{\partial m_o} - \frac{\partial N_o}{\partial m_o} = \overline{\kappa} \\[3em] \delta \kappa = \Pi \ \overline{\frac{\partial \delta m}{\partial m_o}} \ \frac{\partial N}{\partial m} + \Pi \ \frac{\partial \delta N}{\partial m} = \overline{\delta \kappa} . \end{cases}$$

Formulae $(6.52)_1$ and $(6.53)_1$ lend themselves to successive linearizations with respect to $\frac{\partial V}{\partial m_o}$ 25. For $(6.53)_1$ in particular, we achieve this objective by linearizing the formulae :

$$\overline{N} \ dm = 0, \quad \overline{N} \ N = 1 .$$

Non-linear deformation analyses, starting from states which are pre-stressed or not, employ the same methods as those described in chapter V.

As for the <u>conditions of compatibility between γ and κ when these are non-linear</u>, these are obtained by writing that Σ_m in the deformed state, remains embedded in E_3, the latter having zero curvature.

By writing that the Riemann-Christoffel curvature tensor of E_3 is zero, we easily obtain the relations[24] :

$$\hat{d}_1 \frac{\partial N}{\partial m} d_2 m - \hat{d}_2 \frac{\partial N}{\partial m} d_1 m = 0, \quad \forall d_1 m, \ d_2 m \in \vec{E}_2$$

which is the Mainardi-Coddazzi vectorial relation in $\vec{E}_2$, and :

$$\left[\hat{d}_1, \hat{d}_2\right]\left[\text{IIV}\right] = R(\text{IIV})(d_1 m)(d_2 m) = \frac{\partial N}{\partial m}\left[d_1 m \ \overline{d_2 m} - d_2 m \ \overline{d_1 m}\right] \frac{\partial N}{\partial m} \text{IIV}$$

$$\forall \text{IIV}, \ d_1 m, \ d_2 m \in \vec{E}_2, \ \text{with } d_1 d_2 - d_2 d_1 = 0.$$

The latter relation, of Gauss, provides a scalar relation. This is the Riemann-Christoffel curvature tensor, in a vector form.

These two relations are expressed by means of γ and κ, γ being used in particular to calculate the Christoffel symbols, as already mentioned, on Σ_m in the deformed state.

In the natural basis, these general relations are relatively complicated and little used. In linearized form, they give the conditions of compatibility already found in paragraph 3.4.

4. – <u>STRESSES</u>[24]

4.1 – <u>Indeterminate stresses in the case of hypothesis (wo)</u>

In the case of hypothesis (wo), deformations, in the same way as displacements, are subject to a constraint of the type :

$$(6.54) \qquad u = \underset{\sim}{u}\left(\frac{\overline{\partial M}}{\partial M_0} \ \frac{\partial M}{\partial M_0}\right) = 0$$

where u is a scalar function of the Cauchy tensor, assumed to be differentiable. If we decompose the stress C into two parts :

$$C = C_1 + C_2$$

C_2 will be an indeterminate stress if it executes zero virtual work for a displacement δM, satisfying the internal constraint (6.54), namely :

$$T_r\left(C_2 \frac{\partial \delta M}{\partial M}\right) = 0, \quad \forall \delta M \ \text{such that} \ u = 0,$$

$$(6.55) \quad \left[\begin{array}{l} T_r(C_2 \ \delta D_L) = 0 \quad \forall \delta D_L \text{ such that } u = 0 \\[2mm] \text{with :} \\[2mm] \delta D_L = \frac{1}{2} \left[\frac{\partial \delta M}{\partial M} + \overline{\frac{\partial \delta M}{\partial M}} \right], \end{array} \right.$$

namely :

$$\forall \delta D_L \ . \ \left| \ \delta u = \frac{\partial u}{\partial K} \ \partial K = 0, \quad K = \overline{\frac{\partial M}{\partial M_o}} \ \frac{\partial M}{\partial M_o} \ .$$

Now :

$$\frac{\partial u}{\partial K} \ \delta K = \frac{\partial u}{\partial K} \ \overline{\frac{\partial M}{\partial M_o}} \ \delta D_L \ \frac{\partial M}{\partial M_o} = 0 \ .$$

Now $\exists$ an endomorphism Z of E_3 such that [see (A.9)] :

$$\frac{\partial u}{\partial K} \ \delta K = T_r \ (Z \ \overline{\frac{\partial M}{\partial M_o}} \ \delta D_L \ \frac{\partial M}{\partial M_o}) = 0$$

$$= T_r \ (\frac{\partial M}{\partial M_o} \ Z \ \overline{\frac{\partial M}{\partial M_o}} \ \delta D_L) = 0 \ .$$

Therefore (6.55) is written :

$$(6.56) \qquad T_r(C_2 \ \delta D_L) = 0, \ \forall \delta D_L \ | \ T_r \ (\frac{\partial M}{\partial M_o} \ Z \ \overline{\frac{\partial M}{\partial M_o}} \ \delta D_L) = 0 \ .$$

Therefore $\exists$ a scalar Lagrangian multiplier λ, such that [see (A.101)] :

$$C_2 = \lambda \ \frac{\partial M}{\partial M_o} \ Z \ \overline{\frac{\partial M}{\partial M_o}} \ .$$

In $\vec{E}_6$, C_2 is orthogonal to δD_L and parallel to $\frac{\partial M}{\partial M_o} \ Z \ \overline{\frac{\partial M}{\partial M_o}}$, and λ has an arbitrary value. C is therefore defined, within an additive stress proportional to $\frac{\partial M}{\partial M_o} \ Z \ \overline{\frac{\partial M}{\partial M_o}}$, normal to the manifold $u = 0$ of E_6.

Let us apply this result to shells, in the case of hypothesis (wo), taking for example a unitary basis T' of E_3, we obtain :

$$\overline{T'} \ T' = 1_{R^3} \quad \text{with } T'_3 = N \ .$$

We can define a Hermitian endomorphism basis in $\vec{E}_6$:

$$\frac{1}{2}\left[\,T'_i\,\overline{T'_j}\,+\,T'_j\,\overline{T'_i}\,\right], \quad i,j = 1,2,3 \; ; \quad i \leqslant j$$

As in the case of hypothesis (wo), the projections of δD_L on $\frac{1}{2}\left[\,T'_i\,\overline{T'_3}\,+\,T'_3\,\overline{T'_i}\,\right]$ are zero, δD_L belonging to the supplementary orthogonal vector space, and C_2 to the space subtended by these vectors. The components of C_2 on these vectors are indeterminate, as we have just seen. If in addition $\delta z = 0$, the component of C_2 on vector $T'_3\,\overline{T'_3}$ of $\vec{E}_6$ is also indeterminate. In conclusion, in this case σ_{13}, σ_{23} and σ_{33} are a priori indeterminate in the shell. We shall use this result in due course.

4.2 – <u>Symmetrical stresses in the case of hypothesis (wo)</u>

The kinematic Kirchhoff–Love hypotheses can be used to introduce symmetrical generalized surface stresses.

We shall make this calculus on the assumption that $\delta z = 0$, without altering generality. In this case, we could write the virtual deformation energy, taking account of (6.12) (see paragraph 3.3), as follows :

$$(6.57) \qquad \delta w = \iint_{\Sigma_m} T_r\left(\overline{\Pi\,\mathbf{N}}\;\Pi\,\frac{\partial\,\delta m}{\partial m} - \overline{i_2\,\mathbf{M}}\;\Pi\,\frac{\partial\,\delta N}{\partial m}\right)d\Sigma_m .$$

It is easy to exhibit the symmetrical curvature deformation $(6.37)_2$, and the symmetrical stress (6.30) :

$$\delta w = \iint_{\Sigma_m} T_r\left(\left[\overline{\Pi\,\mathbf{N}} + i_2\mathbf{M}\,\frac{\partial N}{\partial m}\right]\Pi\,\frac{\partial\,\delta m}{\partial m} - i_2\mathbf{M}\,\frac{\partial N}{\partial m}\;\Pi\frac{\partial\,\delta m}{\partial m} - \overline{i_2\,\mathbf{M}}\;\Pi\frac{\partial\,\delta N}{\partial m}\right)d\Sigma_m .$$

$$= \iint_{\Sigma_m} T_r\left(\left[\overline{\Pi\,\mathbf{N}} + i_2\mathbf{M}\,\frac{\partial N}{\partial m}\right]\Pi\,\frac{\partial\,\delta m}{\partial m} - i_2\mathbf{M}\left[\frac{\partial N}{\partial m}\,\Pi\,\frac{\partial\,\delta m}{\partial m} + \overline{\Pi\,\frac{\partial\,\delta N}{\partial m}}\right]\right)d\Sigma_m .$$

Thus :

$$(6.58) \qquad \left|\;\; \delta w = \iint_{\Sigma_m} T_r\left(\mathbf{n}\,\delta\gamma + \mathbf{m}\delta\kappa\right)\,d\Sigma_m \right.$$

with :

$$n = \overline{\Pi\,\mathbf{N}} + i_2\,\mathbf{M}\,\frac{\partial N}{\partial m} = \overline{n} \;\in\; \mathcal{L}\,(\vec{E}_2,\vec{E}_2)$$

$$m = -\,\frac{i_2\,\mathbf{M} + \overline{i_2\,\mathbf{M}}}{2} = m \;\in\; \mathcal{L}(\vec{E}_2,\vec{E}_2)$$

(6.58)
cont$\underline{\mathrm{d}}$

$$\delta\gamma = \frac{1}{2}\left[\Pi\,\frac{\partial\,\delta m}{\partial m} + \overline{\Pi\,\frac{\partial\,\delta m}{\partial m}}\,\right] = \overline{\delta\gamma} \;\in\; \mathcal{L}\,(\vec{E}_2,\vec{E}_2)$$

$$\delta\kappa = \overline{\Pi\,\frac{\partial\,\delta N}{\partial m}} + \frac{\partial N}{\partial m}\,\Pi\,\frac{\partial\,\delta m}{\partial m} = \overline{\delta\kappa} \;\in\; \mathcal{L}(\vec{E}_2,\vec{E}_2)$$

and

$$\delta N = -\,\overline{\frac{\partial\,\delta m}{\partial m}}\,N \quad (\text{hypothesis (wo)})\;.$$

We can also write the equilibrium equations according to generalized symmetrical stresses **n** and **m**. In fact we obtain :

$$\delta w = \iint_{\Sigma_m} T_r\Big(\,\mathbf{n}\,\Pi\,\frac{\partial\,\delta m}{\partial m} + \mathbf{m}\Big[\frac{\partial N}{\partial m}\,\Pi\frac{\partial\,\delta m}{\partial m} + \Pi\,\frac{\partial\,\delta N}{\partial m}\Big]\Big)\,d\Sigma_m$$

$$= \iint_{\Sigma_m} T_r\Big(\Big[\,\mathbf{n} + \mathbf{m}\,\frac{\partial N}{\partial m}\Big]\,\Pi\,\frac{\partial\,\delta m}{\partial m} + \mathbf{m}\,\Pi\,\frac{\partial\,\delta N}{\partial m}\,\Big)\,d\Sigma_m$$

$$= \iint_{\Sigma_m}\Big[-\,\widehat{\mathrm{div}}\Big[\,\mathbf{n} + \mathbf{m}\,\frac{\partial N}{\partial m}\Big]\,\delta m + \widehat{\mathrm{div}}\;\mathbf{m}\;\overline{\Pi\frac{\partial\,\delta m}{\partial m}}\,N\,\Big]\,d\Sigma_m$$

$$\qquad -\int_{C_m}\overline{\delta m}\Big[\,\mathbf{n} + \frac{\partial N}{\partial m}\,\mathbf{m}\Big]i_2\;dm \;-\;\int_{C_m}\overline{\delta N}\,\mathbf{m}\;i_2\;dm$$

$$= \iint_{\Sigma_m}\Big[-\,\widehat{\mathrm{div}}\Big[\,\mathbf{n} + \mathbf{m}\,\frac{\partial N}{\partial m}\Big]\,\delta m + \widehat{\mathrm{div}}\Big[\,\overline{N\,\widehat{\mathrm{div}}\,\mathbf{m}\,.\,\delta m}\,\Big] -$$

$$\qquad -\,\widehat{\mathrm{div}}\Big[\,\overline{N\,\widehat{\mathrm{div}}\,\mathbf{m}}\,\Big]\,\delta m\,\Big]\,d\Sigma_m\;-$$

$$\qquad -\int_{C_m}\Big[\,\overline{\delta m}\Big[\,\mathbf{n} + \frac{\partial N}{\partial m}\,\mathbf{m}\Big] + \overline{\delta N}\,\mathbf{m}\,\Big]i_2 dm\;.$$

$$(6.59) \qquad \delta w = - \iint_{\Sigma_m} \widehat{\operatorname{div}} \left[\mathbf{n} + \mathbf{m}\, \frac{\partial N}{\partial m} + N\, \overline{\widehat{\operatorname{div}}\, \mathbf{m}} \right] \delta m \; d\Sigma_m -$$

$$- \int_{C_m} \left[\overline{\delta m} \left[\mathbf{n} + \frac{\partial N}{\partial m}\, \mathbf{m} + N\, \widehat{\operatorname{div}}\, \mathbf{m} \right] + \overline{\delta N}\, \mathbf{m} \right] i_2 dm .$$

Thus to find the new equilibrium equations, it is merely necessary to replace $\overline{\Pi N}$ by $\mathbf{n} + \mathbf{m}\,\dfrac{\partial N}{\partial m}$, and $- \overline{i_2 M}$ by $\mathbf{m}$, in (6.28). It becomes :

$$(6.60) \quad
\begin{cases}
\widehat{\operatorname{div}} \left[\mathbf{n} + \mathbf{m}\, \dfrac{\partial N}{\partial m} + N\, \overline{\widehat{\operatorname{div}}\, \mathbf{m}} \right] + \overline{R_m} + \widehat{\operatorname{div}} \left[\mathbf{G}_m N \right] = 0 \ \text{sur} \ \Sigma_m \\[2ex]
\Sigma_k \left[\overline{t}\, \mathbf{m}\, \nu + \overline{t}\, \mathbf{G}_L t \right]_{-}^{+} \overline{N}\delta m = 0 \ \text{ at the points of discontinuity on } C_m \\[2ex]
- \overline{\Pi\delta m} \left[\mathbf{n} + 2\, \dfrac{\partial N}{\partial m}\, \mathbf{m} \right] \nu + \overline{N}\delta m \left[- \widehat{\operatorname{div}}\, \mathbf{m} + \dfrac{\partial}{\partial m} \left[\overline{t}\, \mathbf{m}\, \nu \right] i_2 \right] \nu + \\[2ex]
\qquad\qquad\qquad + \overline{\nu}\, \mathbf{m}\, \nu\, \dfrac{\partial \overline{N}\delta m}{\partial \nu} = \\[2ex]
= \overline{\Pi\delta m} \left[\Pi\, R_L + \dfrac{\partial N}{\partial m}\, \mathbf{G}_L \right] t + \overline{N}\delta m \left[\mathbf{G}_m\, i_2 + \overline{N}\, R_L + \dfrac{\partial}{\partial m} \left[\overline{t}\, \mathbf{G}_L\, t \right] \right] t \\[2ex]
\qquad\qquad\qquad - \overline{\nu}\mathbf{G}_L\, t\, \dfrac{\partial \overline{N}\delta m}{\partial \nu} , \ \text{on} \ C_m, \ \forall \delta m \ \text{and} \dfrac{\partial \overline{N}\delta m}{\partial \nu} , \ K.A.
\end{cases}$$

To conclude, it should be pointed out that the constitutive laws are written either as functions of the first generalized stresses, and primitive generalized deformations, or as functions of the new stresses and symmetrical deformations.

4.3 – Stress functions in the case of hypothesis (wo)

Using a direct method, and as in the three-dimensional case, we shall establish stress functions from which the generalized stresses will be derived, in the case of internally homogeneous equilibrium.

1 – Non-symmetrical stresses

The equation of internally homogeneous equilibrium is written [see (6.28$_1$)] :

$$(6.61) \qquad \widehat{\mathrm{div}} \left[\overline{\Pi \mathbf{N}} - \overline{\mathbf{N} \ \widehat{\mathrm{div}} \ \overline{i_2 \mathbf{M}}} \right] = 0.$$

Furthermore, we know that the stresses satisfy the relation of symmetry (6.29) :

$$(6.62) \qquad \mathrm{T_r}\left(\left[\Pi \ \mathbf{N} - i_2 \mathbf{M} \ \frac{\partial \mathbf{N}}{\partial m} \right] i_2 \right) = 0.$$

Now with (6.42) we saw that the mapping $\left[\Pi \mathrm{N} - \mathbf{N} \ \widehat{\mathrm{div}} \ \overline{i_2 \mathbf{M}} \right] i_2$ which satisfies (6.61) is a co-edge, and by application of Poincaré's theorem, there exists in an arbitrary simply connected open set of Σ_m a vector field $\Omega^* \epsilon \vec{E}_3$ such that :

$$(6.63) \qquad \Pi \ \mathbf{N} \ i_2 - \mathbf{N} \ \widehat{\mathrm{div}} \ \overline{i_2 \mathbf{M}} \ i_2 = \frac{\partial \Omega^*}{\partial m}.$$

This formula can be split into tangential and normal components :

$$\left[\begin{array}{l} \Pi \ \mathbf{N} \ i_2 = \Pi \ \dfrac{\partial \Omega^*}{\partial m}. \\[2em] - \ \mathrm{div} \ \overline{i_2 \mathbf{M}} \ i_2 = \bar{\mathrm{N}} \ \dfrac{\partial \Omega^*}{\partial m}, \end{array} \right.$$

or otherwise :

$$(6.64) \qquad \left[\begin{array}{l} \Pi \ \mathbf{N} \ i_2 = \Pi \ \dfrac{\partial \Omega^*}{\partial m} \\[2em] - \ \widehat{\mathrm{div}} \ \mathbf{M} = \bar{\mathrm{N}} \ \dfrac{\partial \Omega^*}{\partial m}. \end{array} \right.$$

Vector field Ω^* is calculated by curvilinear integration, to within an arbitrary constant vector, and we can define an operation A* such that :

$$(6.65) \qquad \mathrm{A}^* = \mathbf{M} \ i_2 + i(\Omega^*)\Pi \quad \epsilon \ \mathcal{L}(\vec{E}_2, \vec{E}_3)$$

which gives the components :

$$(6.66) \quad \begin{cases} \Pi A^* = \Pi \, \mathbf{M} \, i_2 + i_2 \, \bar{N} \, \Omega^* \\[2ex] - \, i_2 \overline{A^*} \, N = \Pi \Omega^*. \end{cases}$$

Let us remember that we have demonstrated (6.45), namely :

$$(6.67) \quad \widehat{\mathrm{div}} \left[i_2 \, i(\Omega) \right] \Pi \; = - \, \bar{N} \, \frac{\partial \Omega}{\partial \mathbf{m}}.$$

$(6.64)_2$ therefore gives, with (6.67) :

$$\widehat{\mathrm{div}} \left[i_2 \, i_2 \overline{\mathbf{M}} \right] \Pi \; = - \, \widehat{\mathrm{div}} \left[i_2 \, i(\Omega^*) \right] \Pi$$

or

$$\widehat{\mathrm{div}} \left[i_2 \left[i_2 \overline{\mathbf{M}} + \Pi \, i(\Omega^*) \right] \right] \Pi \; = 0,$$

namely, with definition (6.65) :

$$(6.68) \quad \widehat{\mathrm{div}} \left[i_2 \, \overline{A^*} \right] \Pi = 0.$$

Let us now use (6.62), where $\Pi \mathbf{N}$ is taken from $(6.64)_1$ and $i_2 \mathbf{M}$ from $(6.66)_1$; we obtain, in succession :

$$T_r (\Pi \, \frac{\partial \Omega^*}{\partial \mathbf{m}} - \left[\Pi A^* i_2 + \bar{N} \Omega^* \Pi \right] \frac{\partial N}{\partial \mathbf{m}}) = 0.$$

$$T_r (\Pi \, \frac{\partial}{\partial \mathbf{m}} \left[\Pi \Omega^* + N \bar{N} \Omega^* \right] - \Pi A^* i_2 \frac{\partial N}{\partial \mathbf{m}} - \bar{N} \Omega^* \frac{\partial N}{\partial \mathbf{m}}) = 0$$

$$T_r (\Pi \, \frac{\partial \Pi \Omega^*}{\partial \mathbf{m}} - \Pi A^* i_2 \frac{\partial N}{\partial \mathbf{m}}) = 0.$$

Taking $\Pi \Omega^*$ from $(6.66)_2$:

$$T_r (- \, \Pi \, \frac{\partial}{\partial \mathbf{m}} \left[i_2 \, \overline{A^*} \, N \right] + i_2 \, \overline{\Pi A^*} \frac{\partial N}{\partial \mathbf{m}}) = 0$$

which is written in the natural basis S :

$$\alpha_S^{-1} \, i_2 \left[\partial_\alpha \, \overline{A^*} \, N - \overline{\Pi A^*} \, \partial_\alpha N + \overline{\Pi A^*} \, \partial_\alpha N \right] = 0 \, , \quad \alpha = 1,2,$$

Namely :

$$(6.69) \qquad \widehat{\mathrm{div}} \left[i_2 \, \overline{A^*} \right] N = 0 .$$

Grouping (6.68) and (6.69), we obtain :

$$\widehat{\mathrm{div}} \left[i_2 \, \overline{A^*} \right] = 0 .$$

Consequently by application of Poincaré's theorem, there exists in a simply connected open set of Σ_m, a vector field $V^* \in \vec{E}_3$ such that, to within an arbitrary constant vector :

$$(6.70) \qquad A^* = \frac{\partial V^*}{\partial m} .$$

But then $(6.66)_2$ enables to calculate $\Pi \Omega^*$ as a function of V^* :

$$(6.71) \qquad \Pi \Omega^* = - i_2 \, \frac{\overline{\partial V^*}}{\partial m} \, N .$$

Next $(6.64)_1$ enables us to calculate $\Pi \mathbf{N}$ as a function of Ω^*, therefore as a function of V^* and scalar $\overline{N}\Omega^*$, and finally $(6.66)_1$ can be used to calculate $\mathbf{M}$ as a function of these two quantities. We obtain :

$$(6.71) \qquad \begin{cases} \Pi \, \mathbf{N} = i_2 \, \dfrac{\partial}{\partial m} \left[\dfrac{\overline{\partial V^*}}{\partial m} \, N \right] i_2 - \overline{N}\Omega^* \, \dfrac{\partial N}{\partial m} \, i_2 \\[2ex] \mathbf{M} = - \, \Pi \, \dfrac{\partial V^*}{\partial m} \, i_2 \, \neg \, \overline{N} \, \Omega^* . \end{cases}$$

We need therefore a vector stress function $V^* \in \vec{E}_3$, and a scalar stress function $\overline{N}\Omega^*$, to calculate the generalized stresses $\Pi \mathbf{N}$ and $\mathbf{M}$, which interact only in the case of hypothesis (wo).

However, if we calculate the symmetrical stresses $\mathbf{n}$ and $\mathbf{m}$ defined by $(6.58)_2$ and $(6.58)_3$, we find that the scalar function $\overline{N}\Omega^*$ is eliminated. We then have :

$$(6.72) \qquad \begin{cases} \mathbf{n} = i_2 \, \dfrac{\partial}{\partial m} \left[\dfrac{\overline{\partial V^*}}{\partial m} \, N \right] i_2 - \dfrac{\partial N}{\partial m} \, i_2 \, \dfrac{\overline{\partial V^*}}{\partial m} \, i_2 \\[2ex] \mathbf{m} = \dfrac{1}{2} \, i_2 \left[\Pi \, \dfrac{\partial V^*}{\partial m} + \overline{\Pi \, \dfrac{\partial V^*}{\partial m}} \right] i_2 . \end{cases}$$

Thus the symmetrical stresses follow from only three scalar stress functions, and not four.

2 - Symmetrical stresses

We can obtain this result directly. Let us take the homogeneous equilibrium equation $(6.60)_1$:

$$(6.73) \qquad \widehat{\text{div}}\left[\mathbf{n} + \mathbf{m}\,\frac{\partial N}{\partial \mathbf{m}} + \overline{N\,\widehat{\text{div}}\,\mathbf{m}} \right] = 0$$

with :

$$\mathbf{n} = \bar{\mathbf{n}} \;,\; \mathbf{m} = \bar{\mathbf{m}} \;,\; \in \mathcal{L}(\vec{E}_2,\vec{E}_2)\,.$$

The operator between brackets is a co-cycle, and $\exists$ in a simply connected open set of Σ_m, a vector field $\Omega^* \in \vec{E}_3$, defined to within an arbitrary constant vector, such that [see (6.42)] :

$$(6.74) \quad \left[\begin{array}{l} \left[\mathbf{n} + \dfrac{\partial N}{\partial \mathbf{m}}\mathbf{m} + N\,\widehat{\text{div}}\,\mathbf{m} \right] i_2 = \dfrac{\partial \Omega^*}{\partial \mathbf{m}} \\[2em] \text{or :} \\[1em] \left[\mathbf{n} + \dfrac{\partial N}{\partial \mathbf{m}}\,\mathbf{m} \right] i_2 = \Pi\,\dfrac{\partial \Omega^*}{\partial \mathbf{m}} \\[2em] \widehat{\text{div}}\,\mathbf{m}\; i_2 = \bar{N}\,\dfrac{\partial \Omega^*}{\partial \mathbf{m}}\; . \end{array}\right.$$

We are led to define, as before, an operator $A^* \in \mathcal{L}\,(\vec{E}_2,\,\vec{E}_3)$ by :

$$(6.75) \quad \left[\begin{array}{l} A^* = i_2\,\mathbf{m}\,i_2 + i(\Omega^*)\Pi \\[1.5em] \Longleftrightarrow \left[\begin{array}{l} \Pi A^* = i_2\,\mathbf{m}\,i_2 + i_2\,\bar{N}\Omega^* \\[1.5em] \Pi\Omega^* = -\,i_2\,\overline{A^*}\,N\,. \end{array}\right. \end{array}\right.$$

Now we have formula (6.67) which has already been demonstrated :

$$(6.76) \qquad \widehat{\text{div}}\left[i_2\,i(\Omega^*) \right]\Pi \;=\; -\,\bar{N}\,\frac{\partial \Omega^*}{\partial \mathbf{m}}\,.$$

From that time $(6.74)_3$ and (6.76) give :

$$\widehat{\operatorname{div}} \left[\overline{i_2 m} + i_2\, i(\Omega^*) \right] \Pi = 0$$

and with $(6.75)_1$:

$$(6.77) \qquad \widehat{\operatorname{div}} \left[i_2\, \overrightarrow{A^*} \right] \Pi = 0.$$

Furthermore :

$$\mathbf{n} = \overline{\mathbf{n}} ,$$

can be written :

$$T_r\, (\mathbf{n}\, i_2) = 0 .$$

Replacing $\mathbf{n}$ by its value taken from $(6.74)_2$, we have :

$$T_r(\Pi\, \frac{\partial \Omega^*}{\partial m} - \frac{\partial N}{\partial m}\, \mathbf{m}\, i_2) = 0.$$

By a calculation similar to that which gave us (6.69), we again find :

$$(6.78) \qquad \widehat{\operatorname{div}} \left[i_2\, \overline{A^*} \right] N = 0 .$$

Thus (6.77) and (6.78) once more give :

$$\widehat{\operatorname{div}} \left[i_2\, \overline{A^*} \right] = 0$$

which is the condition for local closure of the vectorial form A^* of degree 1. In the simply connected open set of Σ_m, $\exists$ a vector field $V^* \in \overrightarrow{E}_3$ defined to within an arbitrary constant vector, such that :

$$A^* = \frac{\partial V^*}{\partial m} .$$

(6.75) and (6.74) then easily give formulae (6.72).

It is worth noting that the definition of A^* in (6.75) is slightly different from that of (6.65).

4.4 - <u>Stress functions in the case of hypothesis (w)</u>

In this hypothesis, stress functions can be found, but as with non-symmetrical stresses in the preceding case, four scalar stress functions are required[24].

Let us consider the homogeneous equilibrium in the interior, as given by $(6.26)_1$ and $(6.26)_2$:

$$(6.79) \quad \begin{cases} \widehat{\operatorname{div} \bar{\mathbf{N}}} = 0 \\[2ex] \widehat{\operatorname{div} \bar{\mathbf{M}}} + \bar{\mathbf{N}} \mathbf{N} \, i_2 = 0 . \end{cases}$$

The presence of transverse shear introduces supplementary forces in our calculations.

Furthermore, the generalized stresses meet the relation of symmetry (6.29) :

$$(6.80) \quad T_r \left(\left[\Pi \mathbf{N} - i_2 \mathbf{M} \frac{\partial N}{\partial m} \right] i_2 \right) = 0 .$$

Thus $(6.79)_1$ provides the existence, in a simply connected open set of Σ_m, of a vector field $\Omega^* \in \vec{E}_3$ such that :

$$(6.81) \quad \begin{cases} \mathbf{N} \, i_2 = \dfrac{\partial \Omega^*}{\partial m} \\[3ex] \Leftrightarrow \begin{cases} \Pi \mathbf{N} i_2 = \Pi \dfrac{\partial \Omega^*}{\partial m} \\[3ex] \bar{\mathbf{N}} \mathbf{N} i_2 = \bar{\mathbf{N}} \dfrac{\partial \Omega^*}{\partial m} . \end{cases} \end{cases}$$

With $(6.79)_2$, $(6.81)_3$ is written :

$$- \widehat{\operatorname{div} \mathbf{M}} = \bar{\mathbf{N}} \frac{\partial \Omega^*}{\partial m} .$$

(6.81) therefore becomes :

$$(6.82) \quad \begin{cases} \Pi \mathbf{N} i_2 = \Pi \dfrac{\partial \Omega^*}{\partial m} \\[3ex] - \widehat{\operatorname{div} \mathbf{M}} = \bar{\mathbf{N}} \dfrac{\partial \Omega^*}{\partial m} . \end{cases}$$

As Ω^* can be calculated, let us put :

$$(6.83) \quad \left[\Leftrightarrow \left[\begin{array}{l} A^* = \mathbf{M}\, i_2 + i(\Omega^*)\Pi \\[2ex] \Pi A^* = \mathbf{M}\, i_2 + \bar{N}\, \Omega^* i_2 \\[2ex] \Pi \Omega^* = -\, i_2\, \overline{A^*}\, N\,. \end{array} \right. \right.$$

In this case, it should be noted that :

$$\frac{\Pi A^* - \overline{\Pi A^*}}{2} \neq i_2\, \bar{N}\, \Omega^*$$

as $M \neq \bar{M}$.

As before, $(6.82)_2$ gives, with (6.76) :

$$\widehat{\mathrm{div}}\left[i_2 \left[i_2 \bar{\mathbf{M}} + \Pi i(\Omega^*) \right] \right] \Pi = 0,$$

hence with $(6.83)_1$:

$$(6.84) \quad \widehat{\mathrm{div}}\left[i_2\, \overline{A^*} \right] \Pi = 0\,.$$

Then (6.80) is written, with $(6.81)_2$ and $(6.83)_2$ as before :

$$(6.85) \quad \widehat{\mathrm{div}}\left[i_2\, \overline{A^*} \right] N = 0\,.$$

Thus (6.84) and (6.85) indeed give again :

$$\widehat{\mathrm{div}}\left[i_2\, \overline{A^*} \right] = 0$$

and in the simply connected open set of Σ_m, there exists a vector field $V^* \in \vec{E}$ such that :

$$(6.86) \quad A^* = \frac{\partial V^*}{\partial m}\,.$$

Whence :

$$\Omega^* = \Pi\Omega^* + N\,\bar{N}\,\Omega^* = -\,i_2\,\frac{\overline{\partial V^*}}{\partial m}\,N + N\left[\bar{N}\,\Omega^*\right]$$

$(6.83)_1$ then gives **M**, and $(6.81)_1$ gives **N**, namely :

$$(6.87)\quad \left[\begin{array}{l} \mathbf{N} = -\,\dfrac{\partial\Omega^*}{\partial m}\,i_2\ \text{with}\ \ \Omega^* = -\,i_2\,\dfrac{\overline{\partial V^*}}{\partial m}\,N + N\left[\bar{N}\,\Omega^*\right] \\[2em] \mathbf{M} = -\,\Pi\,\dfrac{\partial V^*}{\partial m}\,i_2 - \Pi\,\bar{N}\,\Omega^* \end{array}\right.$$

In contrast to the preceding case, we observe that it is no longer possible to derive the generalized stresses from only three scalar functions, corresponding to the components of vector V*, a fourth function $\overline{N\Omega^*}$ being required.

Case of multiconnected shells[24,60]

In this case, it is necessary to introduce global closure conditions to ensure the univoque existence of the stress functions. These conditions are found by application of de Rham's theorem, by integration on the non-connected edges $C_{m\ell}$ which constitute C_m [see (A.100)].

$(6.79)_1$ gives immediately, with (6.86), and for p edges :

$$(6.88)\quad \left[\begin{array}{l} \displaystyle\int_{C_{m\ell}} \mathbf{N}\,i_2\,dm = 0 \\[2em] \displaystyle\int_{C_{m\ell}} A^*\,dm = 0 \end{array}\right\} \quad \forall\,\ell = 2,3,\ \ldots,\ p.$$

For simplification's sake, let us assume the variables δm and δN to be independent on C_m, the conditions for equilibrium on the edge are written :

$$(6.89)\quad \left[\begin{array}{l} \mathbf{N}\,i_2\,t = \mathbf{N'}\,i_2\,t \\[1em] \mathbf{M}\,i_2\,t = \mathbf{M'}\,i_2\,t, \end{array}\right\} \quad \text{on}\ C_m$$

where **N'** and **M'** are given. (6.88) also gives, with (6.89) and $(6.83)_1$:

$$(6.90) \quad \left[\begin{array}{l} \int_{C_{m\ell}} \mathbf{N'} \, i_2 \, dm = 0 \\[2em] \int_{C_{m\ell}} \left[\mathbf{M'} \, i_2 dm + i(\Omega^*) dm \right] = 0 \end{array} \right\} \quad \ell = 2,3, \ldots, p.$$

Integration of $(6.90)_2$ by parts easily gives :

$$\int_{C_{m\ell}} \left[\mathbf{M'} \, i_2 \, dm + i(m) \, \frac{\partial \Omega^*}{\partial m} \, dm \right] = 0,$$

and using $(6.81)_1$ and $(6.89)_1$:

$$\int_{C_{m\ell}} \left[\mathbf{M'} \, i_2 \, dm + i(m) \, \mathbf{N'} \, i_2 \, dm \right] = 0, \, \forall \ell = 2,3, \ldots, p.$$

Finally, the supplementary global closure conditions are written :

$$(6.91) \quad \left[\begin{array}{l} \int_{C_{m\ell}} \mathbf{N'} \, i_2 \, dm = 0 \\[2em] \int_{C_{m\ell}} \left[\mathbf{M'} + i(m) \, \mathbf{N'} \right] i_2 \, dm = 0 \end{array} \right\} \quad \forall \ell = 2,3, \ldots, p.$$

Interpretation of this result is easy : it represents cancellation of the resultant force and resultant moment of the given external forces on the edge of each hole. The external forces satisfying the global equilibrium of the shell, we find again that conditions (6.91) need only be satisfied on p-1 disconnected edges.

If the displacements are only given on a part of $C_{m\ell}$, and forces on the complementary part, the conditions must be modified in consequence[24].

5. – <u>VARIATIONAL PRINCIPLES</u>

Using the results of paragraph 2, and those of chapter III, we will give the surface translation of the principle of virtual work and of the principle of virtual stresses, taking for example the case of hypothesis (wo). We assign index d to the given kinematic quantities. We shall again assume $\delta z = 0$.

As will be remembered, the principle of virtual displacements is written :

$$(6.92) \quad \iint_{\Sigma_m} T_r(\mathbf{n}\,\delta\gamma + \mathbf{m}\,\delta\kappa)\,d\Sigma_m - \iint_{\Sigma_m} \left[\overline{\delta m}\,\mathbf{R}_m - \overline{N}\,\frac{\partial \delta m}{\partial m}\,\mathbf{G}_m \right] d\Sigma_m$$

$$- \int_{C_{mF}} \left[\overline{\delta m}\,\mathbf{R}_L - \overline{N}\,\frac{\partial \delta m}{\partial m}\,\mathbf{G}_L \right] dm = 0, \quad \forall \delta m \ \text{K.A.}$$

where C_{mF} represents the part of C_m where the forces are given. In the case of hypothesis (w), δw is given by (6.23).

The principle of virtual stresses is written :

$$(6.93) \quad \left[\begin{array}{l} \displaystyle \iint_{\Sigma_m} T_r(\delta \mathbf{n}\ \gamma + \delta \mathbf{m}\ \kappa)\,d\Sigma_m \\[2em] \displaystyle - \int_{C_{mV}} \left[\overline{\Pi V_d}\left[- \delta\cdot\mathbf{n} - 2\,\frac{\partial N}{\partial m}\,\delta\mathbf{m} \right]\nu + \overline{N}V_d\left[- \widehat{div}\,\delta\mathbf{m} + \frac{\partial \overline{t}\delta\mathbf{m}\nu}{\partial m}\,i_2 \right]\nu \right. \\[2em] \displaystyle \left. + \overline{\nu}\delta\mathbf{m}\ \nu \left[\frac{\partial \overline{N}V}{\partial \nu} \right]_d \right] ds = 0 \\[2em] \forall \delta\mathbf{n}, \delta\mathbf{m}\quad \text{S.A.H.} \end{array} \right.$$

To write down this principle (6.93), we have used equation $(6.28)_3$.

These two formulations can be used in numerical analyses, after discretization using the finite element or finite difference method. The first of these principles can be used in the non-linear case.

It should be noted that if we put the conditions for kinematic compatibility, using a Lagrangian multiplier, in (6.92), we find the stress function in the form of this multiplier.

Conversely, introducing the stress functions in (6.93), we find the conditions for kinematic compatibility.

In the first case, the principle also gives the global conditions without difficulty[24,60].

Other variational formulations can be found, in particular mixed and non-linear[61].

6. - LINEAR CONSTITUTIVE LAWS[7]

It goes without saying that a theory of shells can only be used as such, insofar as the theory is associated with surface constitutive laws. Now these laws naturally depend on the simplifying hypotheses, formulated beforehand to establish the theory. It can even be said that these laws form part of the theory itself.

The conditions required for formulating these laws, are firstly that the laws must be consistent with the theory used, and furthermore they must supply a correct description of the actual behaviour of the shell. The latter condition means that the pre-formulated simplifying hypotheses must themselves be properly representative of the static or kinematic behaviour of the shell.

This type of behaviour is generally characterized by the fact that the normal deformation energy is negligible. We have seen that this can be translated in practice, either by neglecting the normal deformation, which is translated in virtual terms by $\delta z = 0$, or by neglecting the normal stress.

We will consider these two hypotheses in succession, together with their consequences on the constitutive laws in the linearized case. In both cases, we will start with a linear, hyperelastic three-dimensional behaviour, and expose Naghdi's elegant method[7].

1) Zero normal stress

The problem consists in finding a two-dimensional constitutive law, which takes account of a set of kinematic and dynamic hypotheses. The preceding

theory was established on the basis of hypothesis (w), with transverse
shear, and assuming $\delta z = 0$ for example. This led us to an indeterminate
normal stress (see § 4.1), and remains partially valid for the same kinematic hypotheses, with the exception of those relating to normal deformation. For example, we can assume $\delta z \neq 0$, and zero normal stress, this
hypothesis being contained in that of indeterminate normal stresses. The
two types of hypothesis (w) and $\sigma_{33} = 0$, are therefore not incompatible.
But we have also seen that transverse shear deformations are modifed when
$\delta z \neq 0$.

This being so, the simultaneous introduction of these hypotheses, can,
and should therefore be accompanied by a certain type of interpolation of
transverse shear deformations, provided certain equations of the problem
are satisfied in the mean, using a weak formulation. As it is a question
of satisfying the constitutive laws in the mean, it is natural to utilize
the three-field principle of Hu-Washizu, given by (3.9) or (3.7). Furthermore, the deformation and the stress are independent unknown fields in this principle, and lend themselves to hypotheses of simultaneous interpolation.

This interpolation for the transverse shear deformation, is accompanied, in
linear elasticity, by an interpolation of the same type for transverse
shear stresses, by reason of the linear relations linking the two.

We therefore select a polynominal interpolation in the thickness direction, namely in z, to obtain finally the surface laws.

For transverse shear stress, it is natural to adopt the form :

$$(6.94) \qquad \Pi C \, N\overline{N} = \frac{3}{2h} \mu Q \left[1 - \left[\frac{z}{h/2} \right]^2 \right] \overline{N}$$

which cancels out for $z = \pm \frac{h}{2}$, which is compatible with the usual case that
the surface Σ^+ and Σ^- are free from transverse shear forces. Generalized stress
$Q \in \vec{E}_2$ will be transverse shear force on Σ_m ($z = 0$).

Furthermore, we have seen that the generalized deformations, other than
the normal deformation, are given by (6.33), so that when $\delta z = 0$:

$$(6.95) \quad \begin{cases} 2\,D_1 = 2\Pi\,D_L\,\Pi = [\gamma' + z\kappa']\,\mu^{-1} + \mu^{-1}\,[\gamma' + z\kappa'] \\[2ex] 2\,D_2 = 2\Pi\,D_L\,N\overline{N} = \mu^{-1}\,\gamma_3\,\overline{N}\ ,\ \left(\delta\gamma_3 = \dfrac{\overline{\partial\delta m}}{\partial m}\,N + \delta N\ \epsilon\ \vec{E}_2\right) \\[2ex] 2\,D_3 = 2\,N\overline{N}\,D_L\,\Pi = 2\,\overline{D_2} \\[2ex] 2\,D_4 = 2\,N\overline{N}\,D_L\,N\overline{N} = 0. \end{cases}$$

When the normal deformation is zero, the transverse shear deformation in independent of z. It is therefore natural to adopt for that deformation D_2, an interpolation of the same type as (6.94), which cancels out with the stress, giving for z = 0, the preceding deformation D_2 on Σ_m. This leads to :

$$(6.96) \qquad D_2 = \mu^{-1}\,\gamma_3\,\overline{N}\left[1 - \left[\frac{z}{h/2}\right]^2\right] \times \frac{5}{8}\ .$$

The numerical coefficients have been selected to simplify the coefficients resulting from subsequent integrations on z.

It is then merely necessary to introduce hypotheses (w) (6.94) and (6.96) in the three-field principle, to obtain, as we know, all the equations of the problem including the boundary conditions.

It is clear that integration in z will give the equilibrium equations, and the surface boundary conditions, as also the deformation/displacement relations [see (3.8)] , as in the utilization of the primal principle for the equilibrium. On the other hand, we also obtain the surface constitutive laws resulting from (3.8)$_1$, averaged in thickness, giving for a linear hyperelastic body :

$$(6.97) \qquad \int_\Omega \left[\frac{\partial\alpha}{\partial D_L} - \tilde{C}\right]\,\delta D_L\ d\Omega = 0\ ,\ \forall\delta D_L\ .$$

A calculation indentical to that for § 2.1, obviously gives :

$$\int_\Omega \tilde{C}\ \delta D_L\ d\Omega = \int_{\Sigma_m} T_r(\overline{\Pi\mathbf{N}}\ \delta\gamma' - \overline{i_2\mathbf{M}}\ \delta\kappa')d\Sigma_m +$$

$$+ \int_\Omega T_r(\Pi C N \overline{N}\ \overline{\delta D_2})d\Omega$$

$$= \int_{\Sigma_m} T_r(\overline{\Pi\mathbf{N}}\ \delta\gamma' - \overline{i_2\mathbf{M}}\ \delta\kappa')d\Sigma_m +$$

$$+ \int_{\Sigma_m} d\Sigma_m \overline{Q}\ \delta\gamma_3 \int_{-\frac{h}{2}}^{+\frac{h}{2}} \frac{15}{16h}\left[1 - \left[\frac{z}{h/2}\right]^2\right]^2 dz \ .$$

(6.97) is therefore written :

(6.98)
$$\int_\Omega \frac{\partial\alpha}{\partial D_L}\ \delta D_L\ d\Omega = \int_{\Sigma_m} \left[T_r(\overline{\Pi\mathbf{N}}\ \delta\gamma' - \overline{i_2\mathbf{M}}\ \delta\kappa') + \overline{Q}\ \delta\gamma_3\right]d\Sigma_m$$

$$\forall\ \delta\gamma',\ \delta\kappa',\ \delta\gamma_3$$

with (6.95).

It is then sufficient to calculate the first member, averaged on z, and to identify the terms in $\delta\gamma'$, $\delta\kappa'$ and $\delta\gamma_3$ respectively, to obtain the generalized stresses $\overline{\Pi\mathbf{N}}$, $- \overline{i_2\mathbf{M}}$ and Q as functions of the coefficients of α.

As we are concerned with obtaining detailed expressions, as functions of the components in the natural bases of E_3 and E_2, the calculations are made in these bases.

We shall assume that the shell has symmetrical mechanical properties with respect to the middle surface Σ_m, by a unitary transformation with respect to the tangent plane $\vec{E}_2$. This transformation is equal to $\Pi - N\overline{N}$. For the three-dimensional law having this symmetry, and assumed to be heterogeneous and anisotropic [*], we find :

---

[*] In all the sequel, the Greek indices will take the values 1 and 2. The indexed notations used are those of Naghdi.

$$(6.99) \quad \begin{cases} \sigma^{\alpha\beta} = C^{\alpha\beta\gamma\delta} \, e_{\gamma\delta} + C^{\alpha\beta33} \, e_{33} \\[2mm] \sigma^{\alpha3} = 2 \, C^{\alpha z \gamma 3} \, e_{\gamma 3} \\[2mm] \sigma^{33} = C^{33\gamma\delta} \, e_{\gamma\delta} + C^{3333} \, e_{33} \\[2mm] \alpha = \dfrac{1}{2} \, \sigma^{ij} \, e_{ij} \quad (i,j = 1,2,3). \end{cases}$$

with :

$$(6.100) \quad \begin{cases} \sigma^{i}_{\ j} = C^{ik}_{\ j\ell} \, e^{\ell}_{k} \qquad\qquad i,j,k,\ell = 1,2,3 \\[2mm] C^{\alpha\gamma}_{\ \beta 3} = C^{\gamma\alpha}_{\ 3\beta} = C^{\alpha 3}_{\ 33} = C^{3\alpha}_{\ 33} = 0 \quad \text{(symmetrical with respect to } \vec{E}_2). \end{cases}$$

In (6.99), the linear deformation D_L is expressed in covariant components e_{ij}, in other words D_L is expressed in the natural basis T for def(D_L), and in the supplementary basis T' for val(D_L), such that :

$$(6.101) \quad \begin{cases} T' = \overline{T^{-1}} \\[3mm] \text{where :} \\[2mm] T = [\mu S \ N], \quad T_3 = N = \overline{T^{-1}}_3 - = T'_3 \\[3mm] \overline{T} \, T = \begin{bmatrix} \overline{S} \ \mu^2 S & 0 \\[4mm] 0 & 1 \end{bmatrix} = G_T, \quad \overline{T}_i \, T_j = {}^{i}G_{T_j} = g_{T^{ij}} \\[4mm] T^{-1} = G_T^{-1} \, \overline{T} \, , \quad \overline{T^{-1}} = T \, G_T^{-1} = T' \\[3mm] D_L = T'_j \, e_{ij} \, {}^{j}T^{-1} \, , \quad e_{ij} = \overline{T}_i \, D_L \, T_j \, . \end{cases}$$

In this basis T', tensor $\underset{\sim}{\alpha}$ has components $C^{ijk\ell}$, such that :

$$\begin{aligned}
c^{\alpha\beta\gamma\delta} &= \underset{\sim}{\alpha}(T'_\alpha \, \overline{T'_\beta})(T'_\gamma \, \overline{T'_\delta}) = \overline{c}^{\,\alpha'\beta'\gamma'\delta'} \; m^\alpha_{\alpha'} \, m^\beta_{\beta'} \, m^\gamma_{\gamma'} \, m^\delta_{\delta'}
\end{aligned}$$

where :

$$m = \mu^{-1} = \left[1_{E_2} +, z \frac{\partial N}{\partial m}\right]^{-1}, \quad m^\alpha_{\alpha'} = {}^\alpha S^{-1} \, \mu^{-1} \, S_{\alpha'}$$

and :

$$\overline{c}^{\,\alpha\beta\gamma\delta} = \overline{c}^{\,\alpha\beta}_{\gamma'\delta'} \; g^{\gamma\gamma'} \, g^{\delta\delta'}$$

$$\overline{c}^{\,\alpha\beta}_{\gamma\delta} = \underset{\sim}{\alpha}(S_\gamma \, {}^\alpha S^{-1})(S_\delta \, {}^\beta S^{-1})$$

$$G = \overline{S}\, S, \quad {}^\gamma G_{\gamma'} = g_{\gamma\gamma'}, \quad {}^\gamma G^{-1}_{\gamma'} = g^{\gamma\gamma'}$$

(Remark : $\overline{c}^{\,\alpha\beta}_{\gamma\delta} \neq \overline{c}^{\,\beta\alpha}_{\gamma\delta} \neq \dots$)

Then :

$$c^{\alpha 3\beta 3} = \underset{\sim}{\alpha}(T'_\alpha \, \overline{N})(T'_\beta \, \overline{N}) = \overline{c}^{\,\alpha'3\beta'3} \; m^\alpha_{\alpha'} \, m^\beta_{\beta'} \, .$$

(6.102)

The preceding formulae offer the advantage of being functions of the coefficients of $\underset{\sim}{\alpha}$ in S, which depend only of point $m \in \Sigma_m$ if the properties of the material are homogeneous in z.

Let us now introduce the condition :

$$\sigma_{33} = 0 .$$

By $(6.99)_3$, we obtain :

$$c^{33\gamma\delta} \, e_{\gamma\delta} + c^{3333} \, e_{33} = 0 .$$

Whence eliminating e_{33}, we have :

$$\alpha = \frac{1}{2} \, \sigma^{ij} \, e_{ij}$$

$$\alpha = \frac{1}{2} \, A^{\alpha\beta\gamma\delta} \, e_{\gamma\delta} \, e_{\alpha\beta} + 2 \, A^{\alpha 3\beta 3} \, e_{\gamma 3} \, e_{\alpha 3}$$

with :

$$A^{\alpha\beta\gamma\delta} = c^{\alpha\beta\gamma\delta} - \frac{c^{33\gamma\delta} \, c^{\alpha\beta 33}}{c^{3333}}$$

(6.103)

$$(6.103) \text{ continued} \quad \begin{cases} A^{\alpha\beta\gamma\delta} = A^{\beta\alpha\gamma\delta} = A^{\alpha\beta\delta\gamma} = A^{\gamma\delta\alpha\beta} \\[2mm] A^{\alpha3\beta3} = A^{\gamma3\alpha3} = A^{\alpha3\gamma3} = A^{\gamma3\alpha3} \end{cases}$$

The contravariant components of A are expressed in basis S, as those of C, by means of the same formulae (6.102) and barred contravariant coefficients $\overline{A}$.

Let us then consider the first member of (6.98), we have [see (6.7)] :

$$\int_\Omega \frac{\partial\alpha}{\partial D_L} \delta D_L \, d\Omega = \int_{\Sigma_m} d\Sigma_m \int_{-\frac{h}{2}}^{+\frac{h}{2}} \det(\mu) \frac{\partial\alpha}{\partial D_L} \delta D_L \, dz .$$

$$(6.104) \qquad \int_\Omega \frac{\partial\alpha}{\partial D_L} \delta D_L \, d\Omega = \int_{\Sigma_m} d\Sigma_m \int_{-\frac{h}{2}}^{+\frac{h}{2}} \det(\mu) \, \sigma^{ij} \, \delta e_{ij} \, dz .$$

Now in the bases $T' = T^{-1}$ and T, (6.95) gives for covariant components :

$$(6.105) \quad \begin{cases} 2\,e_{\alpha\beta} = \mu_\alpha^{\alpha'} \left[\gamma'_{\alpha'\beta} + z\kappa'_{\alpha'\beta} \right] + \mu_\beta^{\beta'} \left[\gamma'_{\beta'\alpha} + z\kappa'_{\beta'\alpha} \right] \\[3mm] 2e_{\alpha3} = \gamma_{\alpha3} \left[1 - \left[\frac{z}{h/2} \right]^2 \right] \cdot \frac{5}{4} : \end{cases}$$

(6.104) is then written, with (6.105) and (6.103) :

$$(6.106) \quad \begin{aligned} &\int_\Omega \frac{\partial\alpha}{\partial D_L} \delta D_L \, d\Omega = \int_{\Sigma_m} d\Sigma_m \int_{-\frac{h}{2}}^{+\frac{h}{2}} \det(\mu) A^{\alpha\beta\gamma\delta} \mu_\alpha^{\alpha'} \mu_\gamma^{\gamma'} \left[\gamma'_{\gamma'\delta} + z\kappa'_{\gamma'\delta} \right] \times \\[3mm] &\hspace{4cm} \left[\delta\gamma'_{\alpha'\beta} + z\delta\kappa'_{\alpha'\beta} \right] dz \\[3mm] &\quad + \int_{\Sigma_m} d\Sigma_m \int_{-\frac{h}{2}}^{+\frac{h}{2}} \det(\mu) A^{\alpha3\beta3} \gamma_{\beta3} \delta\gamma_{\alpha3} \left[1 - \left[\frac{z}{h/2} \right]^2 \right]^2 \cdot \frac{25}{16} \, dz . \end{aligned}$$

The second member of (6.98) is also written as functions of the components in the same bases :

$$(6.107) \qquad \int_{\Omega} T_r(C\, \delta D_L)\, d\Omega = \int_{\Sigma_m} \left[N^{\beta\alpha}\, \delta\gamma'_{\alpha\beta} + M^{\beta\alpha}\, \delta\kappa'_{\alpha\beta} + Q^{\alpha}\, \delta\gamma_{\alpha 3} \right] d\Sigma_m$$

where $N^{\beta\alpha}$, $M^{\beta\alpha}$, Q^{α} represent the contravariant components of $\overline{\overline{\Pi}\,\mathbf{N}}$, $-\overline{i_2\mathbf{M}}$ and Q in basis S, such that :

$$\begin{cases} N^{\beta\alpha} = {}^{\beta}S^{-1}\ \overline{\overline{\Pi\mathbf{N}}}\ {}^{\overline{\alpha}_S{}^{-1}} \\[2ex] M^{\beta\alpha} = -\ {}^{\beta}S^{-1}\ \overline{\overline{i_2\mathbf{M}}}\ {}^{\overline{\alpha}_S{}^{-1}} \\[2ex] Q^{\alpha} = {}^{\alpha}S^{-1}\ Q\ . \end{cases}$$

Finally, (6.106) and (6.107) give, by identification :

$$(6.108) \qquad \begin{cases} N^{\beta\alpha} = B_0^{\alpha\beta\gamma\delta}\ \gamma'_{\gamma\delta} + B_1^{\alpha\beta\gamma\delta}\ \kappa'_{\gamma\delta} \\[2ex] M^{\beta\alpha} = B_1^{\alpha\beta\gamma\delta}\ \gamma'_{\gamma\delta} + B_2^{\alpha\beta\gamma\delta}\ \kappa'_{\gamma\delta} \\[2ex] Q^{\alpha} = \dfrac{5}{6}\, B_0^{\alpha 3\beta 3}\ \gamma_{\beta 3}, \end{cases}$$

with, taking account of (6.102)

$$(6.109) \qquad \begin{cases} B_n^{\alpha\beta\gamma\delta} = \overline{A}^{\alpha\beta'\gamma\delta'} \displaystyle\int_{-\frac{h}{2}}^{+\frac{h}{2}} \mathrm{d\acute{e}t}(\mu)\ m_{\beta'}^{\beta},\ m_{\delta'}^{\delta},\ z^n dz \quad (n = 0,1,2) \\[4ex] B_0^{\alpha 3\beta 3} = \dfrac{15}{8}\, \overline{A}^{\alpha'3\beta'3} \displaystyle\int_{-\frac{h}{2}}^{+\frac{h}{2}} \mathrm{d\acute{e}t}(\mu)\ m_{\alpha'}^{\alpha},\ m_{\beta'}^{\beta},\ \left[1 - \left[\dfrac{z}{h/2}\right]^2\right]^2 dz, \end{cases}$$

assuming the material to be homogeneous in z.

Furthermore, the relations of symmetry (6.100) give :

$$B_n^{\alpha\beta\gamma\delta} = B_n^{\gamma\delta\alpha\beta}, \quad B_o^{\alpha 3\beta 3} = B_o^{\beta 3\alpha 3}$$

But :

$$B_n^{\alpha\beta\gamma\delta} \neq B_n^{\beta\alpha\gamma\delta} \neq B_n^{\beta\alpha\delta\gamma}$$

because of (6.102).

Moreover, the coefficients $B_n^{\alpha\beta\gamma\delta}$ must satisfy certain relations result-ing from the relation of symmetry (6.30), always satisfied by the general-ized stresses, namely, in contravariant components :

$$N^{\beta\alpha} + b_\beta^\beta, \, N^{\beta'\alpha} = N^{\alpha\beta} + M^{\alpha\beta'} \, b_\beta^\beta,$$

where, let us remind it :

$$b_{\beta'}^\beta = - \, {}^\beta S^{-1} \, \frac{\partial N}{\partial m} \, S_{\beta'}.$$

This gives :

$$B_o^{\alpha\beta\gamma\delta} \, \gamma_{\gamma\delta}' + B_1^{\alpha\beta\gamma\delta} \, \kappa_{\gamma\delta}' + b_{\beta'}^\beta \left[B_1^{\alpha\beta'\gamma\delta} \, \gamma_{\alpha\delta}' + B_2^{\alpha\beta'\gamma\delta} \, \kappa_{\gamma\delta}' \right],$$

$$= B_o^{\alpha\beta\gamma\delta} \, \gamma_{\gamma\delta}' + B_1^{\beta\alpha\gamma\delta} \, \kappa_{\gamma\delta}' + b_{\beta'}^\beta \left[B_1^{\beta'\alpha\gamma\delta} \, \gamma_{\alpha\delta}' + B_2^{\beta'\alpha\gamma\delta} \, \kappa_{\gamma\delta}' \right],$$

namely :

$$(6.110) \quad \begin{cases} B_o^{\alpha\beta\gamma\delta} + b_{\beta'}^\beta \, B_1^{\alpha\beta'\gamma\delta} = B_o^{\beta\alpha\gamma\delta} + b_{\beta'}^\beta \, B_1^{\beta'\alpha\gamma\delta} \\[2em] B_1^{\alpha\beta\gamma\delta} + b_{\beta'}^\beta \, B_2^{\alpha\beta'\gamma\delta} = B_1^{\beta\alpha\gamma\delta} + b_{\beta'}^\beta \, B_2^{\beta'\alpha\gamma\delta}. \end{cases}$$

Approximate surface constitutive laws

The laws (6.108) and (6.109) can be the subject of approximate calcul-ations. In (6.109) :

$$(6.111) \quad \begin{cases} m = \mu^{-1} = \dfrac{Adj(\mu)}{det(\mu)} \quad \text{with} \quad \mu = 1_{E_2} + z\,\dfrac{\partial N}{\partial m} \\[2em] det(\mu) = 1 - 2\,zH + z^2 K \quad \text{with} \quad \begin{cases} 2H = T_r\left(\dfrac{\partial N}{\partial m}\right) \\[1.5em] K = det\left(\dfrac{\partial N}{\partial m}\right) \end{cases} \\[2em] Adj(\mu) = 1_{E_2} - z\left[\dfrac{\partial N}{\partial m} - 2H.\,1_{E_2}\right]. \end{cases}$$

Making a limited expansion of μ^{-1} with respect to z, we find, with all calculations made[7] :

$$(6.112) \quad \begin{cases} B_o^{\alpha\beta\gamma\delta} = h\bar{A}^{\alpha\beta'\gamma\delta'}\left[{}^{\beta}|_{\beta'}\; {}^{\delta}|_{\delta'} + \dfrac{h^2}{12}\, I^{\beta\delta}_{\beta'\nu}\, b^{\nu}_{\delta'} + O\!\left(\dfrac{h^4}{R^4}\right)\right] \\[2em] B_1^{\alpha\beta\gamma\delta} = h\bar{A}^{\alpha\beta'\gamma\delta'}\left[\dfrac{h^2}{12}\, I^{\beta\delta}_{\beta'\delta'} + O\!\left(\dfrac{h^2}{R^2}\right)\right] \\[2em] B_2^{\alpha\beta\gamma\delta} = h\bar{A}^{\alpha\beta'\gamma\delta'}\left[\dfrac{h^2}{12}\, {}^{\beta}|_{\beta'}\; {}^{\delta}|_{\delta'} + O\!\left(\dfrac{h^2}{R^2}\right)\right] \\[2em] B_o^{\alpha 3\,\beta 3} = h\bar{A}^{\alpha'3\beta'3}\left[{}^{\alpha}|_{\alpha'}\; {}^{\beta}|_{\beta'} + \dfrac{h^2}{28}\, I^{\alpha\beta}_{\alpha'\nu}\, b^{\nu}_{\beta'} + O\!\left(\dfrac{h^4}{R^4}\right)\right] \\[2em] \text{where } \bar{A} \text{ is given by formulae (6.102) and (6.103), h is the} \\ \text{thickness of the shell, R the smallest local radius of} \\ \text{curvature of } \Sigma_m, \text{ and where :} \\[2em] I^{\beta\delta}_{\beta'\delta'} = {}^{\beta}|_{\beta'}\, b^{\delta}_{\delta'} + b^{\beta}_{\beta'}\, {}^{\delta}|_{\delta'} - 2H\, {}^{\beta}|_{\beta'}\; {}^{\delta}|_{\delta'} \\[1.5em] I^{\beta\delta}_{\beta'\nu}\, b^{\nu}_{\delta'} = b^{\beta}_{\beta'}\, {}^{\delta}|_{\delta'} - K\, {}^{\beta}|_{\beta'}\; {}^{\delta}|_{\delta'} \\[1.5em] \text{and :} \\[1em] {}^{\alpha}|_{\beta} = \begin{cases} 0 \text{ if } \alpha \neq \beta \\[1em] 1 \text{ if } \alpha = \beta \end{cases} \qquad\qquad \text{[see appendix § 2]}. \end{cases}$$

Then assuming $\left[\dfrac{h}{R}\right]^2 \ll 1$, (6.108) becomes, with (6.112) :

$$(6.113)\begin{cases} N^{\beta\alpha} = h\bar{A}^{\alpha\beta'\gamma'\delta'}\left[\beta|_{\beta'}\ \gamma'_{\gamma'\delta'} + \dfrac{h^2}{12}\ I^{\beta\nu}_{\beta'\delta'}\left[b^{\delta}_{\nu}\ \gamma'_{\gamma'\delta'} + \kappa'_{\gamma'\nu}\right]\right] \\[3ex] M^{\beta\alpha} = \dfrac{h^3}{12}\ \bar{A}^{\alpha\beta'\gamma'\delta'}\left[\beta|_{\beta'}\ \kappa'_{\gamma'\delta'} + I^{\beta\nu}_{\beta'\delta'}\ \gamma'_{\gamma'\nu}\right] \\[3ex] Q^{\alpha} = \dfrac{5}{6}\ h\bar{A}^{\alpha'3\beta'3}\left[\alpha|_{\alpha'}\ \beta|_{\beta'} + \dfrac{h^2}{28}\ I^{\alpha\beta}_{\alpha'\nu}\ b^{\nu}_{\beta'}\right]\gamma_{\beta3}\ . \end{cases}$$

<u>Isotropic case</u>

The isotropic three-dimensional law corresponding to (6.99), is obtained from (1.41) which gives here :

$$C^{ijk\ell} = \mu\left[\dfrac{2\nu}{1-2\nu}\ g_T^{ij}\ g_T^{k\ell} + g_T^{jk}\ g_T^{i\ell} + g_T^{j\ell}\ g_T^{ik}\right]$$

where ν is the Poisson coefficient of the material, and μ the shear modulus.

In fact (1.41) gives :

$$^i\sigma_j = \lambda\ ^k\varepsilon_k\cdot\ ^i|_j + 2\mu\ ^i\varepsilon_j$$

$$= \lambda\ \varepsilon^{\ell}_k\ ^k|_{\ell}\cdot\ ^i|_j + 2\mu\ \varepsilon^{\ell}_k\ ^k|_j\cdot\ ^i|_{\ell}$$

$$= \left[\lambda\ ^k|_{\ell}\ ^i|_j + 2\mu\ ^k|_j\ ^i|_{\ell}\right]\varepsilon^{\ell}_k$$

$$\sigma^{ij} = \left[\lambda\ ^k|_{\ell}\ ^i|_{j'}\ g_T^{j'j} + 2\mu\ ^k|_{j'}\ ^i|_{\ell}\ g_T^{j'j}\right]\varepsilon^{\ell}_k$$

$$= \left[\lambda\ ^k|_{\ell}\ ^i|_{j'}\ g_T^{j'j}\ g_T^{\ell\ell'} + 2\mu\ ^k|_{j'}\ ^i|_{\ell}\ g_T^{j'j}\ g_T^{\ell'\ell}\right]\varepsilon_{\ell k}$$

$$\sigma^{ij} = \left[\lambda\ g_T^{ij}\ g_T^{k\ell} + 2\mu\ g_T^{kj}\ g_T^{i\ell}\right]\varepsilon_{\ell k},$$

whence $C^{ijk\ell}$, taking account of $\varepsilon_{\ell k} = \varepsilon_{k\ell}$ and $\lambda = 2\mu\ \dfrac{\nu}{1-2\nu}$.

The surface quantities overlined in (6.102) then become :

$$\bar{C}^{\alpha\beta\gamma\delta} = \mu \left[\frac{2\nu}{1-2\nu} g^{\alpha\beta} g^{\gamma\delta} + g^{\alpha\gamma} g^{\beta\delta} + g^{\alpha\delta} g^{\beta\gamma} \right],$$

with :

$$g_{\alpha\beta} = {}^{\alpha}G_{\beta} = \overline{S_{\alpha}} \, S_{\beta} \, , \quad g^{\alpha\beta} = {}^{\alpha}G^{-1}{}_{\beta},$$

$$\bar{C}^{\alpha3\beta3} = \mu g^{\alpha\beta},$$

from which we take, as in (6.103) :

$$(6.114) \quad \left[\begin{array}{l} \bar{A}^{\alpha\beta\gamma\delta} = \dfrac{E}{1-\nu^2} \left[\nu g^{\alpha\beta} g^{\gamma\delta} + \dfrac{1-\nu}{2} \left[g^{\alpha\delta} g^{\beta\gamma} + g^{\alpha\gamma} g^{\beta\delta} \right] \right] \\[2em] \bar{A}^{\alpha3\beta3} = \mathbf{G} g^{\alpha\beta} . \end{array} \right.$$

In this formula, $\mathbf{E}$ is the Young's modulus, and $\mathbf{G} = \mu$ is the shear modulus of the material, and ν the Poisson coefficient.

If we take the <u>principal lines of curvature</u> as the coordinate lines of Σ_m, and calling R_1 and R_2 the principal radii of curvature at point m, we find :

$$I^{\beta\delta}_{\beta'\delta'} = \left[\begin{array}{ll} \dfrac{1}{R_2} - \dfrac{1}{R_1} & \text{for} \quad \beta = \beta' = \delta = \delta' = 1 \\[1.5em] \dfrac{1}{R_1} - \dfrac{1}{R_2} & \text{for} \quad \beta = \beta' = \delta = \delta' = 2 \\[1.5em] 0 & \text{for the other index values} . \end{array} \right.$$

In this particular basis S, taking account of (6.114), law (6.113) becomes :

$$
\begin{aligned}
N^{11} &= \frac{Eh}{1-\nu^2}\left[\gamma'_{11} + \nu\gamma'_{22} + \frac{h^2}{12}\left[\frac{1}{R_1} - \frac{1}{R_2}\right]\left[\frac{\gamma'_{11}}{R_1} - \kappa'_{11}\right]\right] \\[2ex]
N^{22} &= \frac{Eh}{1-\nu^2}\left[\gamma'_{22} + \nu\gamma'_{11} + \frac{h^2}{12}\left[\frac{1}{R_2} - \frac{1}{R_1}\right]\left[\frac{\gamma'_{22}}{R_2} - \kappa'_{22}\right]\right] \\[2ex]
N^{12} &= \frac{Eh}{2\left[1+\nu\right]}\left[\gamma'_{12} + \gamma'_{21} + \frac{h^2}{12}\left[\frac{1}{R_1} - \frac{1}{R_2}\right]\left[\frac{\gamma'_{21}}{R_1} - \kappa'_{21}\right]\right] \\[2ex]
N^{21} &= \frac{Eh}{2\left[1+\nu\right]}\left[\gamma'_{12} + \gamma'_{21} + \frac{h^2}{12}\left[\frac{1}{R_2} - \frac{1}{R_1}\right]\left[\frac{\gamma'_{12}}{R_2} - \kappa'_{12}\right]\right] \\[2ex]
M^{11} &= \frac{Eh^3}{12\left[1-\nu^2\right]}\left[\kappa'_{11} + \nu\kappa'_{22} + \left[\frac{1}{R_2} - \frac{1}{R_1}\right]\gamma'_{11}\right] \\[2ex]
M^{22} &= \frac{Eh^3}{12\left[1-\nu^2\right]}\left[\kappa'_{22} + \nu\kappa'_{11} + \left[\frac{1}{R_1} - \frac{1}{R_2}\right]\gamma'_{22}\right] \\[2ex]
M^{12} &= \frac{Eh^3}{24\left[1+\nu\right]}\left[\kappa'_{12} + \kappa'_{21} + \left[\frac{1}{R_2} - \frac{1}{R_1}\right]\gamma'_{21}\right] \\[2ex]
M^{21} &= \frac{Eh^3}{24\left[1+\nu\right]}\left[\kappa'_{12} + \kappa'_{21} + \left[\frac{1}{R_1} - \frac{1}{R_2}\right]\gamma'_{12}\right] \\[2ex]
Q^1 &= \frac{5}{6}\,Gh\left[1 + \frac{h^2}{28R_1}\left[\frac{1}{R_2} - \frac{1}{R_1}\right]\gamma_{13}\right] \\[2ex]
Q^2 &= \frac{5}{6}\,Gh\left[1 + \frac{h^2}{28R_2}\left[\frac{1}{R_1} - \frac{1}{R_2}\right]\gamma_{23}\right].
\end{aligned}
\qquad (6.115)
$$

The following formulae will be associated with the above :

$$
\begin{aligned}
&\alpha_1 = \left[g_{11}\right]^{\frac{1}{2}},\ \alpha_2 = \left[g_{22}\right]^{\frac{1}{2}} \\[2mm]
&v^{\alpha} = \left[\Pi V\right]^{(\alpha)},\ \bar{N}V = w\ ,\ \left[\Delta N\right]^{(\alpha)} = \beta^{(\alpha)} \\[2mm]
&\gamma'_{11} = \frac{1}{\alpha_1}\left[v^{(1)}_{,1} - \frac{v^{(2)}}{\alpha_2}\,\alpha_{1,2}\right] + \frac{w}{R_1}\ ;\ \gamma'_{22} = \frac{1}{\alpha_2}\left[v^{(2)}_{,2} + \frac{v^{(1)}}{\alpha_1}\,\alpha_{2,1}\right] + \frac{w}{R_2} \\[2mm]
&\gamma'_{12} = \frac{1}{\alpha_2}\left[v^{(1)}_{,2} - \frac{v^{(2)}}{\alpha_1}\,\alpha_{2,1}\right]\ ;\ \gamma'_{21} = \frac{1}{\alpha_1}\left[v^{(2)}_{,1} - \frac{v^{(1)}}{\alpha_2}\,\alpha_{1,2}\right] \\[2mm]
&\kappa'_{11} = \frac{1}{\alpha_1}\left[\beta^{(1)}_{,1} - \frac{\beta^{(2)}}{\alpha_2}\,\alpha_{1,2}\right]\ ;\ \kappa'_{22} = \frac{1}{\alpha_2}\left[\beta^{(2)}_{,2} + \frac{\beta^{(1)}}{\alpha_1}\,\alpha_{2,1}\right] \\[2mm]
&\kappa'_{12} = \frac{1}{\alpha_2}\left[\beta^{(1)}_{,2} - \frac{\beta^{(1)}}{\alpha_1}\,\alpha_{2,1}\right]\ ;\ \kappa'_{21} = \frac{1}{\alpha_1}\left[\beta^{(2)}_{,1} - \frac{\beta^{(1)}}{\alpha_2}\,\alpha_{1,2}\right] \\[2mm]
&\gamma_{13} = \beta^{(1)} + \frac{1}{\alpha_1}\,w_{,1} - \frac{v^{(1)}}{R_1}\ ;\ \gamma_{23} = \beta^{(2)} + \frac{1}{\alpha_2}\,w_{,2} - \frac{v^{(2)}}{R_2}\ .
\end{aligned}
\tag{6.116}
$$

It is clear that (6.115) can be simplified further with $\frac{h^2}{R^2} \ll 1$, but this leads to non-invariant laws. Reference 7 will be consulted for a historical discussions concerning these different approximations.

The preceding approximate law (6.115) complies with the approximation of Flügge (1934), Lur'e (1940) and Byrne (1944). It has been expressed in its general form by Naghdi (1963). However, it should be noted that the presence of a non-zero normal deformation affects the tangent deformations of quantities which have in fact been ignored[26,24].

2 - Zero normal and transverse shear deformation

This case corresponds strictly to the kinematic Kirchhoff-Love, namely hypothesis (wo) with $\delta z = 0$.

The laws are deduced from the preceding laws, cancelling out the corresponding energy terms. No interpolation is necessary. It is natural to consider the symmetrical stresses defined in § 4.2, and also the symmetrical deformations γ and κ, the only useful deformations. We then find :

$$(6.117) \qquad \int_{\Sigma_m} \left[\mathbf{n}^{\alpha\beta} \, \delta\gamma_{\alpha\beta} + \mathbf{m}^{\alpha\beta} \, \delta\kappa_{\alpha\beta} \right] d\Sigma_m =$$

$$= \int_{\Sigma_m} \left[D_o^{\alpha\beta\gamma\delta} \, \gamma_{\gamma\delta} \, \delta\gamma_{\alpha\beta} + D_1^{\alpha\beta\gamma\delta} \, \kappa_{\gamma\delta} \, \delta\gamma_{\alpha\beta} + D_1^{\alpha\beta\gamma\delta} \, \gamma_{\gamma\delta} \, \delta\kappa_{\alpha\beta} \right.$$

$$\left. + D_2^{\alpha\beta\gamma\delta} \, \kappa_{\gamma\delta} \, \delta\kappa_{\alpha\beta} \right] d\Sigma_m \, .$$

Whence :

$$(6.118) \qquad \left[\begin{array}{l} n^{\alpha\beta} = D_o^{\alpha\beta\gamma\delta} \, \gamma_{\gamma\delta} + D_1^{\alpha\beta\gamma\delta} \, \kappa_{\gamma\delta} \; ; \quad n^{\alpha\beta} = \alpha_S^{-1} \, \mathbf{n}^{\overline{\beta_S - 1}} \\[3em] m^{\alpha\beta} = D_1^{\alpha\beta\gamma\delta} \, \gamma_{\gamma\delta} + D_2^{\alpha\beta\gamma\delta} \, \kappa_{\gamma\delta} \; ; \quad m^{\alpha\beta} = \alpha_S^{-1} \, \mathbf{m}^{\overline{\beta_S - 1}} \end{array} \right.$$

with :

$$(6.119) \qquad \left[\begin{array}{l} D_o^{\alpha\beta\gamma\delta} = B_o^{\alpha\beta\gamma\delta} + b_{\delta'}^{\delta} \, B_1^{\alpha\beta\gamma\delta'} + b_{\delta'}^{\alpha} \, B_1^{\gamma\delta\beta\delta'} + b_{\alpha'}^{\alpha} \, b_{\gamma'}^{\gamma} \, B_2^{\beta\alpha'\delta\gamma'} \\[2.5em] D_1^{\alpha\beta\gamma\delta} = \dfrac{1}{2} \left[B_1^{\alpha\beta\gamma\delta} + B_1^{\alpha\beta\delta\gamma} \right] + \dfrac{1}{2} b_{\alpha'}^{\alpha} \left[B_2^{\gamma\delta\beta\alpha'} + B_2^{\beta\alpha'\gamma\delta} \right] \\[2em] \qquad\qquad\qquad\qquad + \dfrac{1}{2} b_{\alpha'}^{\alpha} \left[B_2^{\delta\gamma\beta\alpha'} + B_2^{\beta\alpha'\gamma\delta} \right] \\[2.5em] D_2^{\alpha\beta\gamma\delta} = \dfrac{1}{4} \left[B_2^{\alpha\beta\gamma\delta} + B_2^{\beta\alpha\gamma\delta} + B_2^{\alpha\beta\delta\gamma} + B_2^{\beta\alpha\delta\gamma} \right] \, . \end{array} \right.$$

We see that :

$$D_n^{\alpha\beta\gamma\delta} = D_n^{\beta\alpha\gamma\delta} = D_n^{\alpha\beta\delta\gamma} = D_n^{\gamma\delta\alpha\beta} \quad (n = 0,1,2) \ .$$

Approximated constitutive law in hypothesis (wo) and symmetrical stresses

Following the Koiter's ideas, partially adopted by Naghdi, we can pursue the following argument :

Equilibrium equation $(6.26)_2$ shows that the transverse shear stress takes the form :

$$0 \ (\frac{h}{L} \ \overline{i_2 \mathbf{M}})$$

where L is the shortest wavelength of the deformed surface Σ_m, and where $\overline{i_2 \mathbf{M}}$ is the generalized bending stress operator. Therefore to neglect the transverse shear energy implies an error of :

$$0\left[|\frac{h}{L} \ i_2 \mathbf{M}|^2\right] = 0 \ \frac{h^2}{L^2} \ \mathcal{E}_f$$

where $\mathcal{E}_f$ is the bending deformation energy.

With (6.119) and (6.112) we then find :

$$(6.120) \quad \begin{cases} D_0^{\alpha\beta\gamma\delta} = h\overline{A}^{\alpha\beta\gamma\delta}\left[1 + 0(\frac{h^2}{R^2})\right] \\[2em] D_1^{\alpha\beta\gamma\delta} = \overline{A}^{\alpha\beta\gamma\delta} \ 0(\frac{h^3}{R}) \\[2em] D_2^{\alpha\beta\gamma\delta} = \frac{h^3}{12} \ \overline{A}^{\alpha\beta\gamma\delta}\left[1 + 0(\frac{h^2}{R^2})\right] . \end{cases}$$

According to these results, (6.117) supplies approximate values for the energy errors :

$$(6.121) \quad \int_{\Sigma_m}\left[n^{\alpha\beta} \ \delta\gamma_{\alpha\beta} + m^{\alpha\beta} \ \delta\kappa_{\alpha\beta}\right] d\Sigma_m = \delta\int_{\Sigma_m}\left[\mathcal{E}_e\left[1 + 0(\frac{h^2}{R^2})\right] + \right.$$
$$\left. + \overline{A}^{\alpha\beta\gamma\delta} \ \gamma_{\alpha\beta} \ \kappa_{\gamma\delta} \ 0(\frac{h^3}{R}) + \mathcal{E}_f\left[1 + 0(\frac{h^2}{R^2})\right]\right] d\Sigma_m, \forall \delta\gamma, \delta\kappa \ ,$$

where $\mathcal{E}_e$ and $\mathcal{E}_f$, representing membrane and bending energies, are positive definite formes in γ and κ respectively :

$$\left[\begin{array}{l} \mathcal{E}_e = \dfrac{1}{2}\, h\overline{A}^{\alpha\beta\gamma\delta}\, \gamma_{\alpha\beta}\, \gamma_{\gamma\delta} \\[2ex] \mathcal{E}_f = \dfrac{1}{2}\, \dfrac{h^3}{12}\, \overline{A}^{\alpha\beta\gamma\delta}\, \kappa_{\alpha\beta}\, \kappa_{\gamma\delta} \, . \end{array}\right.$$

Therefore using the Schwartz inequality (Koiter, 1960) :

$$\left[\overline{A}^{\alpha\beta\gamma\delta}\, \gamma_{\alpha\beta}\, \kappa_{\gamma\delta}\right]^2 \leqslant \left[\overline{A}^{\alpha\beta\gamma\delta}\, \gamma_{\alpha\beta}\, \gamma_{\gamma\delta}\right]\left[\overline{A}^{\alpha\beta\gamma\delta}\, \kappa_{\alpha\beta}\, \kappa_{\gamma\delta}\right] \, .$$

Thus :

$$\left|\dfrac{h^3}{R}\, \overline{A}^{\alpha\beta\gamma\delta}\, \gamma_{\alpha\beta}\, \kappa_{\gamma\delta}\right| \leqslant \left|\dfrac{h}{R}\right|\left[\mathcal{E}_e\, \mathcal{E}_f\right]^{1/2}$$

$$\leqslant \left|\dfrac{h}{R}\right|\, \mathcal{E}_e \quad \text{if}\quad \mathcal{E}_e \geqslant \mathcal{E}_f$$

$$\leqslant \left|\dfrac{h}{R}\right|\, \mathcal{E}_f \quad \text{if}\quad \mathcal{E}_e \leqslant \mathcal{E}_f \, .$$

The term of highest order in (6.12) can now be neglected, if we admit an error of $0(\dfrac{h}{R}\,\mathcal{E}_e)$ or $0(\dfrac{h}{R}\,\mathcal{E}_f)$ on the deformation energy. This error is admissible if $\dfrac{h}{R} \ll 1$. Furthermore, if we neglect the transverse shear in hypothesis (wo), we have seen that we then admit an energy error of $0(\dfrac{h^2}{L^2}\,\mathcal{E}_f)$. Finally, Koiter also demonstrated that to neglect normal stress σ_{33} produces an error of $0(\dfrac{h}{R}\,\mathcal{E}_e)$.

Thus if we accept these approximations, we can reduce (6.118), with the aid of (6.120), to :

$$(6.122)\qquad \left[\begin{array}{l} n^{\alpha\beta} = h\overline{A}^{\alpha\beta\gamma\delta}\, \gamma_{\gamma\delta} \\[2ex] m^{\alpha\beta} = \dfrac{h^3}{12}\, \overline{A}^{\alpha\beta\gamma\delta}\, \kappa_{\gamma\delta} \end{array}\right.$$

In the normal basis (principal curvatures), we find, in the isotropic case :

$$(6.123) \quad
\begin{cases}
n^{11} = \dfrac{Eh}{1-\nu^2}\left[\gamma_{11} + \nu\gamma_{22}\right] = N^{11} - \dfrac{1}{R_1} M^{11} \\[2em]
n^{22} = \dfrac{Eh}{1-\nu^2}\left[\gamma_{22} + \nu\gamma_{11}\right] = N^{22} - \dfrac{1}{R_2} M^{22} \\[2em]
n^{12} = \dfrac{Eh}{1-\nu}\gamma_{12} = {}^{21} = N^{12} - \dfrac{1}{R_2} M^{21} = N^{21} - \dfrac{1}{R_1} M^{12} \\[2em]
m^{11} = \dfrac{Eh^3}{12\left[1-\nu^2\right]}\left[\kappa_{11} + \nu\kappa_{22}\right] = M^{11} \\[2em]
m^{22} = \dfrac{Eh^3}{12\left[1-\nu^2\right]}\left[\kappa_{22} + \nu\kappa_{11}\right] = M^{22} \\[2em]
m^{12} = \dfrac{Eh^3}{12\left[1+\nu\right]}\kappa_{12} = m^{21} = \dfrac{1}{2}\left[M^{12} + M^{21}\right]
\end{cases}$$

where furthermore we have :

$$(6.124) \quad
\begin{cases}
\gamma_{11} = \gamma'_{11} \ ; \ \gamma_{22} = \gamma'_{22} \ ; \ \gamma_{12} = \dfrac{1}{2}\left[\gamma'_{12} + \gamma'_{21}\right] = \gamma_{21} \\[2em]
\kappa_{11} = \kappa'_{11} + \dfrac{\gamma'_{11}}{R_1} \ ; \ \kappa_{22} = \kappa'_{22} + \dfrac{\gamma'_{22}}{R_2} \\[2em]
\kappa_{12} = \kappa_{21} = \kappa'_{12} + \dfrac{\gamma'_{12}}{R_2} = \kappa'_{21} + \dfrac{\gamma'_{21}}{R_1}
\end{cases}$$

associated with (6.116).

The preceding constitutive laws should be compared with those given by Koiter (1960), Sanders (1959), or also partially by Novozhilow (1959).

7. – <u>SHELLS OF REVOLUTION</u>

A point M on a shell of revolution will be parameterized conventionally by the position of a point m of Σ_m on a meridian, that is to say by its curvilinear abscissa s, the angle θ made by the meridian plane with a reference plane passing through the axis of revolution ZZ', and the normal co-ordinate $|z| = |\overrightarrow{mM}|$, with z > 0 in outward direction.

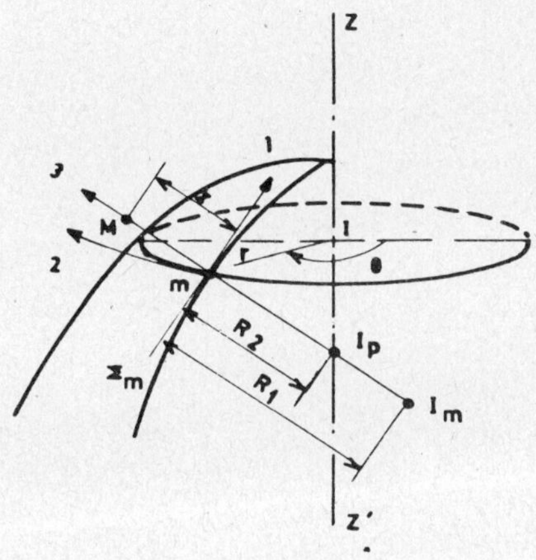

Therefore :

$$X = \begin{bmatrix} s \\ \theta \end{bmatrix} \qquad y = \begin{bmatrix} s \\ \theta \\ z \end{bmatrix} = \begin{bmatrix} ^1y \\ ^2y \\ ^3y \end{bmatrix}$$

We again call h the thickness of the shell. We also call r the radius of the parallel with centre I, passing through m :

$$r = |\overrightarrow{Im}| \ .$$

Let I_p be the intersection of the normal to Σ_m with the axis, this being the centre of curvature corresponding to the tangent to the parallel (Meusnier's theorem), and I_m the centre of curvature of the meridian at m. The two tangents at m, to the meridian and the parallel, constitute the principal directions. We take R_1 and R_2 as the corresponding radii of curvature.

We then have, with $X = \begin{bmatrix} s \\ \theta \end{bmatrix}$

$$\partial_1 m = \frac{\partial m}{\partial s} \ , \quad \partial_2 m = \frac{\partial m}{\partial \theta} = r \quad (d_2 m = r d\theta)$$

$$\left| S_1 \right| = \left| \frac{\partial m}{\partial s} \right| = 1 \ ; \quad \left| S_2 \right| = \left| \frac{\partial m}{\partial \theta} \right| = r$$

$$g_{11}^{1/2} = \alpha_1 = 1 \ , \quad g_{22}^{1/2} = \alpha_2 = r \ , \quad g_{12} = 0 \ .$$

In the following, hypothesis (w) with transverse shear will be adopted, taking $\delta z = 0$, namely zero normal deformation, for sake of simplification. Let V be the displacement of point M, we shall put here in the local basis, in standard notations :

$$V \Rightarrow \begin{bmatrix} v^{(1)} = u \\[2mm] v^{(2)} = v \\[2mm] w \quad\ = w \end{bmatrix} \qquad \Delta N \Rightarrow \begin{bmatrix} \beta^{(1)} = \varphi \\[2mm] \beta^{(2)} = \psi \end{bmatrix}$$

Thus the displacement of point M, namely W, will have the following components :

$$\begin{bmatrix} U^{(1)}(Y) = u(s,\theta) + z\,\varphi(s,\theta) \\[2mm] U^{(2)}(Y) = v(s,\theta) + z\,\psi(s,\theta) \\[2mm] U^{(3)}(Y) = w(s,\theta) \ . \end{bmatrix}$$

Formulae (6.116) supply the generalized deformations :

$$(6.125) \quad \begin{bmatrix} \gamma'_{11} = \dfrac{\partial u}{\partial s} + \dfrac{w}{R_1} & ; & \gamma'_{22} = \dfrac{1}{r}\dfrac{\partial v}{\partial \theta} + \dfrac{u}{r}\dfrac{\partial r}{\partial s} + \dfrac{w}{R_2} \\[4mm] \kappa'_{11} = \dfrac{\partial \varphi}{\partial s} & ; & \kappa'_{22} = \dfrac{1}{r}\dfrac{\partial \psi}{\partial \theta} + \dfrac{\varphi}{r}\dfrac{\partial r}{\partial s} \\[4mm] \gamma_{13} = \dfrac{\partial w}{\partial s} - \dfrac{u}{R_1} + \varphi & ; & \gamma_{23} = \dfrac{1}{r}\dfrac{\partial w}{\partial \theta} - \dfrac{v}{R_2} + \psi \\[4mm] \gamma'_{21} = \dfrac{\partial v}{\partial s} & ; & \gamma'_{12} = \dfrac{1}{r}\dfrac{\partial u}{\partial \theta} - \dfrac{v}{r}\dfrac{\partial r}{\partial s} \end{bmatrix}$$

$$(6.125) \ \text{continued} \quad \left| \quad \kappa'_{21} = \frac{\partial \psi}{\partial s} \quad ; \quad \kappa'_{12} = \frac{1}{r}\frac{\partial \varphi}{\partial \theta} - \frac{\psi}{r}\frac{\partial r}{\partial s} \right.$$

<u>Surface constitutive laws[26]</u>

We shall consider the case of a homogeneous and orthotropic shell of revolution, namely one having mechanical properties which are symmetrical with respect to the local co-ordinate planes at m, and independent of angle θ.

Under these conditions, with all calculations made and using standard notations, law (6.108) is written :

$$(6.126) \quad
\begin{Bmatrix} N^{11} \\ N^{22} \\ M^{11} \\ M^{22} \\ Q^{1} \\ Q^{2} \\ N^{12} \\ N^{21} \\ M^{12} \\ M^{21} \end{Bmatrix}
=
\begin{bmatrix}
C_1 & C_{12} & \Delta_1 & 0 & 0 & 0 & 0 & 0 & 0 & 0 \\
 & C_2 & 0 & \Delta_2 & 0 & 0 & 0 & 0 & 0 & 0 \\
 & & D_1 & D_{12} & 0 & 0 & 0 & 0 & 0 & 0 \\
 & & & D_2 & 0 & 0 & 0 & 0 & 0 & 0 \\
 & & & & F_1 & 0 & 0 & 0 & 0 & 0 \\
 & & & & & F_2 & 0 & 0 & 0 & 0 \\
 & & & & & & I_2 & I_{12} & \Gamma_1 & 0 \\
 & & & & & & & I_2 & 0 & \Gamma_2 \\
 & & & & & & & & J_1 & 0 \\
 & & & & & & & & & J_2
\end{bmatrix}
\begin{Bmatrix} \gamma'_{11} \\ \gamma'_{22} \\ \kappa'_{11} \\ \kappa'_{22} \\ \gamma_{13} \\ \gamma_{23} \\ \gamma'_{21} \\ \gamma'_{12} \\ \kappa'_{21} \\ \kappa'_{12} \end{Bmatrix}$$

with :

$$\left[\begin{array}{l}
C_1 = \dfrac{E_1 h}{1-\nu_{12}\nu_{21}} \ , \quad C_2 = \dfrac{E_2 h}{1-\nu_{12}\nu_{21}} \ , \quad C_{12} = C_1 \nu_{12} = C_2 \nu_{21} \\[3mm]
D_1 = \dfrac{h^2}{12} C_1 \ , \quad D_2 = \dfrac{h^2}{12} C_2 \ , \quad D_{12} = D_1 \nu_{12} = D_2 \nu_{21} \\[3mm]
\Delta_1 = - D_1 \left[\dfrac{1}{R_1} - \dfrac{1}{R_2} \right] , \quad \Delta_2 = - D_2 \left[\dfrac{1}{R_2} - \dfrac{1}{R_1} \right] \\[3mm]
F_1 = \dfrac{5}{6} h\, G_{13} \qquad\quad , \ F_2 = \dfrac{5}{6} h\, G_{23} \\[3mm]
\qquad\qquad I_{12} = h\, G_{12} \\[3mm]
J_1 = J_2 = J_{12} = I_{12} \dfrac{h^2}{12} \\[3mm]
\Gamma_1 = - J_1 \left[\dfrac{1}{R_1} - \dfrac{1}{R_2} \right] , \quad \Gamma_2 = - \Gamma_1 \\[3mm]
I_1 = I_{12} - \dfrac{\Gamma_1}{R_1} \qquad , \ I_2 = I_{1.2} - \dfrac{\Gamma_2}{R_2} \ .
\end{array}\right. \tag{6.127}$$

In these formulae :

E_1 is the Young modulus in direction s.

E_2 is the Young modulus in direction θ.

ν_{12} is the Poisson coefficient of contraction in direction θ, due to a
traction in direction s.

ν_{21} is the Poisson coefficient of contraction in direction s, due to a
traction in direction θ.

We also have the relation in addition :

$$E_2 \, \nu_{21} = E_1 \, \nu_{12}$$

already visible in the first line of (6.127), and due to relations of symmetry (6.110).

G_{12} is the tangential shear modulus.

G_{13} is the transverse shear modulus in plane (s,z).

G_{23} is the transverse shear modulus in plane (θ,z).

Solution in the form of Fourier series

In the case of an arbitrary static or dynamic loading, including in the case of a freely vibrating shell, the loading surface density (resultant and resultant moment) can be expanded in a Fourier series, in the angle θ, varying from 0 to 2Π. The solution is also represented in a linear problem, in a Fourier basis, in the form :

$$
\begin{aligned}
u &= \Sigma_n \left[U_{1n} \cos n\theta + U_{2n} \sin n\,\theta \right] \\[2ex]
v &= \Sigma_n \left[V_{1n} \cos n\theta + V_{2n} \sin n\,\theta \right] \\[2ex]
w &= \Sigma_n \left[W_{1n} \cos n\theta + W_{2n} \sin n\,\theta \right] \\[2ex]
\varphi &= \Sigma_n \left[\Phi_{1n} \cos n\theta + \Phi_{2n} \sin n\,\theta \right] \\[2ex]
\psi &= \Sigma_n \left[\Psi_{1n} \cos n\theta + \Psi_{2n} \sin n\,\theta \right] .
\end{aligned}
$$

Putting these values into the principle of virtual work [see (6.23)], expressed as functions of displacements, and separating the terms in $\cos n\theta$ and $\sin n\theta$ in the obtained Fourier series, we obtain, with all calculations made and the Fourier basis being orthonormal, equilibrium equations having the form :

$$
\Sigma_n \underset{\sim}{P_{1n}} \,(U_{1n},\ V_{2n},\ W_{1n},\ \phi_{1n},\ \psi_{2n})\, \cos n\theta +
$$

$$
+\ \Sigma_n \underset{\sim}{P_{2n}} \,(U_{2n},\ V_{1n},\ W_{2n},\ \phi_{2n},\ \psi_{1n})\, \sin n\theta = 0
$$

The system degenerates for each value of n, into two identical problems, the second being deduced from the first by putting $n\Theta' = n\Theta + \dfrac{\Pi}{2}$. The

problems are decoupled for each value of n, and it is merely necessary to
find solutions having the form :

$$
\begin{aligned}
u &= U_{1n}\ \cos n\theta \\[1em]
v &= V_{2n}\ \sin n\theta \\[1em]
w &= W_{1n}\ \cos n\theta \\[1em]
\varphi &= \Phi_{1n}\ \cos n\theta \\[1em]
\psi &= \Psi_{2n}\ \sin n\theta.
\end{aligned}
$$

8. - <u>DISCRETIZATION</u>

The variational principles described above can be used to calculate
shells of arbitrary shape, by application of classical discretization methods,
the Ritz or Galerkin methods, the finite difference methods, or also the
finite element method, the latter being particularly suitable for the
calculation of complex structures[27,28]. For static or dynamic problems, we
can apply the principles described in chapters II, III and V. The principal
lines of curvature are generally adopted. The displacements method, which
is the most frequently used, has given birth to many programmes, although
multifield mixed elements have already found many applications.

It is appropriate to distinguish between two types of approach, in the
analysis of shells using the finite element method : the first uses flat
elements, and was described in chapter II ; the second introduces curved
or even isoparametric finite elements, and although more complex, is
frequently found to be indispensable. Obviously it is possible to use
both types of element simultaneously, provided the problems of assembly
(see chapter III) are solved as well as possible. The use of curved
elements allows more accurate modelling of the structure, thus avoiding
artificial geometrical imperfections in problems of buckling, as already
mentioned (see chapter V).

Excellent syntheses have already appeared on the subject of plate and
shell elements, in which we find an abundant bibliography. Let us quote
those of Jones and Strom[29], Gallagher[30], Bushnell[28], Clough and Wilson[31]

and, very recently, Gallagher[32].

The greatest difficulty encountered in the displacement method, lies
in the simultaneous satisfaction of the conditions of conformity, zero
deformation for rigid body displacements, constant deformation, and finally arbitrary geometry. Besides, approximations result from the modelling, the type of interpolation adopted, the relatively good conformity of
the elements, and finally from the hypotheses inherent in the theory of
shells used.

As concerns flat elements, their formulation ranges from the simplest
interpolations, generally cubic for bending, to the most refined, quintic[33]
or even bicubic[34]. Mention should be made of the interesting work of
Hermann[35] and Pian[36], concerning mixed or hybrid finite elements, using
different variational principles (see chapter III) adapted to plates.
Flat elements of the displacement or stress type have been developed by
Sander[37], by application of the static/geometrical analogy, or of the
mixed type[38].

Difficulties concerning continuity in the idealization of shells using
flat elements, have already been mentioned in chapter II, but his technique nevertheless remains in high favour by reason of its simplicity.

Curved elements of a wide variety have been developed up to the present
time, by numerous authors. Some introduce transverse shear and zero normal
stress, or zero normal deformation, while others do not. The elements
range from shallow shells to elements of any shape, for example isoparametric, either linear or non-linear.

The difficulties already mentioned in connection with flat elements,
exist to an amplified degree in the formulation of curved elements.

A discussion of the different aspects of the theory of shells, is
undertaken in this connection by Krauss[39], among others.

In reference 32, Gallagher mentions the special studies which have been
made, concerning the practical geometrical representation of curved
surfaces, in particular by the use of isoparametric elements[40,41],
stressing that bad idealization can give rise to more serious errors than
those originating from mediocre interpolation, as has already been

mentioned about buckling.

The majority of curved elements are based on the displacement method. A difficulty resulting from the search for conformity, can perhaps be solved by the use of intrinsic methods. Furthermore, this condition can be obtained in the form of progressively finer approximation, by the use of increasing degree interpolation.

The zero deformation condition for rigid body displacements, has been studied by Cantin[42], with modification of the initial displacement field, a method also used by Fonder and Clough[43].

The choice of the interpolation degree appears to be guided by the fact that this degree should be less than the maximum order of the derivations existing in the deformation energy. Nevertheless, certain authors recommend an order which is at least identical, and even at least equal to 5, so as to obtain a good degree of accuracy on the stresses. Others recommend the same degree for all unknown fields, and use the third order[44], superabundant terms allowing closer satisfaction of the condition of constant deformation. For example, Cowper, Lindberg and Olson[45] produce a refined conforming triangular element, of the 5th degree in w, and the third degree in u and v, such that the condition of rigid body displacement is satisfied as closely as possible. We should also mention the "Sheba" conforming element of Argyris and Sharpf[46], where the three components u, v and w are interpolated to the 5th degree. The condition of zero deformation for rigid body displacements is fully satisfied. This costly element is particularly accurate.

Certain authors have interpolated displacements in rational fraction form[47,41].

All these sophisticated elements are frequently very costly to use, and are preferred to simpler elements, which are less accurate though more economic, only for particular cases.

The Lagrange multiplier method (see chapter III) has been used to satisfy continuity in the mean[48,49]. Accuracy of the displacements is improved, but it has not been proved that the accuracy of the stresses is improved.

In chapter III, we described the mixed principle of Hellinger-Reissner, and the hybrid principle of Pian. The use of these principles, where the number of unknown fields is increased, allows very simple, linear or quadratic interpolations, and C°-continuity is taken to be sufficient[50-53].

It should be mentioned that the introduction of transverse shear makes it possible to achieve conformity easily, although this introduction is the source of large stiffnesses relative to this deformation, which can make certain static or dynamic problems ill-conditioned. The situation can be remedied by applying hypothesis (wo) at certain nodes of the element[40].

Numerous axisymmetrical elements have also been proposed[54-56], and in general multi-layered elements[57,58,62-64], taking account of transverse shear, which is important in the case of sandwich materials, and achieving inter-layer continuity.

To conclude, let us mention that many users employ isoparametric three-dimensional elements for the calculation of thin shells, even multi-layered, without coming up against problems of ill-conditioning, by reason of the substantial progress achieved with modern computers, in the matter of accuracy and rapidity. As for dynamic vibration problems specific to shells, a remarkable synthesis of this question, together with numerous results, will be found in reference 59.

APPENDIX
NOTATIONS AND FORMULAE [1-5]

The present appendix is limited to indications concerning the mathematical notations, definitions, formulae and results used in the various chapters of this Course.

1 - MAPPINGS - MULTIPLE MAPPINGS

- def (A) : domain of definition of mapping A.

- val (A) : domain of value of mapping A.

- A(X)(Y)(Z) : value of multiple mapping A (in this case triple), such that :

$$Z \in \text{def}\left(A(X)(Y)\right) \quad , \quad [A(X)(Y)](Z) \in \text{val}(A(X)(Y)),$$

$$Y \in \text{def}(A(X)) \quad , \quad [A(X)](Y) \in \text{val}(A(X)),$$

$$X \in \text{def}(A) \quad , \quad A(X) \in \text{val}(A).$$

If the various spaces are respectively vectorial, the mappings A, A(X), A(X)(Y) can be linear. In this case, A is linear, bilinear, trilinear, etc.

- A regular $\iff$ A one-to-one mapping.

- $\vec{E}$: vector space, possibly associated with a linear space E.

- I_E : identical mapping on $\vec{E}$, or E, or more simply $\underline{I}$ if there is no ambiguity.

- λ_E : scalar endomorphism of E, or $\vec{E}$, or $\lambda . I_E$, or more simply $\underline{\lambda}$ if there is no ambiguity

- Co-vector of $\vec{E}$: element of $\vec{E}^*$, dual of $\vec{E}$.

- $\pounds(\vec{E}, \vec{E}')$ set of linear mappings from $\vec{E}$ in $\vec{E}'$.

291

2 - <u>MATRIX KEYS</u>

- Matrix keys of $\mathbb{R}^3$:

$$|_1 = \begin{bmatrix} 1 \\ 0 \\ 0 \end{bmatrix} \quad , \quad |_2 = \begin{bmatrix} 0 \\ 1 \\ 0 \end{bmatrix} \quad , \quad |_3 = \begin{bmatrix} 0 \\ 0 \\ 1 \end{bmatrix}$$

$$^1| = \begin{bmatrix} 1 & 0 & 0 \end{bmatrix} \quad , \quad ^2| = \begin{bmatrix} 0 & 1 & 0 \end{bmatrix} \quad , \quad ^3| = \begin{bmatrix} 0 & 0 & 1 \end{bmatrix}$$

Keys $|_i$, $^i|$, (i=1,2,3), are the basis rows and columns of $\mathbb{R}^3$.

Einstein's convention is used, unless specified otherwise.

$$|_i \cdot {}^i| = |_{\mathbb{R}^3} = \begin{bmatrix} 1 & 0 & 0 \\ 0 & 1 & 0 \\ 0 & 0 & 1 \end{bmatrix} \quad , \quad {}^i||_j = {}^i|_j = \begin{cases} 0 \; , \; i \neq j \\ \\ 1 \; , \; i = j \end{cases}$$

- $\forall$ column X $\in \mathbb{R}^3$:

$$X = |_{R^3} X = |_i \cdot {}^i| X = |_i \cdot {}^i X \; , \; {}^i X \in \mathbb{R}.$$

- $\forall$ line L $\in \mathbb{R}^{3*}$:

$$L = L \cdot |_{R^3} = L |_i \cdot {}^i| = L_i \cdot {}^i| \; , \; L_i \in \mathbb{R}$$

- $\forall$ matrix M of $\mathbb{R}^3$, M $\in \mathcal{L}(\mathbb{R}^3, \mathbb{R}^3)$:

$$M = |_i \cdot {}^i| M |_j \cdot {}^j| = |_i \cdot {}^i M_j \cdot {}^j| \; , \; (i,j = 1,2,3) , {}^i M_j \in \mathbb{R}.$$

- More generally :

$$\vec{E} = \vec{E}_1 \times \vec{E}_2 \times \dots \vec{E}_n \; , \; \vec{E}' = \vec{E}'_1 \times \vec{E}'_2 \times \dots \times \vec{E}'_n , ,$$

The spaces $\vec{E}_i$, $\vec{E}'_j$ are real vector spaces :

$$|_i \in \mathcal{L}\left(\vec{E}_i, \vec{E}\right) \; , \; {}^i| \in \mathcal{L}\left(\vec{E}, \vec{E}_i\right) \; , \; i = 1,2, \dots,n.$$

$$|_j \in \dot{\mathcal{L}}\left(\vec{E}'_i, \vec{E}'\right) \;, \quad {}^j| \in \mathcal{L}\left(\vec{E}', \vec{E}'_j\right) \;, \quad j = 1,2,\ldots,n'.$$

$$|_i \cdot {}^i| = {}^1_E \;, \quad |_j \cdot {}^j| = {}^1_{E'} \,,$$

$$
{}^i|_k = \left\{ \begin{array}{l} o \;, \quad i \neq k \\[1em] {}^1_{E_i} \;, \quad i = k \end{array} \right.
\;, \quad
{}^j|_\ell = \left\{ \begin{array}{l} 0 \;, \quad j \neq \ell \\[1em] {}^1_{E'_j} \;, \quad j = \ell \end{array} \right.
$$

We then have a generalization of the formulae relating to rows, columns and matrices, and in particular :

$$\forall M \in \mathcal{L}(\vec{E}, \vec{E}')$$

$$ {}^j|M|_i = {}^j M_i \in \mathcal{L}(\vec{E}_i, \vec{E}'_j) \,. $$

3 – BASES – REPRESENTATION OF LINEAR MAPPINGS

Let E be a vector space, real and with finite dimension n (for example), by definition :

$$[S \text{ is a basis of } \vec{E}] \Longleftrightarrow [S \in \mathcal{L}\,(\mathbb{R}^n, \vec{E}) \text{ and } S \text{ bijective i.e. a one}$$
$$\text{to one mapping from } \mathbb{R}^n \text{ onto } \vec{E}]$$

Furthermore :

$$S^{-1} \in \mathcal{L}\left(\vec{E}, \mathbf{R}^n\right) \text{ is a } \underline{\text{cobasis}}.$$

Example : n = 3 :

$$S = \begin{bmatrix} S_1 & S_2 & S_3 \end{bmatrix} \;, \quad S^{-1} = \begin{bmatrix} {}^1S^{-1} \\ {}^2S^{-1} \\ {}^3S^{-1} \end{bmatrix}.$$

Vectors S_i are the basis vectors, and covectors ${}^iS^{-1}$ are the basis co-vectors.

We then have :

$$S\,X = V, \quad X \in \mathbb{R}^3, \quad V \in \vec{E}, \quad X = S^{-1}V, \quad {}^iX = {}^iS^{-1}V \,.$$

- Let $A \in \mathcal{L}(\vec{E}, \vec{E}')$, $\dim(\vec{E}) = n$, $\dim(\vec{E}') = n'$.

  S a basis of $\vec{E}$, and S' a basis of $\vec{E}'$:

$$\forall V \in \vec{E}, \quad AV = W \in \vec{E}'$$

$$V = SX, \quad X \in \mathbb{R}^n, \quad W = S'X', \quad S' \in \mathbb{R}^{n'}$$

$$ASX = S'X' \implies X' = S'^{-1} ASX$$

$$X' = MX \text{ with } M = S'^{-1} AS \in \mathcal{L}(R^n, R^n)$$

$$A = S' M S^{-1}$$

$$^i M_j = {}^i S'^{-1} AS_j, \quad i = 1, 2, \ldots, n', \quad j = 1, 2, \ldots, n.$$

M is the matrix representing A in S and S'.

If $A \in \mathcal{L}(\vec{E}, \vec{E})$:

$$A = SMS^{-1}, \quad M = S^{-1}AS \in \mathcal{L}(\mathbb{R}^n, \mathbb{R}^n).$$

- Let $C \in \vec{E}^*$, $\forall V \in \vec{E} : V = SX$

$$CV = CSX = CS_i \, {}^i X$$

$$CS = L \in \mathbb{R}^{n*}, \quad CS_i = L_i$$

$$C = LS^{-1} = L_i \, {}^i S^{-1}$$

Line L represents C in S.

Change of bases

Let : $S' = SK, \quad K \in \mathcal{L}(\mathbb{R}^n, \mathbb{R}^n)$

$$K = S^{-1}S'$$

$$V = S'X' = SX \implies X' = S'^{-1}SX = K^{-1}X$$

$$C = L'S'^{-1} = LS^{-1} \implies L' = LS^{-1}S' = LK$$

$$A = S'M'S'^{-1} = SMS^{-1} \implies M' = S'^{-1}SMS^{-1}S' = K^{-1}MK.$$

4. – <u>TENSORS</u>

Given a vector space $\vec{E}$ with dimension n, and its dual $\vec{E}*$, a tensor A of order p + q is a multilinear mapping with scalar values, such that :

$$A \in \mathcal{L}\left(\vec{E}^{*q} \times \vec{E}^{p}, \mathbb{R}\right).$$

<u>Examples</u> :

a) $A \in \mathcal{L}(\vec{E} \times \vec{E}, \mathbb{R})$, A is a second order covariant tensor :

$$\forall v_1, v_2 \in \vec{E}, \ v_1 = SX_1, \ v_2 = SX_2.$$

$$A\left(v_1\right)\left(v_2\right) = A\left(SX_1\right)\left(SX_2\right) = A\left(S_i \ {}^i X_1\right)\left(S_j \ {}^j X_2\right)$$

$$= A\left(S_i\right)\left(S_j\right) \cdot {}^i X_1 \cdot {}^j X_2$$

$$= A_{ij} \ {}^i X_1 \ {}^j X_2, \ \text{with} \ A_{ij} = A\left(S_i\right)\left(S_j\right)$$

A_{ij} components of A in S.

b) $A \in \mathcal{L}(\vec{E}* \times \vec{E}, \mathbb{R})$, A is a second order mixed tensor :

$$\forall C \in \vec{E}*, \ \forall V \in \vec{E}, \ C = LS^{-1}, \ L \in \mathbb{R}^{n*}, \ C = L_i \ {}^i S^{-1}$$

$$V = SX, \ X \in \mathbb{R}^n, \ V = S_i \ {}^i X$$

$$A(C)(V) = A\left(L_i \ {}^i S^{-1}\right)\left(S_j \ {}^j X\right) = A\left({}^i S^{-1}\right)\left(S_j\right) \cdot L_i \cdot {}^j X$$

$$= A^i_j \cdot L_i \cdot {}^j X$$

with $A^i_j = A\left({}^i S^{-1}\right)\left(S_j\right)$, the component of A in S.

It should be remembered that the space of the second order mixed tensors on $\vec{E}$ and $\vec{E}*$, is isomorphic with those of the endomorphisms of $\vec{E}$.

c) $A \in \mathcal{L}(\vec{E}* \times \vec{E}*, \mathbb{R})$, A is a contravariant tensor :

$$\forall\, c_1, c_2 \in \vec{E}^{*}, \quad c_1 = L_1 s^{-1}, \quad c_2 = L_2 s^{-1}$$

$$A\big(c_1\big)\big(c_2\big) = A\big(L_1 s^{-1}\big)\big(L_2 s^{-1}\big) = A\big(L_{1i}\,{}^{i}s^{-1}\big)\big(L_{2j}\,{}^{j}s^{-1}\big)$$

$$= A\big({}^{i}s^{-1}\big)\big({}^{j}s^{-1}\big) L_{1i}\, L_{2i}$$

$$= A^{ij}\, L_{1i}\, L_{2j}$$

with $A^{ij} = A\,({}^{i}s^{-1})\,({}^{j}s^{-1})$, the component of A in S.

These formulae can be used to execute changes of basis without diffi-culty.

5. – <u>GAUGES</u>

A gauge of a real vector space $\vec{E}$, with dimension n, is a non-zero anti-symmetrical multilinear mapping of degree n, with real values, giving the volume n-form.

Example : $n = 3$: $\forall V_1$, V_2, $V_3 \in \vec{E}$, we have the 3-form :

$$\mathrm{vol}_E\big(V_1\big)\big(V_2\big)\big(V_3\big)\,.$$

More frequently, we shall write vol instead of vol_E, where there will be no ambiguity.

It should be remembered that all gauges of $\vec{E}$ are proportional, and that by convention :

$$\mathrm{vol}_{R^3}\big(|_1\big)\big(|_2\big)\big(|_3\big) = \mathrm{vol}_{123} = 1\,.$$

In the case of a two-dimensional space $\vec{E}$, the gauge is called vol_2 to distinguish it from the three-dimensional gauge.

– An <u>alternor</u> is an anti-symmetrical p-linear mapping, with vector value, defined on space $\vec{E}$ with dimension n, and giving an anti-symmetrical p-form (if $p > n$, the alternor is zero.)

Thus $\forall V \in \vec{E}$, $\mathrm{vol}_E(V)$ is an alternor of order $n-1$.

> Reciprocally, $\forall$ alternor A with scalar values, of order n-1,
> giving the n-1-form :
>
> (A.1) $\quad A\big(V_1\big)\big(V_2\big) \cdots \big(V_{n-1}\big), \forall\, V_1, V_2, \ldots, V_{n-1} \in \vec{E}$ gauged
>
> $\exists V$ unique $\in \vec{E}$, such that $A = \mathrm{vol}_E(V)$.

- det(A) : determinant of a linear mapping A, which maps $\vec{E}$ of dimension n in $\vec{E}'$, and which has the same dimension, $\vec{E}$ and $\vec{E}'$, being gauged.

Example n = 3. def(A) = $\vec{E}$, val(A) $\subset \vec{E}'$,

$\forall V_1, V_2, V_3 \in \vec{E}$, by definition :

(A.2) $\qquad \mathrm{vol}_{E'}\big(AV_1\big)\big(AV_2\big)\big(AV_3\big) = \det(A)\; \mathrm{vol}_E\big(V_1\big)\big(V_2\big)\big(V_3\big).$

In particular, let S be a basis of $\vec{E}$:

$$V_1 = SX_1 \;,\quad V_2 = SX_2 \;,\quad V_3 = SX_3$$

$$\mathrm{vol}_E\big(SX_1\big)\big(SX_2\big)\big(SX_3\big) = \det(S)\; \mathrm{vol}_{R^3}\big(X_1\big)\big(X_2\big)\big(X_3\big)$$

$$\mathrm{vol}_E\big(S_1\big)\big(S_2\big)\big(S_3\big) = \det(S)\,\mathrm{vol}_{R^3}\big(I_1\big)\big(I_2\big)\big(I_3\big) = \det(S).$$

- Adj(A) : adjoint of a linear mapping A, mapping $\vec{E}$ in $\vec{E}'$ having the same dimension n, and gauged.

Example : n = 3, $\forall V_2, V_3 \in \vec{E}, \forall V_1 \in \vec{E}'$, by definition :

(A.3) $\qquad \mathrm{vol}_{E'}\big(V_1\big)\big(AV_2\big)\big(AV_3\big) = \mathrm{vol}_E\big(\mathrm{Adj}(A).V_1\big)\big(V_2\big)\big(V_3\big).$

- $T_r(A)$: trace of an _endomorphism_ A of $\vec{E}$, gauged and having dimension n.

Example : n = 3 : $\forall V_1, V_2, V_3 \in \vec{E}$, by definition :

(A.4) $\qquad T_r(A)\, \mathrm{vol}_E\big(V_1\big)\big(V_2\big)\big(V_3\big) = \mathrm{vol}_E\big(AV_1\big)\big(V_2\big)\big(V_3\big) + \mathrm{vol}_E\big(V_1\big)\big(AV_2\big)\big(V_3\big)$

$$+\; \mathrm{vol}_E\big(V_1\big)\big(V_2\big)\big(AV_3\big).$$

In particular :

(A.5) $T_r(V\Gamma) = \Gamma V, \forall\, V \in \vec{E}, \forall\, \Gamma \in \vec{E}^*$,

   $-\forall\, A \in \mathcal{L}(\vec{E},\vec{E}'), \forall\, B \in \mathcal{L}(\vec{E}',\vec{E}),\ \dim(\vec{E}) = n,$

   $\dim(\vec{E}') = n',\quad AB \in \mathcal{L}(\vec{E}',\vec{E}'),\ BA \in \mathcal{L}(\vec{E},\vec{E})$

(A.6) $T_r(AB) = T_r(BA)$.

In particular, let S be a basis of $\vec{E}$, $A \in \mathcal{L}(\vec{E},\vec{E})$:

   $A = SMS^{-1}$.

(A.8) $T_r(A) = T_r(M) = {}^iS^{-1}AS_i = {}^iM_i$.

   $-$ Let $\mathcal{T} = \mathcal{L}(\vec{E},\vec{E})$, $\forall\, A \in \mathcal{L}(\mathcal{T}, \mathbb{R})$,

                          $\forall\, B \in \operatorname{def}(A) \subset \mathcal{T}$,

   $\exists$ an endomorphism $C \in \mathcal{T}$, unique, representing A, such that :

(A.9) $A(B) = T_r(C\,B)$.

   This theorem is generalized in the case where $\mathcal{T} = \mathcal{L}(\vec{E}, \vec{E}')$,
   then $C \in \mathcal{L}(\vec{E}',\vec{E})$.

6. $-$ <u>SCALAR PRODUCT</u> ·

   $-\ g_E$: fundamental metric tensor of a real, Euclidian space $\vec{E}$. g_E is
bilinear, symmetrical and regular. Thus :

   $\forall\, V, W \in \vec{E},\ g_E(V)(W) = (V,W)$: scalar product of V,W.

   $-\ \overline{A}$: transpose of a linear mapping A, mapping $\vec{E}$ in $\vec{E}'$ both Euclidian and of finite dimension. We have :

   $\forall\, V \in \vec{E},\ V' \in \vec{E}'$

   $g_E(V')(AV) = g_E(\overline{A}V')(V),\ \overline{\overline{A}} = A$

Thus identifying $\vec{E}$ to $\mathcal{L}(\mathbb{R},\vec{E})$, we have :

$$(A.10) \qquad (V,W) = g_E(V)(W) = \overline{V}\,W = \overline{W}\,V, \quad \forall V,\ W \in \vec{E}$$

$$\overline{V},\overline{W} \in \vec{E}^*$$

- If $\vec{E}$ is complex, we still have :

$$(\overline{V},W) = \overline{V}W, \quad \forall V,W \in \vec{E} \ ; \quad \overline{V} \in \vec{E}^*$$

$$\text{and} \ \ \overline{zV} = \overline{z}\,\overline{V}, \quad \forall z \in \mathbb{C}, \ V \in \vec{E}, \ \text{where} \quad z = x+iy, \ \ \overline{z} = x-iy, \ (x,y \in \mathbb{R})$$

$$\text{and} \ \overline{i} = -i$$

Furthermore, if $\vec{E} = \mathbb{R}^n$:

$$\overline{|_i} = {}^i| \ , \ \overline{{}^i|} = |_i \ , \quad i = 1,2,\ldots,n .$$

If E is real, Euclidian and gauged, we have $\forall A \in \mathcal{L}(\vec{E},\vec{E})$:

$$(A.11) \qquad \begin{cases} \det(\overline{A}) = \overline{\det(A)} = \det(A) \\[2ex] \mathrm{Adj}(\overline{A}) = \overline{\mathrm{Adj}(A)} \\[2ex] T_r(\overline{A}) = \overline{T_r(A)} = T_r(A) . \end{cases}$$

- If $A \in \mathcal{L}(\vec{E},\vec{E}')$. $\vec{E},\vec{E}'$ of the same dimension n and Euclidian :

$$(A.12) \qquad A \text{ regular} \Longleftrightarrow \overline{A} \text{ regular} \Longrightarrow \overline{A^{-1}} = \overline{A}^{-1}.$$

In particular, let S be a basis of E :

$$\overline{S^{-1}} = \overline{S}^{-1} = \text{supplementary basis of S.}$$

- Gram matrix of a basis S. This is the matrix :

$$G = \overline{S}\,S$$

the components of which are :

$${}^iG_j = {}^i\overline{S}S_j = \overline{S_i}\,S_j = g_E\big(S_i\big)\big(S_j\big) = g_{ij} = g_{ji} .$$

Let $S' = \overline{S^{-1}}$ be the supplementary basis of S :

$$G = \overline{S}S \Rightarrow \overline{S} = GS^{-1}, \quad \overline{S^{-1}} = SG^{-1}$$

$$\forall V \in \vec{E} : V = SX \Rightarrow X = S^{-1} V \Rightarrow {}^iX = {}^iS^{-1} V, \quad (i = 1,\ldots,n).$$

But we can define the "covariant components" of V by :

$$X_i = \overline{S_i}\, V = {}^i\overline{S}\, V = {}^iGS^{-1} V = {}^iG_{i'}\, {}^{i'}S^{-1} V = g_{ii'}\, {}^{i'}X.$$

If $A \in \mathcal{L}(\vec{E},\vec{E})$, with S a basis of E :

$$M = S^{-1} AS \Rightarrow {}^iM_j = {}^iS^{-1} AS_j, \quad (i,j = 1,2, \ldots, n).$$

We can define the "covariant components" of A by :

$$M_{ij} = \overline{S_i}\, AS_j = {}^i\overline{S}AS_j = {}^iGS^{-1} AS_j = {}^iG_{i'}\, {}^{i'}S^{-1} AS_j$$

(A.13) $$M_{ij} = g_{ii'}\, {}^{i'}M_j.$$

We can also define the "contravariant components" of A by :

$$M_{ij} = \overline{S'_i}\, AS'_j, \quad (\text{where } S' = \overline{S^{-1}})$$

$$= {}^i\overline{S'}\, AS'_j = {}^iS^{-1}\, \overline{AS^{-1}}_j$$

$$= {}^iS^{-1}ASG^{-1}_j = {}^iS^{-1}AS_{j'}\, {}^{j'}G^{-1}_j.$$

(A.14) $$M^{ij} = {}^iM_j\, g^{j'j}, \quad \text{with } g^{j'j} = {}^{j'}G^{-1}_j$$

- <u>Change of tensor variance on an Euclidian space</u>

The preceding formulae are generalized to tensors. Taking $\vec{E}$ as a real Euclidian space of dimension n, we have seen (A.10) that :

$$\forall V \in \vec{E}, \quad \overline{V} \in \vec{E}^*$$

$$\forall C \in \vec{E}^*, \quad \overline{C} \in \vec{E}.$$

Let A be a mixed tensor on $\vec{E}$ and $\vec{E}^*$, for example of second order, we can define a covariant tensor B on $\vec{E}$, such that :

$$A(C)(V) = B(\overline{C})(V) \quad , \quad \forall C \in \vec{E}^*, \quad \forall V \in \vec{E}.$$

Let S be a basis of $\vec{E}$, we have :

$$B_{ij} = B\left(s_i\right)\left(s_j\right) = A\left(\overline{s_i}\right)\left(s_j\right) = A\left(\overline{s_i}\ s\ s^{-1}\right)\left(s_j\right)$$

$$= A\left(\overline{s_i}\ s_\alpha{}^\alpha s^{-1}\right)\left(s_j\right) = A\left({}^\alpha s^{-1}\right)\left(s_j\right)\overline{s_i}\ s_\alpha$$

$$B_{ij} = A_{\cdot j}^{\alpha}\ g_{i\alpha} .$$

It is frequently put $B_{ij} = A_{ij}$, quantities called the "covariant components" of A. The result is indeed consistent with (A.13), and with the isomorphism between mixed tensors and endomorphisms of $\vec{E}$. We also have :

$$A_{\cdot j}^{i} = A\left({}^i s^{-1}\right)\left(s_j\right) = B\left(\overline{{}^i s^{-1}}\right)\left(s_j\right) = B\left(s\ s^{-1} . \overline{{}^i s^{-1}}\right)\left(s_j\right)$$

$$= B\left(s_\alpha{}^\alpha s^{-1}\ \overline{{}^i s^{-1}}\right)\left(s_j\right) = B\left(s_\alpha\right)\left(s_j\right){}^\alpha s^{-1}\ \overline{{}^i s^{-1}}$$

$$= B_{\alpha j}\ {}^\alpha s^{-1}\ \overline{s^{-1}}_{\cdot j} = B_{\alpha j}\ {}^\alpha G^{-1}{}_{\cdot i} .$$

We can also define a contravariant tensor from a mixed tensor, and vice versa.

The above considerations apply to the fundamental metric tensor g_E. For the components of the mixed tensor g', we find :

$$g_{\cdot j}^{i} = g'\left({}^i s^{-1}\right)\left(s_j\right) = g_E\left(\overline{{}^i s^{-1}}\right)\left(s_j\right) = \overline{{}^i s^{-1}}\ s_j = {}^i s^{-1} s_j = {}^i|_{\cdot j},$$

and for those of the contravariant tensor g" :

$$g^{ij} = g''\left({}^i s^{-1}\right)\left({}^j s^{-1}\right) = g_E\left(\overline{{}^i s^{-1}}\right)\left(\overline{{}^j s^{-1}}\right) = \overline{{}^i s^{-1}}\ \overline{s^{-1}}_{\cdot j}$$

$$g^{ij} = {}^i s^{-1}\ \overline{s^{-1}}_{\cdot j} = {}^i G^{-1}{}_{\cdot j} .$$

Thus for the mixed tensor A (of second order)

$$(A.15) \qquad A_{ij} = A_{\cdot j}^{i'}\ g_{i'i} \quad , \quad A_{\cdot j}^{i} = A_{i'j}\ g^{i'i} .$$

- $\vec{E}_3$: real vector space, Euclidian, gauged, oriented, of dimension 3, with the gauge vol.

- $\vec{E}_2$: real vector space, Euclidian, gauged, oriented, of dimension 2, with the gauge vol_2.

- An endomorphism A of $\vec{E}$ will be said to be Hermitian if $A = \overline{A}$.

(A.16)

> Fundamental theorem (Sylvester). Let E be an Euclidian space with dimension n (which can be complex or hyperbolic), and $\vec{F}$ be a subspace of $\vec{E}$, with dimension $p \leqslant n$:
>
> $\forall A = \overline{A} \in \mathcal{L}(\vec{E},\vec{E})$, $\exists$ a basis S' of F, such that :
>
> $\overline{S'} \, A \, S' = D$, real diagonal matrix of $\mathbb{R}^p$.
>
> (The proof is given by recurrence).

From the above, we deduce that if $\vec{F}$ is positive, there exist unitary bases of $\vec{F}$, namely S', such that $\overline{S'} \, S' = 1_{\mathbb{R}^p}$. Furthermore, if $\vec{E}$ is positive, A is represented by the diagonal matrix of its real eigenvalues in the basis of its orthonormalized eigenmodes, with possible repetition in the case of multiple eigenvelues associated with eigensubspaces of dimension > 1.

- <u>Vector product</u> in $\vec{E}_3$ and $\vec{E}_2$

i : anti-symmetrical bilinear mapping on $\vec{E}_3$, with values in $\vec{E}_3$, giving the vector product of two vectors.

$$\forall V_1, \ V_2 \in \vec{E}_3, \text{ by definition :}$$

(A.17)
$$V_1 \times V_2 = i\left(V_1\right)\left(V_2\right) = \overline{\text{vol}\left(V_1\right)\left(V_2\right)} \in \vec{E}_3 \ .$$

Whence the mixed product $\forall V_1, \ V_2, \ V_3 \in \vec{E}_3$:

(A.18)
$$\overline{i\left(V_1\right)\left(V_2\right)} \cdot V_3 = \text{vol}\left(V_1\right)\left(V_2\right)\left(V_3\right) \ .$$

i(V) : anti-Hermitian endomorphism of $\vec{E}_3$, $\forall V \in \vec{E}_3$ (from the definition of i) :

(A.19)
$$\overline{i(V)} = - \, i(V) \ .$$

Reciprocally, we demonstrate with (A.1) that $\forall$ the anti-Hermitian endomorphism A of $\vec{E}_3$, $\exists$ an unique vector $V \in E_3$, such that :

(A.20) $A = i(V)$, if $A = -\,\overline{A} \in \mathcal{L}(\vec{E}_3,\vec{E}_3)$.

- Double vector product : $\forall V_1, V_2 \in \vec{E}_3$

(A.21)
$$i\left(v_1\right) \cdot i\left(v_2\right) = v_2\,\overline{v_1} - \overline{v_2}\,v_1 \cdot 1_{E_3}$$
$$i\left(i\left(v_1\right)\left(v_2\right)\right) = v_2\,\overline{v_1} - v_1\,\overline{v_2}\ .$$

i_2 : anti-Hermitian endomorphism of $\vec{E}_2$, giving the "vector product of a vector". $\forall V \in \vec{E}_2$, by definition :

(A.22) $i_2\,V = \overline{vol_2(V)}$.

Hence the "mixed product", $\forall V_1, V_2 \in \vec{E}_2$:

(A.23) $\overline{i_2\,V_1}\,V_2 = vol_2\left(v_1\right)\left(v_2\right)$.

As a result of the definition :

(A.24) $\overline{i_2} = -\,i_2, \quad i_2^2 = -\,1_{E_2},$

i_2 is therefore the operator of rotation of $+\,\dfrac{\Pi}{2}$ in $\vec{E}_2$.

If λ is a real scalar, λi_2 is an anti-Hermitian endomorphism of $\vec{E}_2$. Reciprocally, $\forall$ anti-Hermitian endomorphism A of $\vec{E}_2$, $\exists$ an unique real scalar λ, such that :

(A.25) $A = \lambda i_2, \quad \lambda \cdot 1_{E_2} = -\,i_2 A = -\,A i_2$, if $A = -\,\overline{A} \in \mathcal{L}\left(\vec{E}_2,\vec{E}_2\right)$.

- Let S be a basis of $\vec{E}_3$, $\forall V_1, V_2 \in \vec{E}_3$, $V_1 = SX_1$, $V_2 = SX_2$:

(A.26) $i\left(v_1\right)\left(v_2\right) = \overline{Adj(S)}\ i\left(x_1\right)\left(x_2\right)$.

$$\text{If } X = \begin{bmatrix} {}^1X \\ {}^2X \\ {}^3X \end{bmatrix}, \text{ applying (A.18) to } \mathbb{R}^3, \text{ we obtain :}$$

$$(A.27) \qquad i(X) = \begin{bmatrix} 0 & -{}^3X & {}^2X \\ {}^3X & 0 & -{}^1X \\ -{}^2X & {}^1X & 0 \end{bmatrix}.$$

7. – DIFFERENTIALS

Let E and E' be two normed linear spaces of finite dimensions, with an open set of E, and F a differentiable field applying $\mathcal{O}$ on E', such that:

$$\forall M \in \mathcal{O}, \; F(M) = V \in E',$$

$\dfrac{\partial V}{\partial M}$ is the derivative of the mapping F at point M, or otherwise :

$$\frac{\partial V}{\partial M} = D'F)(M) = F'(M).$$

$\forall$ the field f mapping $\mathcal{O}$ on $\vec{E}$, such that :

$$f(M) = dM \in \vec{E}.$$

d is the "derivation" associated with f.

One defines the differential dV by :

$$(A.28) \qquad dV = \frac{\partial V}{\partial M} dM = F'(M)(dM) \in \vec{E}'.$$

If $\vec{E}'$ is Euclidian, one shows that :

$$(A.29) \qquad d\overline{V} = \overline{dV}.$$

– If $E = \mathbb{R}^3$, $M = X \in \mathcal{O} \subset \mathbb{R}^3$, we consider the derivation ∂_i such that :

$$\partial_i X = |_i, \quad i = 1,2,3.$$

Thus :

$$(A.30) \qquad \partial_i V = \frac{\partial V}{\partial X} \partial_i X = \frac{\partial V}{\partial X} \Big|_i = \frac{\partial V}{\partial {}^i X}.$$

$$(A.31) \qquad dV = \frac{\partial V}{\partial X}dX = \frac{\partial V}{\partial X}\Big|_i \; {}^i dX = \begin{bmatrix} \partial_1 V & \partial_2 V & \partial_3 V \end{bmatrix} \begin{bmatrix} d^1X \\ d^2X \\ d^3X \end{bmatrix}.$$

If $E = E' = \mathbb{R}^3 : V = Y \in \mathbb{R}^3 :$

$$\partial_i Y = \partial_i \begin{bmatrix} {}^1Y \\ {}^2Y \\ {}^3Y \end{bmatrix} = \begin{bmatrix} \partial_i \; {}^1Y \\ \partial_i \; {}^2Y \\ \partial_i \; {}^3Y \end{bmatrix}$$

$$\frac{\partial Y}{\partial X} = \begin{bmatrix} \partial_1 \; {}^1Y & \partial_2 \; {}^1Y & \partial_3 \; {}^1Y \\ \partial_1 \; {}^2Y & \partial_2 \; {}^2Y & \partial_3 \; {}^2Y \\ \partial_1 \; {}^3Y & \partial_2 \; {}^3Y & \partial_3 \; {}^3Y \end{bmatrix}.$$

− Let A and ϕ be two differentiable mappings in an open set $\mathcal{O}$ of E, with values in normed linear spaces, such that :

$$F = A(X), \; Y = \phi(X), \quad \forall X \in \mathcal{O},$$

and such that F be a differentiable mapping in an open set of def(F). The derivation of a composition product leads to :

$$Z = F(Y)$$

$$(A.32) \qquad dZ = dF(Y) + F'(Y)(dY), \quad \forall dX \in \vec{E}.$$

If F is linear, $F'(Y) = F \Longrightarrow$

$$(A.33) \qquad dZ = dF(Y) + F(dY), \; F \text{ linear}.$$

Likewise, for a differentiable multilinear mapping F, such that :

$$Z = F(X)(Y), \; X,Y \quad \text{possibly independant}$$

$$(A.34) \qquad dZ = dF(X)(Y) + F(dX)(Y) + F(X)(dY),$$

$\forall$ the fields of differentials of the independent variable.

- Let us take the composition product of two differentiable linear mappings A and B, on the respective open sets of their normal linear domains of definition, we have :

$$(A.35) \qquad d\,[A.B] = dA.B + A.dB, \quad \forall\, dX \in \vec{E}$$

with : $\mathrm{val}(B) \subset \mathrm{def}(A), \ \mathrm{def}(B) = E.$

- d [det(A)]: Let $A : F(M) \in \mathcal{L}\,(\vec{E}_3, \vec{E}_3)$, F differentiable in an open set of E. From (A.2), (A.3), (A.4) and (A.34), we deduce :

$$(A.36) \qquad d[\det(A)] = T_r(\mathrm{Adj}(A).dA), \quad \forall\, dM \in \vec{E}.$$

- <u>Second differentials</u> : Let a field F be twice differentiable in an open set $\mathcal{O}$ of the normed linear space E, and mapping E on E', linear and normed, E and E' having finite dimensions for sake of simplification, and let f_1 and f_2 be two differentiable fields, applying $\mathcal{O}$ on $\vec{E}$, such that :

$$d_1 X = f_1(X), \ d_2 X = f_2(X), \quad \forall\, X \in \mathcal{O}$$

and

$$Y = F(X) \in E'.$$

We have : $d_1 Y = F'(X)\left(d_1 X\right) , \ d_2 Y = F'(X)\left(d_2 X\right)$

$$d_2 d_1 Y = F''(X)\left(d_2 X\right)\left(d_1 X\right) + F'(X)\left(d_2 d_1 X\right)$$

$$d_1 d_2 Y = F''(X)\left(d_1 X\right)\left(d_2 X\right) + F'(X)\left(d_1 d_2 X\right),$$

with :

$$F''(X)\left(d_1 X\right)\left(d_2 X\right) = F''(X)\left(d_2 X\right)\left(d_1 X\right).$$

Whence :

$$(A.37) \qquad \left[d_2 d_1 - d_1 d_2\right] Y = F'(X)\left(\left[d_2 d_1 - d_1 d_2\right] X\right).$$

$d_1 d_2 - d_1 d_2$, a bracket of differentials, defines a differential.

- <u>Taylor's formula</u> : If F is differentiable p times in the open set $\mathbf{O}$ of E, we have :

$$(A.38) \qquad F(X+H) = F(X) + F'(X)(H) + \frac{1}{2!} F''(X)(H)(H) + \ldots$$

$$\ldots + \frac{1}{p!} F^{(p)}(X)(H)\ldots(H) + o\left(|H|^p\right)$$

where $H \in \vec{E}$, and $|H|$ designate the norm of H in E, and $o\left(|H|^p\right)$ a vector of E, such that $\dfrac{o\,|H|^p}{|H|^p}$ tends to zero when $|H|$ tends to zero, with X remaining fixed.

In this formula, $F^{(p)}(X)$ also designates the pth derivative of F, at point X, p times linear and symmetrical. It should be noted that we must have $X+H \in \mathbf{O}$.

8. - <u>EXTERNAL DERIVATIVES</u>

Let E be a normed linear space of dimension n, and an alternor field A of order p on $\vec{E}$, defining a differentiable p-form with vectorial values :

$$A\left(d_1M\right)\left(d_2M\right)\ldots\left(d_pM\right) \,,\, \forall d_iM = f_i(M), \quad M \in \mathbf{O} \subset E, \quad (i=1,2,\ldots,p)$$

(A is an anti-symmetric covariant tensor of order p, with vectorial values).

If A is differentiable in $\mathbf{O}$, we define the external derivative, or co-edge of A, namely ∇A, $\forall dM = f(M)$, by :

$$(A.39) \qquad \nabla A\left(dM\right)\left(d_1M\right)\left(d_2M\right)\ldots\left(d_pM\right) = dA\left(d_1M\right)\left(d_2M\right)\ldots\left(d_pM\right)$$

$$- d_1A(dM)\left(d_2M\right)\ldots\left(d_pM\right)$$

$$- d_2A\left(d_1M\right)(dM)\ldots\left(d_pM\right)$$

$$- \ldots\ldots\ldots\ldots\ldots\ldots$$

$$- d_pA\left(d_1M\right)\left(d_2M\right)\ldots\left(dM\right).$$

∇A is an alternor of order p+1 on $\vec{E}$.

- An alternor is said to be closed in an open-set, if its co-edge is zero in this open set (A is a co-cycle).

- If A is differentiable twice in $\mathcal{O}$, we have :

(A.40) $\nabla\nabla A = 0$.

Reciprocally (Poincaré's theorem), if B is an alternor of order p+1, of class C^1 in $\mathcal{O}$, and if $\mathcal{O}$ is convex or starred, or homeomorph with an open set of this type, and if B is a co-cycle, namely if $\nabla B = 0$ in $\mathcal{O}$, then :

(A.41) $[\nabla B = 0 \text{ in } \mathcal{O}] \Longleftrightarrow \left[\begin{array}{l} \exists \text{ an alternor A of order p, differentiable} \\ \text{twice in } \mathcal{O}, \text{ such that } B = \nabla A. \end{array}\right]$

Under these conditions therefore, a co-cycle is a co-edge (generally a co-cycle is defined to within any co-edge, thus making it possible to define classes of cohomology).

- <u>Particular cases</u> : Let E be a normed linear space, we assume the necessary conditions for derivability in an open set $\mathcal{O}$ of E to be satisfied.

1) <u>A is of order zero</u> : This is scalar $u = F(M), \forall M \in \mathcal{O}$

$$\nabla u = \frac{\partial u}{\partial M} \in \vec{E}*$$

If E is Euclidian :

$$\text{grad } u = \frac{\overline{\partial u}}{\partial M} \in \vec{E}.$$

2) <u>A is of order 1 with scalar values</u> on $\vec{E}$. This is a co-vector $C = F(M) \in \vec{E}*$.

$$\nabla C \left(d_1 M\right)\left(d_2 M\right) = d_1 C d_2 M - d_2 C d_1 M$$

if $E = E_3$, $\forall V = F(M) \in \vec{E}_3$, and $C = \overline{V} \in \vec{E}_3^*$

(A.42) $\nabla \overline{V}\left(d_1 M\right)\left(d_2 M\right) = \overline{d_1 V}\, d_2 M - \overline{d_2 V}\, d_1 M$.

In this case, (E_3), $\nabla\overline{V}$ is an alternor of order 2 on $\vec{E}_3$, therefore [see (A.1)] , there exists a unique vector called rot V, such that, by definition :

(A.43) $\nabla\overline{V} = \text{vol(rot V)}$.

if $V = \text{grad } u$, $u \in \mathbb{R}$, $\overline{V} = \dfrac{\partial u}{\partial M} = \nabla u$.   (A.40) $\Longrightarrow$

(A.44) $\text{rot grad } u = 0$.

Furthermore, it can be shown that :

(A.45) $i(\text{rot } V) = \dfrac{\partial V}{\partial M} - \dfrac{\overline{\partial V}}{\partial M}$, $V \in \vec{E}_3$.

We also put :

$$\text{grad } V = \dfrac{\overline{\partial V}}{\partial M}$$

3) <u>A is of order n-1 with scalar values</u>, on $\vec{E}$, gauged and of dimension n.

Then (A.1) $\Longrightarrow$ $\exists V$ unique $\in \vec{E}$, such that :

$$A = \text{vol}(V).$$

$\nabla[\text{vol}(V)]$ is an alternor of order n, and therefore a gauge of $\vec{E}$. By definition :

(A.46) $\nabla[\text{vol}(V)] = \text{div } V.\text{vol}$.

It is easily shown by (A.39) and (A.4) that :

(A.47) $\text{div } V = T_r\!\left(\dfrac{\partial V}{\partial M}\right)$.

In particular, if $\vec{E} = \vec{E}_3$ and if $V = \text{rot } W$:

$$\nabla\overline{W} = \text{vol}(\text{rot } W)$$

$$\nabla[\text{vol}(\text{rot } W)] = \nabla\nabla\overline{W} = \text{div rot } W.\ \text{vol} = 0.$$

Whence :

(A.48) $\text{div rot } W = 0$.

- By definition, if $u \in \mathbb{R}$, $\text{grad } u = \dfrac{\overline{\partial u}}{\partial M}$:

(A.49) $\text{div grad } u = \nabla u$ (Laplacian of u).

4) <u>Divergence of an endomorphism</u> : Let E be linear and normed, of dimension n, an endomorphism field A of $\vec{E}$, and a vector field V of $\vec{E}$, we define div A by the formula :

$$(A.50) \qquad \text{div}[AV] = T_r\left(\frac{\partial AV}{\partial M}\right) = \text{div } A.V + T_r\left(A\frac{\partial V}{\partial M}\right)$$

Thus :

$$\text{div } A \in \vec{E}^*.$$

We have also :

$$\text{div }[AV] = \text{div } A.V, \quad \forall \text{ constant vector field V.}$$

5) <u>Rotational (curl) of an endomorphism</u> : Given an endomorphism field A of $\vec{E}_3$,

$$(A.51) \qquad \text{rot }[AV] = \text{rot } A.V, \quad \forall \text{ constant vector field V.}$$

$$\text{rot } A \in \mathcal{L} \ (\vec{E}_3, \vec{E}_3).$$

- <u>Diverse formulae</u> : All quantities considered are differentiable fields, or twice differentiable in an open set of E_3.

$$(A.52) \qquad \text{grad}[\lambda u] = \lambda \text{ grad } u + u \text{ grad } \lambda, \ u, \ \lambda \in \mathbb{R} \ .$$

$$(A.53) \qquad \text{rot}[V\lambda] = \text{rot } V.\lambda + i(\text{grad } \lambda)(V), \ \lambda \in \mathbb{R}, \ V \in \vec{E}_3 .$$

$$(A.54) \qquad \text{div}[V\lambda] = \lambda \text{ div } V + \overline{\text{grad } \lambda}.V \qquad , \lambda \in \mathbb{R}, \ V \in E_3 .$$

$$(A.55) \qquad \text{rot}[V \times W] = V \text{ div } W - W \text{ div } V + \overline{\text{grad } V}.W - \overline{\text{grad } W}.V, \ V, \ W \in \vec{E}_3 .$$

$$(A.56) \qquad \text{div}[i(V)] = \overline{\text{rot } V} \quad , \ V \in \vec{E}_3 .$$

$$(A.57) \qquad \text{rot}[\lambda.1_E] = i(\text{grad } \lambda), \quad \lambda \in \mathbb{R}.$$

$$(A.58) \qquad \text{rot}[i(V)] = \frac{\partial V}{\partial M} - \text{div } V.1_{E_3}, \ V \in \vec{E}_3 .$$

$$(A.59) \qquad \text{rot } \frac{\partial V}{\partial M} = \frac{\partial}{\partial M} \text{ rot } V \ , \quad V \in \vec{E}_3 .$$

$$(A.60) \qquad \text{rot } \frac{\overline{\partial V}}{\partial M} = \text{rot grad } V = 0, \ V \in \vec{E}_3 .$$

(A.61) $\nabla V = \text{div grad } V = \text{div } \dfrac{\overline{\partial V}}{\partial M} = \text{grad div } V - \text{rot rot } V, \quad V \in \vec{E}_3$

Let S be a basis of $\vec{E}_3$ in $M \in E_3$:

(A.62) $\text{div } A = {}^i S^{-1} \partial_i A \quad , \; A \in \mathcal{L}\left(\vec{E}_3, \vec{E}_3\right).$

(A.63) $\text{rot } A = i\left(\overline{{}^i S^{-1}}\right).\partial_i A, \; A \in \mathcal{L}\left(\vec{E}_3, \vec{E}_3\right).$

– <u>Representation of external derivatives</u> : An open set $\mathbf{O}$ of a normed linear space E of dimension n, being parameterized by a map ϕ :

$$M = \phi(X) \in \mathbf{O}, \; X \in \phi^{-1}(\mathbf{O}) \; \subset \mathbb{R}^n,$$

Let S be the natural basis at M (see paragraph 9) :

$$S = \frac{\partial M}{\partial X} = \phi'(X) = \left[\partial_1 M \; \partial_2 M \; \cdots \; \partial_n M\right].$$

The representation in S of the various quantities previously introduced, comes from the following theorem :

An alternor A of order p on $\vec{E}$, being represented by the alternor B on $\vec{\mathbb{R}}^n$ in a basis S, we have :

$$A\left(d_1 M\right)\left(d_2 M\right)\cdots\left(d_p M\right) = B\left(d_1 X\right)\left(d_2 X\right)\cdots\left(d_p X\right)$$

with :

$$d_i M = S d_i X, \quad i = 1, 2, \ldots, p.$$

B is the reciprocal image of A.

We deduce from the definition :

(A.64) $\nabla_M A(dM)\left(d_1 M\right)\left(d_2 M\right)\cdots\left(d_p M\right) = \nabla_X B(dX)\left(d_1 X\right)\left(d_2 X\right)\cdots\left(d_p X\right)$

with :

$$dM = S dX.$$

This property is translated in saying that the operation co-edge commutes with the operation reciprocal image.

In particular, if :

$$V = SY$$

we find, by application of (A.64) :

(A.65)
$$\text{div}_M V = \frac{1}{\sqrt{|\det(G)|}}\, T_r\!\left(\frac{\partial}{\partial X}\left[Y\sqrt{|\det(G)|}\right]\right), \text{ with } \bar{S}S = G$$

(A.66)
$$\left[\begin{array}{l}
\text{if } V \in \vec{E}_3 \text{ and S is a direct basis :} \\[2ex]
\underset{M}{\text{rot }} V = SZ \ , \ Z \in \mathbb{R}^3 \\[2ex]
Z = \frac{1}{\sqrt{|\det(G)|}}\ \underset{X}{\text{rot }}[GY] \\[2ex]
\text{with} \\[2ex]
{}^i|_i\left(\underset{X}{\text{rot }}[GY]\right)\Big|_j = {}^i\left|\left[\frac{\partial GY}{\partial X} - \overline{\frac{\partial GY}{\partial X}}\right]\right|_j = \frac{\partial^i GY}{\partial^j X} - \frac{\partial^j GY}{\partial^i X}
\end{array}\right.$$

Let $A \in \mathcal{L}(\vec{E}_3,\vec{E}_3)$, $A = SBS^{-1}$, $B \in \mathcal{L}(\mathbb{R}^3, \mathbb{R}^3)$

$$S = \phi'(X), \quad dS = \phi''(X)(dX) = S\Gamma(dX)$$

where $\Gamma(dX)$ is a matrix of $\mathbb{R}^3$, linear in dX. Still putting (see paragraph 7) :

$$\partial_j X = |_j \quad (j = 1,2,3),$$

we obtain :

$$\partial_j S_k = S \ \Gamma\!\left(\partial_j X\right)\!\left(|_k\right) = S \ \Gamma\!\left(|_j\right)\!\left(|_k\right) = S \ \Gamma_{jk} = S_i \ {}^i\Gamma_{jk}$$

(A.67)
$$\partial_j S_k = \phi''(X)\!\left(|_j\right)\!\left(|_k\right) = S_i \ {}^i\Gamma_{jk} \ , \ (i,j,k = 1,2,3).$$

Scalars ${}^i\Gamma_{jk}$ are the Christoffel symbols relative to E_3, and to the map ϕ considered.   (A.67) $\Longrightarrow {}^i\Gamma_{jk} = {}^i\Gamma_{kj}$.

The quantities $\partial_j S^{-1}$ are deduced from (A.67), by differentiation of $S^{-1}S = 1_{\mathbb{R}^3}$.

We then deduce from (A.62) and (A.67) :

$$\text{div}_M A.S_k = {}^iS^{-1} \partial_i \left[SBS^{-1} \right] S_k$$

(A.68) $$\text{div}_M A.S_k = {}^i\Gamma_{ij} \, {}^jB_k + \partial_i \, {}^iB_k - {}^iB_j \, {}^j\Gamma_{ik}$$

The representation of rot A would also be deduced from (A.67) and (A.63).

9. – <u>DIFFERENTIABLE MANIFOLDS</u>

Let $\mathcal{V}$ be a set and a mapping F mapping an open set $\mathcal{O}$ of $\mathbb{R}^p$ on $\mathcal{V}$:

$$\forall X \in \mathcal{O} , \quad F(X) = M \in \mathcal{V} ,$$

with F regular $\Rightarrow$ F is a pre-mapping of $\mathcal{V}$.

Given two pre-mappings F and G, on two open sets of $\mathbb{R}^p$, such that :

$$M = F(X) = G(X') \in \mathcal{V}, \quad X \in \mathcal{O} = \text{def}(F), \quad X' \in \mathcal{O}' = \text{def}(G)$$

$F^{-1}G$ maps $\mathbb{R}^p$ in $\mathbb{R}^p$.

If $F^{-1}G$ is differentiable on $\mathcal{O}'$, while its inverse is differentiable on $\mathcal{O}$, F and G are said to be coherent (even if $\text{val}(F) \cap \text{val}(G) = \emptyset$).

An atlas A of $\mathcal{V}$ is a set of two by two coherent pre-mapping, the values of which cover $\mathcal{V}$.

An atlas A being chosen, a map of $\mathcal{V}$ is any pre-map coherent with A. An atlas gives $\mathcal{V}$ the structure of a differentiable manifold (two non-coherent atlas give $\mathcal{V}$ two distinct manifold structures). $\mathcal{V}$ has dimension p.

A part Ω of $\mathcal{V}$ is an open set, if Ω is the union of map images $\Rightarrow$ a topology of $\mathcal{V}$.

If A applies $\mathcal{V}$ in a manifold $\mathcal{V}'$, A is continuous if $\forall$ the open set $\Omega' \subset \mathcal{V}'$, $\exists$ an open set $\Omega \subset \mathcal{V}$, such that :

$$A^-(\Omega') = \Omega \cap \text{def}(A).$$

Let $\mathcal{V}$ and $\mathcal{V}'$ be two manifolds, an immersion of $\mathcal{V}$ in $\mathcal{V}'$ is a mapping A of $\mathcal{V}$ to $\mathcal{V}'$, such that A is differentiable and regular, with its derivative regular at any point of def(A) (reciprocal images are used by means of the maps).

A manifold $\mathcal{V}$ of dimension p is immersed in a manifold $\mathcal{V}'$, of dimension n, if the maps of $\mathcal{V}$ are immersions of $\mathbb{R}^p$ in $\mathcal{V}'$.

If $\mathcal{V}$ is immersed in E, linear and of dimension n, there is no difficulty in defining the tangent vector space at any regular point of $\mathcal{V}$, where this tangent plane is defined.

The vector space tangent to $\mathcal{V}$ is a vector subspace of $\vec{E}$, of dimension p (therefore $p \leqslant n$).

If the union $\mathcal{D}$ of the domains of definition of the maps of an atlas is bicompact, $\mathcal{V}$ is bicompact, and $\exists$ an atlas composed of a finite number of maps, but larger than 1.

The edge of $\mathcal{V}$ is the image, by the maps, of the edge of $\overline{\mathcal{D}}$. A pseudo-manifold, or pseudo-edge is a singular manifold.

The canonical orientation of a manifold immersed in E, linear and oriented, is defined without difficulty, by the continuous canonical orientation of the tangent vector space.

It is to be understood in all cases in the following, that the manifolds are canonically oriented, bicompact and connex (o-connexity).

– <u>Linear connection – covariant differentials</u> : Let $\mathcal{V}$ be a differentiable manifold, of class C^2 and dimension p, immersed in a normed linear space E of dimension n ($p \leqslant n$), we define the linear connection of $\mathcal{V}$, by the giving a differentiable field of projectors Π, defined at any point m of the regular part of $\vec{E}$, and projecting any vector of E on the plane tangent to $\mathcal{V}$ at m.

Let F be the map of $\mathcal{V}$ mapping an open set $\mathcal{O}$ of $\mathbb{R}^p$ in $\mathcal{V}$, and the natural basis S of the tangent plane at point $m \in \mathcal{V}$:

$$m = F(X), \quad \forall X \in \mathcal{O} \subset \mathbb{R}^p$$

$$S = F'(X) = \frac{\partial m}{\partial X} = \left[\partial_1 m \; \partial_2 m \; \cdots \; \partial_p m\right] = \left[S_1 \; S_2 \; \cdots \; S_p\right].$$

Let V be a differentiable vector field tangent to $\mathcal{U}$:

$$V = SY, \quad Y \in \mathbb{R}^p, \quad Y = f(X), \quad X \in \mathcal{O}.$$

We call covariant differential of V, the tangent differential vector :

$$\hat{d}V = \Pi dV = \Pi d[SY] = \Pi dSY + \Pi SdY$$

$$= \Pi dSY + SdY$$

$\forall$ the tangent differential vector field at m, dm.

If we call $\Gamma(dX)$ the matrix used to express the line dS in S, we obtain:

(A.69) $\qquad \hat{d}S = \Pi dS = S\,\Gamma(dX), \quad (\Gamma \text{ linear})$

(A.70) $\qquad \hat{d}V = S\,[\Gamma(dX)(Y) + dY].$

If $\partial_i X = |_i$, and $V = S_j = S|_j$ $\quad (i,j = 1,2, \ldots, p), \quad$ (A.70) $\Longrightarrow$

(A.71) $\qquad \hat{\partial}_i S_j = S\,\Gamma\!\left(\partial_i X\right)\!\left(|_i\right) = S\,\Gamma\!\left(|_i\right)\!\left(|_j\right) = S\,\Gamma_{ij} = S_k\,{}^k\Gamma_{ij},$

$$i,j,k = 1,2,\ldots,p.$$

The scalars ${}^k\Gamma_{ij}$ are the Christoffel symbols for the connection associated with the basis S (they are used to define the linear connection of an
abstract manifold).

We deduce :

(A.72) $\qquad \hat{\partial}_i V = S_k\,\left[{}^k\Gamma_{ij}\,{}^jY + \partial_i\,{}^kY\right].$

We can agree that :

(A.73) $\qquad \hat{d}W = dW$, if $W \in \vec{E}$, (respectively $\mathbb{R}^n$), with tangent plane $\vec{E}$
$\qquad\qquad$ (respectively $\mathbb{R}^n$), for linear spaces.

Given a differentiable field of <u>covariant tensors</u> A, of order $\leq p$, defined on the plane $\vec{E}_p$ tangent to $\mathcal{U}$ at point m, we extend A on $\vec{E}$ by a tensor A',

such that, for example with p = 3 :

(A.74) $A'\big(W_1\big)\big(W_2\big)\big(W_3\big) = A\big(\Pi W_1\big)\big(\Pi W_2\big)\big(\Pi W_3\big), \; \forall \; W_1,W_2,W_3 \in \vec{E}.$

For any dm, V_1, V_2, $V_3 \in E_p$, and assuming the fields to be differentiable, we define the covariant differential $\hat{d}A$ by :

$$\hat{d}\Big[A\big(V_1\big)\big(V_2\big)\big(V_3\big)\Big] = d\Big[A\big(V_1\big)\big(V_2\big)\big(V_3\big)\Big] = d\Big[A'\big(V_1\big)\big(V_2\big)\big(V_3\big)\Big]$$

$$= dA'\big(V_1\big)\big(V_2\big)\big(V_3\big) + A'\big(dV_1\big)\big(V_2\big)\big(V_3\big) + A'\big(V_1\big)\big(dV_2\big)\big(V_3\big) +$$
$$+ A'\big(V_1\big)\big(V_2\big)\big(dV_3\big)$$

$$= \Big[\hat{dA}\Big]\big(V_1\big)\big(V_2\big)\big(V_3\big) + A\big(\Pi dV_1\big)\big(V_2\big)\big(V_3\big) + A\big(V_1\big)\big(\Pi dV_2\big)\big(V_3\big) +$$
$$+ A\big(V_1\big)\big(V_2\big)\big(\Pi dV_3\big)$$

Thus :

(A.75) $\hat{d}\Big[A\big(V_1\big)\big(V_2\big)\big(V_3\big)\Big] = \hat{d}A\big(V_1\big)\big(V_2\big)\big(V_3\big) + A\big(\hat{d}V_1\big)\big(V_2\big)\big(V_3\big) + A\big(V_1\big)\big(\hat{d}V_2\big)\big(V_3\big)$
$$+ A\big(V_1\big)\big(V_2\big)\big(\hat{d}V_3\big).$$

Using (A.70), and taking $V_1 = S_p$, $V_2 = S_q$, $V_3 = S_r$, we have :

(A.76) $\Big[\hat{\partial_i A}\Big]_{pqr} = \partial_i A_{pqr} - A_{p'qr}{}^{p'}\Gamma_{ip} - A_{pq'r}{}^{q'}\Gamma_{iq} - A_{pqr'}{}^{r'}\Gamma_{ir}$

Let us note that $\hat{d}A$ is defined as the restriction of dA' to $\vec{E}_p, (dA'|_{\vec{E}_p})$.

In particular, if A is a covector C of $\vec{E}_p$:

$$C'(W) = C(\Pi W), \quad \forall \, W \in \vec{E}.$$

We have :

$$\hat{d}C = dC'\big|_{\vec{E}_p}$$

$$C = LS^{-1}, \; L \in \mathbb{R}^{p*}$$

(A.77) $\Big[\hat{\partial_i C}\Big]_p = \partial_i L_p - L_{p'}{}^{p'}\Gamma_{ip}.$

Furthermore :

$$\hat{\partial}_i C = \left[\hat{\partial}_i C\right]_p {}^P S^{-1},$$

therefore if $C = {}^j S^{-1}$, $L = {}^j|$

(A.78) $\hat{\partial}_i {}^j S^{-1} = - {}^j \Gamma_{ip} {}^P S^{-1}$

(we could also have differentiated $S^{-1} S = 1_{E_p}$).

If A is a field of <u>mixed tensors</u>, for example of order 2, defined on $\vec{E}_p$ and $\vec{E}_p^*$, we extend A on $\vec{E}$ and $\vec{E}^*$ by A', such that :

$$A'(\Gamma)(W) = A(\Gamma|_{E_p})(\Pi W), \quad \forall \Gamma \in \vec{E}^*, \quad \forall W \in \vec{E}.$$

Thus $\forall C \in \vec{E}_p^*$ and $V \in \vec{E}_p$, we extend C on $\vec{E}^*$ as before, by $C' = C\Pi$, and obtain :

$$\hat{d}\left[A(C)(V)\right] = d\left[A(C)(V)\right] = d\left[A'(C')(V)\right]$$

$$= dA'(C')(V) + A'(dC')(V) + A'(C')(dV)$$

(A.79) $\hat{d}\left[A(C)(V)\right] = \hat{d}A(C)(V) + A(\hat{d}C)(V) + A(C)(\hat{d}V)$

where, by definition, $\hat{d}A$ is the restriction of dA' to $\vec{E}_p^*$ and $\vec{E}_p$.

We deduce :

(A.80) $\left[\hat{\partial}_i A\right]{}^P_q = \partial_i A^P_q + A^{P'}_q {}^P\Gamma_{ip'} - A^P_{q'} {}^{q'}\Gamma_{iq}.$

If A is a <u>linear mapping</u> from E on $\vec{E}_p$, the value of a differentiable field in an open set of $\mathcal{V}$, we can either consider the associated mixed tensor, or proceed directly to calculate the covariant differential of A, having the same domain of definition and value as A. Thus :

$$\forall W \in \vec{E}, \quad C \in \vec{E}_p^*, \quad \text{we extend C on } \vec{E}^* \text{ by C'},$$

whence : $\forall dM \in \vec{E}_p$:

(A.81) $\hat{d}[C A W] = d [C A W] = \hat{d}C A W + C \hat{d}A W + C A dW.$

<u>External derivatives</u> : Formulae (A.39) and (A.75) enable us to extend to differentiable manifolds the definition of the co-edge of a differentiable alternor A of order q, on $\vec{E}_p$ (q $\leqslant$ p). Thus $\forall$ the differential fields dm, $d_1 m$, $d_2 m$, ..., $d_q m \in \vec{E}_p$:

$$(A.82) \qquad \hat{\nabla} A(dm) \left(d_1 m\right)\left(d_2 m\right) \ldots \left(d_q m\right) = \hat{d} A \left(d_1 m\right)\left(d_2 m\right) \ldots \left(d_q m\right)$$

$$- \hat{d}_1 A(dm) \left(d_2 m\right) \ldots \left(d_q m\right)$$

$$- \hat{d}_2 A \left(d_1 m\right)(dm) \ldots \left(d_q m\right)$$

$$- \ldots\ldots\ldots\ldots\ldots\ldots$$

$$- \hat{d}_q A \left(d_1 m\right)\left(d_2 m\right) \ldots \left(dm\right) .$$

As in the case of a linear space, the covariant co-edge operation commutes with the reciprocal image operation [see (A.64)] . It should be noted that if q = p, $\hat{\nabla} A = 0$.

We again find in general :

$$(A.83) \qquad \hat{\nabla}\hat{\nabla} A = 0.$$

We can again enunciate Poincaré's theorem, according to which, given an alternor B of order q on $\vec{E}_p$, closed in a simply connected open set of   , namely :

$$\hat{\nabla} B = 0$$

$\exists$ an alternor field A, of order q-1, defined on this open set, such that :

$$B = \hat{\nabla} A.$$

<u>Connection curvature and torsion</u> : $\forall$ the tangent vector fields $d_1 m$, $d_2 m$, $d_3 m \in \vec{E}_p$, we can define the torsion and curvature of the connection, respectively by mappings T and R, with vector values, in $\vec{E}_p$, such that :

$$T \left(d_1 m\right)\left(d_2 m\right) = \left[\hat{d}_1 d_2 - \hat{d}_2 d_1 \right] m - \left[d_1 d_2 - d_2 d_1 \right] m$$

$$R \left(d_1 m\right)\left(d_2 m\right)\left(d_3 m\right) = \left[\hat{d}_1 \hat{d}_2 - \hat{d}_2 \hat{d}_i \right] d_3 m - \left[\hat{d}_1 d_2 - \hat{d}_2 d_1 \right] d_3 m .$$

For a connection defined by a projector Π, we also find :

$$R\left(d_1 m\right)\left(d_2 m\right)\left(d_3 m\right) = \left[d_1 \Pi d_2 \Pi - d_2 \Pi d_1 \Pi\right] d_3 m.$$

Varying a point m, or a tangent vector at m, along a path defined by the differentials $d_1 m$, $d_2 m$ on $\mathcal{V}$, the mappings T and R find a simple interpretation.

We deduce from the definition of T that $T = 0$, for a manifold immersed in affine space E. As a result :

$$(A.84) \qquad {}^k\Gamma_{ij} = {}^k\Gamma_{ji}$$

which would also result from direct differentiation of the natural basis.

10. – <u>RIEMANNIAN MANIFOLDS</u>

A differentiable manifold $\mathcal{V}$ of dimension p, immersed in an Euclidian space E, is called Riemannian if we define the scalar product on $\vec{E}_p$ by introduction of the scaler product of $\vec{E}$ on $\vec{E}_p$, and if its linear connection is defined by a Hermitian projector field $\Pi = \overline{\Pi}$, projecting any vector of $\vec{E}$ on the tangent plane $\vec{E}_p$.

Then let g be the fundamental metric tensor of $\vec{E}_p$ at m, with components g_{ij} in the natural basis S at m, the extension of g on $\vec{E}$ being unique, we can deduce Ricci's lemma :

$$(A.85) \qquad \hat{d}g = 0, \quad \forall\, dm \in \vec{E}_p, \quad g_{ij} = \overline{S_i}\, S_j.$$

It comes as a consequence that :

$$(A.86) \qquad \widehat{d\overline{V}} = \overline{\hat{d}V}, \quad \forall\, V \in \vec{E}_p,$$

and that the connection of a Riemannian manifold is expressed by means of tensor g and Christoffel operator Γ, by the following formulae :

$$(A.87) \qquad \left[\begin{array}{l} {}^k\Gamma_{ij} = g^{k'k}\, \gamma_{k'ij} \\[2em] \gamma_{kij} = \dfrac{1}{2}\left[g_{ki,j} + g_{jk,i} - g_{ij,k}\right], \end{array}\right.$$

which can easily be shown by calculating :

$$\hat{\partial}_k g_{ij} = \partial_k \left[g(S_i)(S_j) \right] .$$

Naturally, we still have the relation of symmetry :

(A.88) $\quad {}^k\Gamma_{ij} = {}^k\Gamma_{ji} .$

Namely the torsion of the connection is zero (relations (A.85) and (A.88) are precisely the axioms of definition for an abstract Riemannian manifold).

The scalar quantities ${}^\gamma_{ijk}$ are called Christoffel symbols of the first kind, and ${}^k\Gamma_{ij}$ Christoffel symbols of the second kind.

If the tangent vector space $\vec{E}_p$ is gauged by a differentiable gauge field vol_p, it can be demonstrated under the preceding conditions :

(A.89) $\quad \hat{d}vol_p = 0, \quad \forall dm \in \vec{E}_p . \Rightarrow \hat{d}i_2 = 0 \quad$ for $\quad p = 2.$

Covariant divergence : Given a Riemannian manifold, gauged, with dimension p and immersed in an Euclidian space E with dimension n. $\vec{E}_p$ is its tangent plane at m, π the Hermitian projector projecting $\vec{E}$ onto $\vec{E}_p$, and vol_p the gauge of $\vec{E}_p$.

Let V be also a tangent vector field, we define without difficulty the (covariant) divergence of V, passing via the co-edge of the alternor $vol_p(V)$ of order p-1, on $\vec{E}_p$. As in (A.46), we have, by definition :

(A.90) $\quad \hat{\nabla}vol_p(V) = \widehat{div}\, V.vol_p, \quad V \in \vec{E}_p ,$

and thanks to (A.89), we again find :

(A.91) $\quad \widehat{div}\, V = T_r \left(\dfrac{\hat{\partial}V}{\partial m} \right) = T_r \left(\Pi \dfrac{\partial V}{\partial m} \right) .$

If S is the natural basis of $\vec{E}_p$ at m, we again find :

(A.92) $\quad \widehat{div}\, V = {}^iS^{-1}\, \hat{\partial}_i V \quad\quad (i = 1,2, \ldots, p),$

and as in (A.65) :

$$\widehat{div}\, [SY] = \frac{1}{\sqrt{|\det(G)|}}\, T_r\left(\frac{\partial}{\partial X} \left[Y\, \sqrt{|\det(G)|} \right] \right).$$

If we consider an endomorphism field A of $\vec{E}_p$, we also define $\widehat{\text{div}}\,A$ by :

$$(A.93) \qquad \widehat{\text{div}}[AV] = \widehat{\text{div}}\,A.V + T_r\left(A\,\widehat{\frac{\partial V}{\partial m}}\right), \quad \widehat{\text{div}}\,A \in \vec{E}_p^*.$$

If now :

$$A = S\,B\,S^{-1},$$

we again find :

$$(A.94) \qquad [\widehat{\text{div}}\,A]S_i = \partial_j\,{}^jB_i + {}^j\Gamma_{kj}\,{}^kB_i - {}^r\Gamma_{ip}\,{}^pB_r .$$

More generally, taking an operator field :

$$A \in \mathcal{L}(\vec{E},\vec{E}_p) ,$$

and a vector field $V \in \vec{E}$, defined on the manifold $\mathcal{V}$, we define $\widehat{\text{div}}\,A$ by :

$$(A.95) \qquad \left[\begin{array}{l} \widehat{\text{div}}[AV] = T_r\left(\widehat{\frac{\partial [AV]}{\partial m}}\right) = \widehat{\text{div}}\,A.V + T_r\left(A\,\frac{\partial V}{\partial m}\right) \\[2em] \widehat{\text{div}}\,A \in \vec{E}.^* \end{array} \right.$$

We again have :

$$(A.96) \qquad \widehat{\text{div}}\,A = {}^iS^{-1}\,\hat{\partial}_iA, \quad A \in \mathcal{L}(\vec{E},\vec{E}_p),$$

This formula can be used for example with (A.81).

We can extend the definition of the covariant divergence to covariant tensors, and for example to the covectors of $\vec{E}_p$, after a change of variance in order to convert a covariant tensor into a mixed tensor, or a covector into a vector. Furthermore, it can be demonstrated that the "change of variance" operation commutes with the covariant "derivation" (because of (A.85)).

11. – <u>MANIFOLD INTEGRAL</u>

Let $\mathcal{V}$ be a manifold of dimension p, compact, gauged, immersed in a linear space E, normed, of dimension n and canonically oriented.

Let also A be an alternor field with scalar values, of order p, each defined on the tangent linear manifold $\vec{E}_p$ at m.

Finally, let S be the natural basis at m, and vol_p the gauge of $\vec{E}_p$; $\mathcal{U}$ is the image by the map of a domain $\overline{\Omega}$ of $\mathbb{R}^p$. $\forall$ the independent tangent differential fields :

$$d_i m = S_i \, d^i x \in \vec{E}_p, \quad {}^i x \in \mathbb{R}, \quad i = 1,2, \ldots, p, \ (i \text{ fixed})$$

$$vol_p \left(d_1 m\right)\left(d_2 m\right) \cdots \left(d_p m\right) = \alpha \left| \det(S) \right| \, d^1 x d^2 x \ldots d^p x.$$

If $\alpha \left| \det(S) \right|$ is summable in $\overline{\Omega}$, we define a manifold integral by :

$$(A.97) \qquad \int p \int_{\mathcal{U}} A\left(d_1 m\right)\left(d_2 m\right) \ldots \left(d_p m\right) = \int p \int_{\overline{\Omega}} \alpha \left| \det(S) \right| d^1 x \, d^2 x \ldots d^p x.$$

In the case of a pseudo-manifold, where the singular points constitute a p-dimensionally negligible set, the integral is taken on the regular part of $\mathcal{U}$.

12. – STOKES FORMULA

Let $\widetilde{\mathcal{U}}$ be a closed manifold (without edge) of the preceding type, and $\mathcal{U}$ a manifold of the same dimension p, embedded in $\mathcal{U}$, but having an edge $\nabla \mathcal{U}$ of dimension p-1, canonically oriented in $\widetilde{\mathcal{U}}$.

Let $d_1 m$, $d_2 m$, $\ldots$, $d_p m$ be differentials tangent to $\mathcal{U}$ at m, and independent. Indices $1,2, \ldots, p$ are taken in order for canonical orientation. If A is a differentiable field of alternors, of order p-1, each defined on $\vec{E}_p$, $\hat{\nabla}A$ will define a p-form on $\vec{E}_p$.

Under relatively general conditions, and using (A.82), it can be demonstrated the Stokes formula :

$$(A.98) \qquad \int p \int_{\mathcal{U}} \hat{\nabla}A\left(d_1 m\right)\left(d_2 m\right) \ldots \left(d_p m\right) = \int p{-}1 \int_{\nabla \mathcal{U}} A\left(d_2 m\right) \ldots \left(d_p m\right).$$

It should be pointed out that orientation of the planes $\vec{E}_p$ in this formula, is such that $d_2 m$, $d_3 m$, $\ldots, d_p m$ constitute a basis canonically oriented in $\vec{E}$ ($d_1 m$ is external to $\vec{E}_{p-1}$). For example, if $\mathcal{V}$ is a manifold Σ_m of dimension 2, immersed in E_3, and C_m its edge, we deduce, for a tangent vector field $V \in \vec{E}_2$:

$$(A.99) \qquad \int_{\Sigma_m} 2 \int \hat{\text{div}}\, V\, \text{vol}_2\big(d_1 m\big)\big(d_2 m\big) = \int_{C_m} \bar{\nu}.V\, ds,$$

where ν is the external normal to C_m in Σ_m, and s is the curvilinear abscissa of C_m, oriented in the direction corresponding to the trigonometrical orientation of $\mathbf{R}^2$. Let us also point out that for Σ_m, if N is the external normal to Σ_m, $d_1 m$, $d_2 m$ and N constitute a basis oriented in the "direct" orientation of $\vec{E}_3$.

13. – <u>DE RHAM'S THEOREM</u>   (Particular case)

Let Σ_m be a manifold (or pseudo-manifold), with an edge (or a pseudo-edge) of dimension 2, immersed in $\vec{E}_3$, gauged. Let C_m be its edge.

Let $\vec{E}_2$ be the plane tangent to Σ_m at m. Given a field A of mappings from $\vec{E}_2$ on $\vec{E}_3$, defined on Σ_m, such that :

$$(A.100) \qquad \int_C Adm = 0, \quad \forall\, \text{circuits C traced on } \Sigma_m.$$

$\exists$ a vector field V defined on Σ_m, such that :

$$V \in \vec{E}_3, \quad A = \frac{\partial V}{\partial m}$$

to within any constant vector.

(It is assumed that the necessary conditions for the integrability of A on these circuits C are satisfied).

In particular, if A is a form of degree 1 of the preceding type, with vector value, closed in any simply connected open set of Σ_m, we only need consider those circuits C which are non-reducable to a point of Σ_m by simple homotopy, by reason of the Stokes formula (A.99), and even all these circuits except one, Σ_m being a part of a closed surface $\overset{\sim}{\Sigma}_m$, by hypothesis.

14. - <u>LAGRANGIAN MULTIPLIERS</u>

From a general theorem of factorization, according to which, given two mappings A and B, and $x \in \text{def}(A) \cap \text{def}(B)$:

$$\left[\left[B(x_1) = B(x_2) = \dots \ . \right] \Rightarrow \left[A(x_1) = A(x_2) = \dots \ , \right]\right] \Leftrightarrow \left[\exists C \mid A = CB\right]$$

and restriction of C to val(B) is unique,

we deduce, when A and B are linear on their common domain of definition :

$$\left[B(x) = 0 \Rightarrow A(x) = 0\right] \Leftrightarrow \left[\exists \Lambda \text{ linear} \mid A = \Lambda B\right]$$

and restriction of Λ to val(B) is unique.

Thus suppose that we have to make stationary the functional :

$$y = \underset{\sim}{y}(x) \text{ when } z = \underset{\sim}{z}(x) = 0, \text{ where } z \in \text{ a linear space E of}$$
$$\text{finite dimension,}$$

we have the necessary condition :

$$\delta y = \frac{\partial y}{\partial x} \ \delta x = 0, \quad \forall \delta x \text{ satisfying } \frac{\partial z}{\partial x} \ \delta x = 0 \ .$$

The preceding theorem gives :

$$\exists \text{ a linear mapping } \lambda, \text{ such that :}$$

$$(A.101) \qquad \left[\begin{array}{l} \dfrac{\partial y}{\partial x} - \lambda \dfrac{\partial z}{\partial x} = 0, \quad \lambda \in L(\vec{E}, \ \mathbb{R}) \\[2em] z = 0. \end{array}\right.$$

(A.101) gives the independence from δx, but not from x. But we can also use the equivalent form where λ is also an independent variable :

$$(A.102) \qquad \delta\left[y - \lambda z\right] = 0, \quad \forall \ \delta x, \ \delta \lambda \ .$$

Naturally, fields x and λ belong to appropriate spaces, for the scalar quantities introduced to be bounded. For example if z is a square summable vector in Ω, λ must be a square summable vector and

$$\overline{\lambda z} = \int_{\Omega} \overline{\underset{\sim}{\lambda}(M)} \ . \ \underset{\sim}{z} \ (\underset{\sim}{x}(M)) d\Omega \ .$$

REFERENCES
CHAPTER I

1   GERMAIN, P., Cours de Mécanique des Milieux Continus. Masson, Paris 1973.

2   MANDEL, J., Cours de Mécanique des Milieux Continus. Gauthier-Villars Paris 1966.

3   TRUEDELL, C. and TOUPIN, R., The classical Field Theory. Handbuch der Physik, V. III. Springer-Verlag, 1960.

4   GERMAIN, P., Théorie Générale des Milieux Continus. Cours de 3ème cycle, Institut H. Poincaré, Paris 1963.

5   CASAL, P., Mécanique des Milieux Continus. Cours de la Faculté des Sciences de Marseille, 1966.

6   PRAGER, W., Introduction to Mechanics of Continua. Ginn and Co., N.Y. 1961.

7   MALVERN, L.E., Introduction to the Mechanics of a Continuous Medium. Prentice Hall, 1969.

8   SEDOV, L.I., A course in Continuum Mechanics V.I. Noordhoff Groningen, 1971.

9   FUNG, Y.C., Foundations of Solid Mechanics. Prentice Hall 1965.

10   ERINGEN, A.C., A unified Theory of Thermomechanical Materials. Int. J. Eng. Sc. V.4, Pergamon Press, 1966, 179-202.

11   GERMAIN, P., La méthode des Puissances Virtuelles en mécanique des Milieux Continus. Journal de Mécanique V.12, No. 2, Juin 1973.

12   VALID, R., La Théorie Lineaire des Coques et son Application aux Calculs Inélastiques. Thèse Poitiers 1973. Publication ONERA No. 147, 1973.

13   SOLOMON, L., Elasticité Linéaire, Masson, Paris 1968.

14   GREEN, A.E. and ZERNA, W., Theoretical Elasticity, Clarendon Press,
      Oxford, 1954.

15   SOKOLNKOF, Mathematical Theory of Elasticity. McGraw Hill, 1956.

16   LANDAU, L. et LIFCHITZ, E., Théorie de l'Elasticité, Ed. Mir, Moscou,
      1967.

17   GERMAIN, P., Mécanique des Milieux Continus. Masson, Paris, 1962.

18   TRUESDELL, C., Invariant and Complete Stress Functions for General
      Continua. Arch. Rat. Mech. Analysis, V.4., No. 1, 1959, 1-29.

19   GURTIN, M.E., A Generalization of the Beltrami Stress-Function in
      Continuum Mechanics. Arch. Rat. Mech. Analysis. V.13, No. 5, 1963,
      321-329.

20   TOUPIN, R.A., Theories of Elasticity with Couple-Stress. Arch. Rat.
      Mech. Analysis. V.17, No. 2, 1964.

21   FRAEIJS de VEUBEKE, B., Stress Function Approach. World Congress in
      Finite Element Methods in Structural Mechanics. J. Robinson ed.,
      Bournemouth, Dorset, England, 1975.

22   SOURIAU, J-M., Une méthode générale de la linéarisation des problèmes
      physiques. P.S.T. Ministère de l'Air, Paris, n° 261 (1952).

REFERENCES
CHAPTER II

1   ARGYRIS, J.H. and KELSEY, S., Energy Theorems and Structural Analysis. Aircraft Eng. Oct. 54 - May 1955 (Reprinted by Butterworth 1960).

2   TURNER, M.J., CLOUGH, R.W., MARTIN, H.C., and TOPP, L.J., Stiffness and Deflection Analysis of Complex Structures. J. Aero, Sc. 23, 1956.

3   ARGYRIS, J.H., On the Analysis of Complex Elastic Structures Appl. Mech. Reviews. July 1958.

4   ROBINSON, J. and REGL, R.R., An automated Matrix Analysis for General Plane Frames ("Rank Technique"). J. of Am. Helicopter Soc. 8(4), 1963.

5   FRAEIJS de VEUBEKE, B., Displacement and Equilibrium Models in the Finite Element Method. Chap. 9 in Stress Analysis, ed. by Zienkiewicz O.C. and Holister G.S., J. Wiley and Sons, 1965.

6   ZIENKIEWICZ, O.C. and CHEUNG, Y.K., The Finite Element in Structural and Continuum Mechanics. McGraw Hill, 1967.
ZIENKIEWICZ, O.C., The Finite Element Method. Third Edition, Mc Graw Hill, 1977.

7   ARGYRIS, J.H., Matrix Analysis in three-dimensional Elastic media ; Small and Large Displacements. J. AIAA. V.3, Jan. 1965.

8   ARGYRIS, J.H., Recent Advances in Matrix Methods of Structural Analysis Pergamon Press, 1964.

9   PRZEMIENIECKI, J.S., Theory of Matrix Structural Analysis. Mc Graw Hill, 1968.

10   RUBINSTEIN, M.F., Structural Systems-Statics ; Dynamics and Stability. Prentice Hall, 1970.

11   ZIENKIEWICZ, O.C., The Finite Element Method in Engineering Sciences. Mc Graw Hill, 1971.

12   SPOONER, J.B., A History of the Finite Element Method. World Congress
      in Finite Element Methods in Structural Mechanics. J. Robinson ed.
      Bournemouth, Dorset, England, Oct. 1975.

13   STRANG, S. and FIX, G.J., An Analysis of the Finite Element Method.
      Prentice Hall, 1973.

14   ZIENKIEWICZ, O.C., The Finite Element Method from Intuition to
      Generality. Appl. Math. Review. V.23, No. 3, March 1970, 249-256.

15   ARANTES e OLIVEIRA, E.R., Completeness and Convergence in the Finite
      Element Method. Proc. 2nd Conf. Math. Meth. Struct. Mech. Wright-
      Patterson AFB. Dayton, Ohio, 1968.

16   ARANTES e OLIVEIRA, E.R., Convergence and Accuracy in the Finite
      Element Method. World Congress on Finite Element Methods in Struc-
      tural Mechanics. J. Robinson ed. Bournemouth, Dorset, England, Oct.
      1975.

17   RAVIART, P.A. et THOMAS, J.M., Cours d'Analyse Numérique-Eléments
      Finis. Université de Paris VI, 1973.

18   ODEN, J.T., Finite Element of non-linear Continua. Mc Graw Hill,
      N.Y. 1972.

19   FRAEIJS de VEUBEKE, B., Sur certaines inégalités fondamentales et leur
      généralisation dans la théorie des bornes supérieures et inférieures
      en élasticité. Revue Universelle des Mines, XVII, No. 5, 1961.

20   FRAEIJS de VEUBEKE, B., Upper and Lower Bounds in Matrix Structural
      Analysis. Agardograph 72, Pergamon Press 1964.

21   CASAL, P., La théorie du second gradient et la capillarité. CRASc.,
      Série A, t 274, 1972, p. 1571-1574.

22   VALID, R., La Théorie des Coques et son Application aux Calculs
      Inélastiques. Thèse , Poitiers 1973, Publ. ONERA no. 147, 1973.

23   GERMAIN, P., La Méthode des Puissances Virtuelles en Mécanique des
     Milieux Continus. J. de Mécanique V.12, No. 2, juin 1973.

24   CLOUGH, R. and WILSON, E.L., Dynamic Finite Element Analysis of
     Arbitrary Thin Shells. IUTAM Symp. on High Speed Comp. of Elastic
     Structures. Fraeijs de Veubeke ed. Liège, Belgium, 1970.

25   GALLAGHER, R.H., Shell Elements. World Congress on Finite Element
     Methods in Structural Mechanics. J. Robinson Ed. Bournemouth,
     Dorset, England. Oct. 1975.

26   GALLAGHER, R.H., Finite Element Representations for Thin Shell
     Instability Analysis. IUTAM Symp. on Buckling of Structures. Harvard
     University, Mass. June 1974.

27   HILL, R., The mathematical Theory of Plasticity. Clarendon Press,
     Oxford, 1960.

28   MANDEL, J., Cours de Mécanique des Milieux Continus, t. 2, Gauthier-
     Villars, Paris, 1966.

29   SON, N.Q., On the Elastic-Plastic Initial Boundary Value Problem and
     its Numerical Integration. Conf. GAMNI, Paris, 22-1-1976.

30   IRONS, B. and RAZZAQUE, A., The Evolution of the Isoparametric
     Elements. World Congress on Finite Element Methods in Structural
     Mechanics. J. Robinson Ed. Bournemouth, Dorset, England, Oct. 1975.

31   ROBINSON, J., Integrated Theory of Finite Element Methods. J. Wiley
     and Sons, London 1973.

32   IRONS, B.M., A frontal Solution Program. Int. J. Num. Meth. Eng., 2,
     5-32, 1970.

33   HINTON, E. and OWEN, D.R., Finite Element Programming, Academic
     Press, London 1977.

34   WILSON, E.L. and CLOUGH, R.W., Dynamic response by step by step
      Analysis. Proc. Symp. on the Use of Computers in Civil Engineering,
      Lisbon, Oct. 1962.

REFERENCES
CHAPTER III

1   TRUESDELL, C. and TOUPIN, R.A., The Classical Field Theories.
Encyclopedia of Physics.　Springer Verlag, Berlin, 1960.

2   Variational Methods in Engineering.　Proc. of an International Conference held at the University of Southampton.　Sept. 1972.　Ed. by
Dept. of Civil Engineering ; Univ. of Southampton 1973, t. 1 and 2.

3   WASHIZU, K., Variational Methods in Elasticity and Plasticity.　Pergamon Press 1975.

4   REISSNER, E., On a variational Theorem in Elasticity.　J. of Math.
and Phys. Vol. 29, 1950.

5   FRAEIJS de VEUBEKE, B., Displacement and Equilibrium Models in the
Finite Element Methods.　Chap. 9 in Stress Analysis.　Ed. O.C.
Zienkiewicz and G.S. Holister. J. Wiley 1965.

6   GERMAIN, P., Mécanique des Milieux Continus, Masson, Paris, 1962.

7   PIAN, T.H.H., Derivation of Element Stiffness Matrices by assumed
Stress Distributions.　AIAA. J. 2 no. 7, 1333-1336, 1964.

8   VALID, R., La Théorie linéaire des Coques et son Application aux
Calculs Inélastiques.　Thèse , Poitiers, 1973, Publ. ONERA no. 147,
1973.

9   TRUESDELL, C., Invariant and Complete Stress-Functions for General
Continua.　Arch. Rat. Mech. Analysis V. 13, no. 5, 1963, 321-329.

10   FRAEIJS de VEUBEKE, B., and ZIENKIEWICZ, O.C., Strain Energy Bounds
in Finite Element Analysis by Slab Analogy.　J. of Strain Analysis
V. 2, 1967.

11   ROBINSON, J., Structural Matrix Analysis for the Engineer.　J. Wiley,
1966.

12   FRAELJS de VEUBEKE, B., Basis of a well-conditioned Force-Program for
      Equilibrium Models via Southwell Slab Analogies. Air Force Report
      AFFDL-TR-67-10, 1967.

13   FRAELJS de VEUBEKE, B., Upper and Lower Bounds in Matrix Structural
      Analysis. Matrix Methods in Structureal Analysis. Agardograph 72.
      Pergamon Press, 1964.

14   SANDER, G. and BECKERS, P., Delinquant Finite Elements for Shell
      Idealization. World Congress on Finite Element Methods in Structural
      Mechanics. J. Robinson Ed. Bournemouth, Dorset, England, Oct. 1975.

15   IRONS, B.M. and RAZZAQUE, A., Experience with the Patch-Test for
      convergence of Finite Elements. Math. Found. of the Finite Element
      Method, Acad. Press 1972, 557-587.

16   COOK, R.D., Concepts and Applications of Finite Element Analysis.
      Prentice Hall N.J. 1975.

17   STRANG, G. and FIX G.F., An Analysis of the Finite Element Method.
      Prentice Hall, N.J. 1973.

18   FRAELJS de VEUBEKE, B., Variational Principles and the Patch test. Int.
      J. Num. Meth. in Eng. V.8, 1974, 783-801.

19   LASCAUX, P. et LESAINT, P., Eléments finis non conformes pour la
      flaxion des plaques minces. Proc. "Journées Eléments Finis". IRIA,
      Univ. de Rennes, 1974.

20   OLSON, M.D., Compatibility. World Congress on Finite Element methods
      in Structural Mechanics. J. Robinson Ed. Bournemouth, Dorset, England,
      Oct. 1975.

21   THOMAS, B., Principe variationnel mixte en élasticité et application au
      calcul d'une plaque en flexion. N.T. ONERA no. 233, 1974.

22   EKELAND, I. et TEMAM, R., Analyse Convexe et problèmes variationnels.
      Dunod, Paris 1974.

23   ODEN, J.T., and REDDY, J.N., An Introduction to the Mathematical Theory
of Finite Elements. John Wiley and Sons (1976).

24   RAVIART, P.A., Mixed Finite Element Methods for solving 2nd order
Elliptic Equations, Paper presented at the Conference on Numerical
Anaysis - Royal Irish Academy, Dublin, July 29th - August 2nd, 1974.

25   VALID, R., An Intrinsic Formulation for the Non-linear Theory of
Shells and some Approximations. Symposium on Future Trends in
Computerized Structural Analysis and Synthesis. Washington D.C.,
Oct. 30th - Nov. 1st, 1978 . Computers and Structures, Vol. 10
pp. 183 - 194, 1979.

REFERENCES
CHAPTER IV

1   SOURIAU, J.M., Calcul linéaire t. 1 et 2, PUF, Paris 1964.

2   VALID, R., L'aéroélasticité et le flottement des avions. Cours E.C.P.
     1970, 1974.

3   DUVAUT, G. et LIONS, J.L., Les Inéquations en Mécanique et en Physique.
     Dunod, Paris, 1972.

4   TEMAM, R., Analyse Numérique Coll. SUP. PUF., Paris, 1970.

5   SCHWARTZ, L., Cours d'analyse t. 2. Hermann, Paris, 1963.

6   DIEUDONNE, J., Eléments d'analyse t. 1. Gauthier-Vilars, 1969.

7   COURANT, R. and HILBERT, D., Methods of Mathematical Physics V. 1,
     Interscience Publishers Inc., N.Y., 1953.

8   STRANG, G. and FIX, G.J., An Analysis of the Finite Element Method.
     Prentice Hall, 1973.

9   HURTY, W.C., On the Dynamic Analysis of Structural Systems using
     Component Modes. 1st AIAA Annual Meeting. Washington D.C., June
     1964. AIAA paper no. 64-487, 1964.

10   CRAIG, R.R. Jr. and BAMBTON, M.C.C., Coupling of Substructures for
     Dynamic Analysis. AIAA Journal V.6 no. 7, July 1968.

11   BAJAN, R.L., FENG, C.C. and JASZLIES, I.J., Vibration Analysis of
     Complex Structural Systems by Modal Substitution. Shock and
     Vibration Bulletin 39 (3), 1969, p. 99-106.

12   HURTY, W.C., COLLINS, J.D. and HART, G.C., Dynamic Analysis of large
     Structures by Modal Synthesis Analysis. Am. Int. J. Computers and
     Structures 1, 4, Dec. 1971, Pergamon Press.

13   KUHAR, E.J., and STAHLE, C.V., Dynamic Transformation Method for
     Modal Synthesis. AIAA J. V. 12, No. 5, May 1974.

14   GUYAN, R.J., Reduction of Stiffness and Mass Matrices. AIAA. J. V.3,
     No. 2, 1965.

15   IRONS, B.M., Structural Eigenvalue Problems. Elimination of variables.
     AIAA Journal V. 3, No. 5, 1965.

16   GERADIN, M., Les grandes méthodes Numériques de l'Analyse Dynamique
     des Structures. Conférence au GAMNI le 22-1-1976, Paris.

17   GERADIN, M., Error Bounds for Eigenvalue Analysis by Elimination of
     Variables. J. of Sound and Vibration. V. 19, no. 1, 1971.

18   GERADIN, M., Analyse Dynamique duale des Structures par la méthode
     des éléments finis. Collection des Publications de la Faculté des
     Sciences Appliquées de Liège, Belgique, no. 36, 1973.

19   KATO, T., On the upper and lower bounds to eigenvalues. J. Phys. Soc.
     Japan, 4, 334, 1949.

20   SCHWARTZ, L., Méthodes mathématiques pour les sciences physiques.
     Hermann, Paris 1961.

21   VALID, R., Les vibrations aléatoires, Cours Postscolaire, ECP 1970.

22   BOURGINE, A., Sur une approche statistiques de la dynamique
     vibratoire des structures. Thèse Paris-Orsay, 1973.

23   MORAND, H., Analyse variationnelle des méthodes de sous-structuration
     dynamique. Publication ONERA (to be published).

24   VALID, R., and OHAYON, R., Influence of sloshing in wing tip tanks on
     the vibration natural modes of an aircraft. La Recherche Aerospatiale,
     N° 1974-5 (E.S.A. translation available).

25   MORAND, H. and OHAYON, R., Substructure variational analysis of
     compressible hydroelastique vibration problems. Finite element results.
     Int. J. for Num. Meth. in Eng. (To appear in 1979).

26   OHAYON, R., Formulation variationnelle symétrique du problème des
     vibrations harmoniques par couplage des principes primal et dual -
     Application aux oscillations de systèmes couplés fluide-structure.
     La Recherche Aerospatiale 79-3 (1979).

27   OHAYON, R. et VALID, R., Principes duaux en vibrations harmoniques.
     Formulations symétriques couplées et sous-structuration (to be
     published).

REFERENCES
CHAPTER V

1   GERMAIN, P., Cours de Mécanique des Milieux Continus t. 1. Masson, Paris 1973.

2   MANDEL, J., Cours de Mécanique des Milieux Continus. t. 1 et 2. Gauthier-Villars, Paris 1970.

3   GERMAIN, P., Théorie Générale des Milieux Continus. Cours de 3ème cycle. Institut H. Poincaré, Paris 1963.

4   PRAGER, W., Introduction to Mechanics of Continua. Ginn and co. N-Y. 1961.

5   MALVERN, L.E., Introduction to the Mechanics of a Continuous Medium. Prentice Hall, 1969.

6   TRUESDELL, C. and NOLL, W., The Non-linear Field Theories of Mechanics. Handbuch der Physik. V. III/3. Springer-Verlag, 1965.

7   ERINGEN, A., Non-linear Theory of Continuous Media. Mc Graw Hill, 1962.

8   NOVOZHILOV, V.V., Foundations of the Non-linear Theory of Elasticity Graylock Press. Rochester N-Y. 2nd printing, 1957.

9   ODEN, J.T., Finite Elements of Non-linear Continua. Mc Graw Hill, 1972.

10   ZIENKIEWICZ,O.C., The Finite Element Method in Engineering Science. McGraw Hill, 1971. The Finite Element Method. Third Edition, McGraw Hill, 1977.

11   WEMPNER, G.W., Discrete Approximation of Elastic-Plastic Bodies by variational Methods. International Conference on Variational Methods in Engineering 25th - 29th Sept. 1972, University of Southampton.

12   BESSELING, J.F., Non-linear Analysis. World Congress on Finite

Element Methods in Structural Mechanics. J.Robinson Ed. Bournemouth, Dorset, England, Oct. 1975.

13 MARCAL, P.V., Effect of Initial Displacement on Problems of Large Deflection and Stability. Tech. Rep. ARPA. E54, Brown University, Nov. 1967.

14 ZIENKIEWICZ, O.C., and CHEUNG, Y.K., The Finite Element in Structural and Continuum Mechanics. Mc Graw Hill, 1967.

15 MARCAL, P.V., A comparative Study of Numerical Methods of Elastic-Plastic Analysis. Proc. AIAA/ASME 8th Structures, Structural Dynamics and Material Conf. March 1967.

16 BOISSENOT, J.M., DUBOIS, M. et LACHAT, J.C., Calcul des Structures Elasto-plastiques par la Méthode des Element finis. Les mémoires du CETIM, (Centre Technique des Industries Mécaniques) Senlis, Dec. 1972

17 HENRY, R., Coques Minces en Grandes Déformations par les Eléménts Finis. Application du Calcul des Contraintes, des Fréquences et Modes de Compression en Rotation. Congrés Français de Mécanique. Sept. 1973, Poitiers.

18 MAU, S.T. and GALLAGHER, R.H., A Finite Element Procedure for Non-linear prebuckling and Initial Post-buckling Analysis. NASA Contrator Rep., NASA CR 1936, Jan. 1972.

19 DUPUIS, G.Q., PFAFFINGER, D. and MARCAL, P.V., Effective use of Incremental Stiffness Matrices in Non-linear Geometric Analysis. IUTAM Symp. on High Speed Computing of Elastic Structures. Liège, Belgium, 1970.

20 SON, N.Q., HAUG, E. et de ROUVRAY, A., Intégration directe en Dynamique non-linéaire. Application au flambage dynamique des structures par éléments finis. Rapport final, contrat DRME. Informatique Internationale. Vol 1 et 2, 1976.

21 BELYTSCHKO, T., Transient Analysis. Structure Mechanics Computer Programs, 1973.

22   COLLATZ, L., The Numerical Treatment of Differential Equations.
     Springer-Verlag, 1966.

23   WILSON, E.L. and CLOUGH, R.W., Dynamic Response by step by step
     Matrix Analysis. Symp. on Use of Computers in Civil Engineering.
     Lisbon Oct. 1962.

24   FARHOOMAND, Non-linear Dynamic Analysis of two-dimensional Systems.
     Ph. D. Thesis. Univ. of California, Berkeley, 1970.

25   NEWMARK, N.M., A Method of Computation for Structural Dynamics. J.
     Eng. Mech. Div. V. 85, EM3 (67-94), 1959.

26   ARGYRIS, J.H., DUNNA, P.C. and ANGELOPOULOS, T., Non-linear oscilla-
     tions using the Finite Element Technique. Comp. Meth. Appl. Mech.
     Eng. V.2.2. 1973.

     Non-linear Dynamics of Structures. Comp. Meth. in Appl. Sc. Eng.
     Proc. IRIA, Paris, 1973.

27   HOUBOLT, J.C., A Recurrence Matrix Solution for Dynamic Response of
     Elastic Aircraft. J. of Aeron. Science. V. 17, Sept. 1950, p. 540-
     550.

28   HILDEBRAND, F.B., Introduction to Numerical Analysis. Mc Graw Hill,
     1956.

29   STRICKLIN, J.A., Geometrically Non-linear Static and Dynamic Analysis
     of Shells of Revolutions. IUTAM Symp. on High Speed Computing of
     Elastic Structures. Fraeijs de Veubeke Ed. 1970.

30   Van der NEUT, A., Post-buckling Behaviour of Structures V. TH. Rep.
     69. Presented to the Agard 6th General Assembly, Brussels, July 1956.

31   TIMOSHENKO, Théorie de la Stabilité Elastique. Lib. Polytechnique Ch.
     Béranger, 1943.

32   ZIEGLER, H., On the Concept of Elastic Stability. Adv. Appl. Mech.,
     V.4., 1956, 351-403.

33 KOITER, W.T., On the Stability of Elastic Equilibrium. Thesis,
    Delft 1945. NASA TTF 10, 1967.

34 BRUHN, E.F., Analysis and Design of Flight Vehicle Structures. Tri-
    State Offset Company. Cincinnati, Ohio, 1965.

35 ROORDA, J., The Buckling Behaviour of Imperfect Structural Systems.
    J. Mech. Phys. Solids. 13, 1965, 267-280.

36 KNOPS, R.J. and WILKES, Theory of Elastic Stability. Handbuch der
    Physik. V. VI a/3. Mechanics of Solids III, 1973.

37 LIAPOUNOV, A.M., Problème général de la Stabilité du Mouvement. Soc.
    Math. de Kharkow 1892. Reprinted by Princeton Univ. Press. (Ann. of
    Maths.) 1949.

38 KOITER, W.T., Stability of Equilibrium of Continuous Bodies Tech.
    Rep. no. 79, Brown University. April 1962.

39 MARTIN, H.C., On the Derivation of Stiffness Matrices for the
    Analysis of Large Deflection and Stability Problems. Proc. of the
    Conf. on Matrix Methods in Structural Mechanics. Wright-Patterson
    A.F.B. Dayton (Ohio), 1965.

40 ALMROTH, B.O. and BROGAN, F.A., Bifurcation Buckling for general
    Shells AIAA/ASME 13th Structures, Structural Dynamics and Material
    Conference. San Antonio (Texas), April 1972, AIAA paper no. 72-352.

41 PRZEMIENIECKI, J.S., Finite Element Structural Analysis of Local
    Stability. AIAA/ASME 13th Structures, Structural Dynamics and Material
    Conference. San Antonio (Texas), April, 1972. AIAA paper no. 72-354.

42 HAISER, W.E., STRIKLIN, J.A. and STEBBINS, F.J., Development and
    Evaluation of Solution Procedures for Geometrically Non-linear
    Structural analysis by the Direct Stiffness Method. AIAA/ASME 12th
    Structures, Structural Dynamics, and Material Conf. Anaheim (Calif.)
    April, 1971.

43   LANSING, W., JENSEN, W.R. and FALBY, W., Matrix Analysis Methods for
      Inelastic Structures. Proc. of the Conf. on Matrix Methods in Struct-
      ural Mechanics. Wright-Patterson A.F.B. Dayton (Ohio), 1965.

44   ODEN, J.T., Formulation and Application of Certain Primal and Mixed
      Finite Element Methods of Finite Deformations of Elastic Bodies.
      Colloque IRIA, Methodes de Calcul Scientifique et Technique.
      Rocquencourt, Décembre 1973.

45   HARRIS, H.G. and PIFKO, A.B., Elastic-Plastic Buckling of Stiffened
      Rectangular Plates. Proc. of Symp. on the Application of Finite
      Element Methods in Civil Engineering. Nashville (Tennessee), Nov.
      1969.

46   THOMSON, J.M.T., A general Theory for the Equilibrium and Stability
      of Discrete Conservative Systems. Zamp. V.20, 1969.

47   THOMSON, J.M.T., and HUNT, G.W., A general Theory of Elastic Stability.
      J. Wiley and Sons, 1973.

48   VALID, R., Déformations non-linéaires et flambage statique. Publication
      ONERA No. 1977-2.

49   VAN der NEUT, A., Mode Interaction with Stiffened Panels. Proc. IUTAM
      Symposium on Buckling of Structures. Harvard University, June 17-21,
      1974. Springer 1974.

50   ALMROTH, B.O., STERN, P., and BROGAN, F.A., Automatic Choice of Global
      Shape Functions in Structural Analysis. AIAA Journal Vol. 16, N°5,
      May 1978.

51   SOURIAU, J-M. Une méthode générale de linéarisation des problèmes
      physiques. P.S.T. Ministère de l'Air, Paris, n° 261 (1952).

REFERENCES
CHAPTER VI

1   LOVE, A.E.H., A Treatise of the Mathematical Theory of Elasticity. Dover Publications N-Y 1964 (4th ed.).

2   GREEN, A.E., and ZERNA W., Theoretical Elasticity. Clarendon Press 1954.

3   FLÜGGE, W., Stress in Shells. Springer-Verlag, 1960.

4   NOVOSHILOV, V.V., The Theory of Thin Shells. P. Noordhoff, 1959.

5   GOL'DENVEIZER, A.L., Theory of Thin Elastic Shells. Pergamon Press, 1961.

6   MUSHTARI, Kh. and GALIMOV, K.Z., Non-linear Theory of Thin Elastic Shells. NASA TTF 62. U.S. Dept. of Commerce. Off. of Technical Services. Washington, 1961.

7   NAGHDI, P.M., Foundations of Elastic Shell Theory. Progress in Solid Mechanics 4, 1-90. North-Holland Publ. Co. Amsterdam, 1963.

8   LUR'E, A.L., On the Static-Geometric Analogue of Shell Theory. Problems of Continuum Mechanics (N.I. Muskhelishvili Anniversary volume). Soc. Ind. Appl. Math. Philadelphia SIAM. Rev. 1961.

9   ROUGEE, P., Equilibre des Coques Elastiques Minces Inhomogène en Théorie Non-linéaires. Thèse Paris, 1969.

10   SANDERS, J.L. Jr., On the Shell Equations in Complex Form. 2nd Symp. of IUTAM on the Theory of Thin Shells. Copenhagen, 1967. Springer-Verlag 1969, 135-154.

11   SANDERS, J.L. Jr., Non-linear Theory of Thin Shells, Quart. of Appl. Math. V. 21, No. 1, April 1963, 21-36-ONR Tech. Rep. No. 10. Harvard Univ., 1961.

12   LEONARD, R.W., Non-linear First Approximation Thin Shell and Membrane
     Theory. NASA Langley Field, Va, 1961.

13   KOITER, W.T., A Consistent First Approximation in the General Theory
     of Thin Elastic Shells. Proc. of the Symp. of the Theory of Thin
     Elastic Shells. Delft, 1959; (North-Holland, Amsterdam, 1960).

14   RUTTEN, H.S., Asymptotic Approximation in the Three-dimensional
     Theory of Thin Shells. Nederlandse Boekdruk Industrie 1971.

15   RUTTEN, H.S., Theory and Design of Shells on the Basis of Asymptotic
     Analysis. Rutten + Kruisman, Netherlands, 1973.

16   KOITER, W.T., On the Foundations of the Linear Theory of Thin Elastic
     Shells. Koninkl. Nederl. Akademi Van Wetenschappen. Amsterdam.
     Proc. Série B 73, No. 3, 1970, p. 169-195.

17   LADEVEZE, P., Comparaison de Modèles de Milieux Continus. Thèse,
     Paris 1975.

18   GREEN, A.E. and NAGHDI, P.M., Shells in the Light of Generalized
     Continua. IUTAM Symp. on the Theory of Thin shells. Copenhague,
     1967, Springer-Verlag, 1969, 39-58.

19   COHEN, H. and De SILVA, C.N., Non-linear Theory of Elastic Surfaces.
     J. of Mathematical Physics, V. 7, No. 2, 1966.

20   KOITER, W.T., On the Non-linear Theory of Thin Elastic Shells. Proc.
     Physical Sciences. Mechanics, Série B, V. 69, No. 1, 1965.

21   SCHWARTZ, L., Cours d'analyse t. 1 et 2, Hermann, Paris 1967.

22   SOURIAU, J.M., Géométrie et Relativité. Hermann, Paris 1964.

23   VALID, R., Notions de Géométrie Différentielle. Cours ESTA, 1969.

24   VALID, R., La Théorie Linéaire des Coques et son Application aux
     Calculs Inélastiques. Thèse , Poitiers, 1973. Publ. ONERA no.147,1973.

25   VALID, R., Déformations non-linéaires et flambage statique. Publication
     ONERA, No. 1977-2.

26   NAGHDI, P.M., On the Theory of Thin Elastic Shells. Quart. Appl. Math.
     V. 14, No. 4, Jan. 1957.

27   FORSBERG, K. and HARTUNG, R., On Evaluation of Finite Difference and
     Finite Element Techniques for General Shells. Proc. IUTAM Symp. on
     High Speed Comp. of Elastic Structures. B. Fraeijs de Veubeke Ed. V. 2
     Liège, 1971.

28   BUSHNELL, D., Thin Shells in Structural Mechanics Computer Programs.
     Surveys, Assessments and Availability. W. Pilkey and Al. Ed. Univ.
     of Virginia Press. Charlottesville, Va, 1974.

29   JONES, R. and STROME, D. Survey of Analysis of Shells by the Dis-
     placement Method. Proc. of Conf. on Matrix Methods on Structural
     Mech. AFFDL-TR-66-80. Dayton (Ohio), 1965, 205-229.

30   GALLAGHER, R.H., Analysis of Plate and Shell Structures. Proc. on
     Application of Finite Element Methods in Civil Engineering. Vanderbilt
     U. 1969, 155-206.

31   CLOUGH, R.W. and WILSON, E.L., Dynamic Finite Element Analysis of
     Arbitrary Thin Shells. Proc. IUTAM Symp. on High Speed Computing
     of Elastic Structures. B. Fraeijs de Veubeke, Ed. Liège, 1971.

32   GALLAGHER, R.H., Shell Elements. World Congress on Finite Element
     Methods in Structural Mechanics. J. Robinson ed. Bournemouth,
     Dorset, England, Oct. 1975.

33   ARGYRIS, J.H. and Al., Some New Elements for the Matrix Displacement
     Method. Proc. 2nd Conf. on Matrix Methods in Structural Mechanics.
     AFFDL-TR-68-150, Dayton (Ohio), 1968, 333-397.

34   GREEN, B.E., JONES, R.E. and STROME, D.R., Dynamic Analysis of Shells
     using Doubly Curved Finite Elements. Proc. 2nd Conf. on Matrix Methods
     in Structural Mechanics. Wright-Patterson AFB Dayton (Ohio), 1968.

35   HERMANN, L.R., Finite Element Bending Analysis for Plates. Proc.
     ASCE J. of Eng. Mech. Div. V. 93, No. EM5, Oct. 1967.

36   PIAN, T.H.H., and TONG, P., Rationalization in deriving Element
     Stiffness Matrix by assumed Stress Approach. Proc. 2nd Conf. on
     Matrix Methods in Structural Mechanics. AFFDL TR-68-150. Dayton
     (Ohio) 1968, 441-469.

37   SANDER, G., Application de la Méthode des Eléments Finis à la Flexion
     des Plaques. Thèse, Liège. Collection des Publications de l'Uni-
     versité des Sciences Appliquées de Liège, No. 15, 1969.

38   SANDER, G., and BECKERS, P., Delinquent Finite Elemnts for Shell
     Idealization. World Congress on Finite Element Methods in Structural
     Mech. J. Robinson ed. Bournemouth, Dorset, England. Oct. 1975.

39   KRAUS, H., Thin Elastic Shells. J. Wiley and Sons, N-Y. 1967.

40   KEY, S. and BEISINGER, Z.E., The Analysis of Thin Shells by the Finite
     Element Method. Proc. of IUTAM Symp. and High Speed Computing of
     Elastic Structures. B. Fraeijs de Veubeke ed. Liège 1971.

41   DUPUIS, G., Application of Ritz's method to Thin Elastic Shell Analy-
     sis. Trans. ASME. J. Appl. Mech. 38, Série E, no. 4, 1971, 987-996.

42   CANTIN, G., Rigid Body Motions in Curved Finite Elements. AIAA J.
     V. 8, no. 7, July 1970, 1252-1255.

43   FONDER, G. and CLOUGH, R.W., Explicit Addition of Rigid Body Motions
     in Curved Finite Elements. AIAA J. V. 11, no. 3, March 1973, 305-312.

44   THOMAS, G.R. and GALLAGHER, R.H., A Triangular Thin Shell Finite
     Element : Linear Analysis. NASA CR-2482, 1975.

45   COWPER, G.R., LINDBERG, G., and OLSON, M.D., Comparison of two High
     Precision Triangular Finite Elements for Arbitrary Deep Shells. Proc.
     3rd Conf. on Matrix Methods in Structural Mech. AFFDL-TR-71-160.
     Dayton (Ohio), 1971, 277-304.

46   ARGYRIS, J.H., and SHARPF, D., The Sheba Family of Shell Elements for
     the Matrix Displacement Method.   The Aeronautical Journal, 1968,
     873-883.

47   ZIENKIEWICZ, O.C., The Finite Element Method in Engineering Science.
     Mc Graw Hill, N-Y. 1971.

48   KIKUCHI, F. and ANDO, Y., Some Finite Element Solutions for Plate
     Bending Problems by Simplified Hybrid Displacement Method.   Nucl. Eng.
     Des. V. 23, 1972, 155-173.

49   KIKUCHI, F. and ANDO, Y., Application of Simplified Hybrid Displace-
     ment Method to Plate and Shell Problems.   Proc. 2nd SMIRT.   Berlin
     1973, V. 5, paper M 5/5.

50   PRATO, C., Shell Finite Elment via Reissner's Principle.   Int. J.
     Solids Struct.   V.5, 1969, 1119-1133.

51   CONNOR, J. and WILL, G., A Mixed Finite Element Shallow Shell Formul-
     ation.   Advances in Matrix Methods of Struct. Analysis and Design.
     R. Gallagher and Al. ed. Univ. of Alabama Press 1969, 105-137.

52   HERMANN, L.R. and MASON, W.E., Mixed Formulations for Finite Element
     Shell Analysis.   Conf. of Computer oriented Analysis of Shell
     Structures.   AFFDL-TR-71-79.   June 1971.

53   EDWARDS, G. and WEBSTER, J.J., Hybrid Cylindrical Shell Elements.   In
     Finite Element Thin Shell Analysis.   D. Ashwell and R. Gallagher ed.
     J. Wiley, 1976.

54   GRAFTON, P. and STROME, D., Analysis of Axisymmetrical Shells by the
     Direct Stiffness Method.   AIAA J. V. 1, no. 10, Oct. 1963, 2342-2347.

55   GOULD, P. and SEN, S., Refined Mixed Finite Elements Methods for
     Shells of Revolution.   Proc. 3rd Conf. on Matrix Methods in Struct.
     Mech. AFFDL-TR-71-160, Dayton (Ohio) 1971, 397-421.

56   AHMAD, S., IRONS, B.M., and ZIENKIEWICZ, O.C., Curved Thich Shell and
     Membrane Elements, with Particular Reference to Axisymmetric Problems.
     Proc. 2nd Conf. on Matrix Methods in Struct. Mech. AFFDL-TR-68-150,
     Dayton (Ohio) 1968, 539-572.

57   ENGRAND, D. and BORDAS, J., Calcul des Coques en Matériaux multi-
     couches et Sandwichs par la Méthode des Eléments Finis. Rech. Aérosp.
     No. 1973-2 (mars-avril), 109-118.

58   ENGRAND, D., Matrix Analysis of Multilayered Shells by the Finite
     Element Method. 10th ICAS Congress Ottawa, Canada, Oct. 1976.

59   LEISSA, A.W., Vibrations of Shells, Doc. NASA, S.P. 288. Scientific
     and Technical Information office, Washington D.C., 1973.

60   VALID, R., Conditions de compatibilité et fonctions de contrainte
     dans le case de coques multiplement connexes. Communication
     présentée au Congrès International de Mécanique Théorique et Appli-
     quée (IUTAM), Delft, 30 Aug.-4 Sept. 1976 (Texte Anglais : T.P. ONERA
     1976-82. E). Rech. Aérosp. no. 1976-5 Sept-Oct, 1976.

61   VALID, R., An Intrinsic Formulation for the Non-linear Theory of
     Shells and some Approximations. Symposium on Future Trends in Com-
     puterized Structural Analysis and Synthesis. Washington D.C., Oct.
     30th - Nov. 1st, 1978.Computers and Structures , Vol.10 pp 183-194, 1979.

62   GIRARD, R., Eléments finis isoparamétriques de coque multicouche
     pour le calcul des structures aérospatiales en matériaux composites.
     La Recherche Aérospatiale no. 1977-2, p. 131-132.

63   ROELANDT, J.M. et VERCHERY, G., Eléments isoparamétriques pour le
     calcul des plaques sandwiches. International Congress on Numerical
     Methods for Engineering. Paris, Nov.27 - Dec.1, 1978.

64   MASSARD, Th., Calcul des contraintes et des déplacements dans une
     structure cylindrique anisotrope stratifiée à l'aide d'une méthode
     variationnelle mixte. International Congress on Numerical Methods for
     Engineering. Paris, Nov.27 - Dec.1, 1978.

REFERENCES
APPENDIX

1   SOURIAU, J.M., Calcul Linéaire t. 1 et 2. PUF. Paris, 1964.

2   SOURIAU, J.M., Géométrie et Relativité. Hermann. Paris, 1964.

3   SCHWARTZ, L., Cours d'Analyse t. 1 et 2. Hermann. Paris, 1967.

4   VALID, R., Méthodes Mathématiques de la Physique. Cours E.C.P. 1969.

5   VALID, R., Notions de Géométrie Différentielle. Cours ESTA, 1969.